# Biostatistics and Microbiology

# Biostatistics and Microbiology:
# A Survival Manual

Daryl S. Paulson
*BioScience Laboratories, Inc.*
*Bozeman, MT, USA*

 Springer

*Author*
Daryl S. Paulson
BioScience Laboratories, Inc.
Bozeman, MT 59715
USA

ISBN: 978-0-387-77281-3          e-ISBN: 978-0-387-77282-0
DOI: 10.1007/978-0-387-77282-0

Library of Congress Control Number: 2008931293

Printed on acid-free paper

9 8 7 6 5 4 3 2 1

springer.com

# Acknowledgements

Researchers do not need to be statisticians to perform quality research, but they do need to understand the basic principles of statistical analysis. This book represents the most useful approaches learned through my years of experience in designing the statistically based studies in microbial death-rate kinetics, skin biotechnology, and clinical trials performed by my company, BioScience Laboratories, Inc.

I am truly grateful to my co-workers at BioScience, who kept the business running while I wrote and rewrote the chapters in this book, in particular, my very capable and supportive wife, Marsha Paulson, Vice-President of BioScience Laboratories, Inc. I am especially indebted to John Mitchell, Director of Quality Assurance, for his many hours of challenging my assumptions, utilizing his vast knowledge of microbiology, and editing this work. Tammy Anderson provided valuable assistance in the development of this book by typing and retyping it, making format changes, editing, creating figures and diagrams, and basically, managing the entire book development process. Her abilities are astounding.

Finally, I thank the staff at Springer for their patience, flexibility, professionalism, and quality concerns.

# Contents

# Chapter 1

## BioStatistics and Microbiology: Introduction

To compete with the many books claiming to demystify statistics, to make statistics easily accessible to the "terrified," or provide an eastern approach purporting to present statistics that do not require computation, as in the "Tao of Statistics," is tough duty, if not utter fantasy. This book does not promise the impossible, but it will enable the reader to access and apply statistical methods that generally frustrate and intimidate the uninitiated. Statistics, like chemistry, microbiology, woodworking, or sewing, requires that the individual put some time into learning the concepts and methods. This book will present in a step-by-step manner, eliminating the greatest obstacle to the learner (not the math, by the way) applying the many processes that comprise a statistical method. Who would not be frustrated, when not only must the statistical computation be made, but an assortment of other factors, such as the $\alpha$, $\beta$, and $\delta$ levels, as well as the test hypothesis, must be determined? Just reading this far, you may feel intimidated. I will counter this fear by describing early in the book a step-by-step procedure to perform a statistical method – a process that we will term "the six-step procedure." All of the testing will be performed adhering to six well-defined steps, which will greatly simplify the statistical process. Each step in the sequence must be completed before moving on to the next step.

Another problem that microbiology and other science professionals often must confront is that most of the training that they have received is "exact." That is, calculating the circumference of a circle tacitly assumes that it is a perfect circle; the weight of a material is measured very precisely to $n$ number of digits; and 50 mL is, all too often, expressed to mean 50.0 mL exactly. This perspective of exactitude usually is maintained when microbiologists employ statistics; however, statistical conclusions deal with long-run probabilities which, by themselves, are nearly meaningless. In the context of microbiology, statistics can be extremely useful in making interpretations and decisions concerning collected data. Statistics, then, is a way of formally communicating the interpretation of clinical or experimental data and is particularly important when a treatment result is not clearly differentiable from another treatment. Yet, and this is the big "yet," the statistic used has much

D.S. Paulson, *Biostatistics and Microbiology*, doi: 10.1007/978-0-387-77282-0_1,
© Springer Science + Business Media, LLC 2008

influence on the conclusions that result. A very "conservative" statistic requires very strong proof to demonstrate significant differences, whereas a "liberal" one requires less. "Yuck," you say already, "I just want to know the answer."

To this, I respond, when in doubt, use a conventional statistical method, one that can be agreed on and accepted by most authorities. These conventional kinds of methods will be presented in this book. As you gain experience, choosing statistical methods will become almost an intuitive process. For example, for problems in which you have little experience, you will be very cautious and conservative. By analogy, this is similar to rafting a river for the first time. If you see rapids in the river, you will be more conservative as you approach them – wearing a life jacket and helmet, and using your paddle to avoid rocks – at least until you have experienced them and developed confidence. You will tend to be more liberal when near a sandy shore in clear, calm, shallow water. For experiments in microbiology in which you have much experience, your microbiological knowledge enables you to be more statistically liberal, as you will know whether the result of statistical analysis is microbiologically rational.

Finally, statistics is not an end-all to finding answers. It is an aid in research, quality control, or diagnostic processes to support critical thinking and decision-making. If the statistical results are at odds with your field knowledge, more than likely, the statistical method and/or the data processed are faulty in some way.

## 1.1 Normal Distribution

Let's get right down to the business of discussing the fundamentals of statistics, starting with the normal distribution, the most common distribution of data. The normal distribution of data is symmetric around the mean, or average value, and has the shape of a bell. For example, in representing humans' intelligent quotients (IQs), the most common, or prevalent IQ value is 100, which is the average. A collection of many individual IQ scores will resemble a bell-shaped curve with the value 100 in the middle (Fig. 1.1).

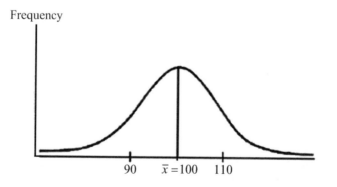

Fig. 1.1 Bell-shaped curve of intelligence quotients

IQ scores that are higher or lower than the mean are symmetrical around it. That is, values ten points above (110) and ten points below (90) the mean are equal distance from the mean, and this symmetrical relationship holds true over the entirety of the distribution. Notice also that, as IQ scores move farther from the mean in either direction, their frequency of occurrence becomes less and less. There are approximately the same number of 90 and 110 IQ scores, but far fewer 60 and 140 IQ scores.

Two values are important in explaining the normal distribution: the mean, or central value, and the standard deviation. When referring to an entire population, these values are referred to as *parameters*. When referring to a sample, they are referred to as *statistics*. The mean (average value) of a population is represented as $\mu$, and the population standard deviation, as $\sigma$. For the most part, the values of the population parameters, $\mu$ and $\sigma$, are unknown for a total "population." For example, the true mean ($\mu$) and standard deviation ($\sigma$) of the numbers of *Staphylococcus aureus* carried in the nasal cavities of humans are unknown, because one cannot readily assess this population among all humans. Hence, the statistical parameters, $\mu$ and $\sigma$, are estimated by sampling randomly from the target population. The sample mean ($\bar{x}$) and the sample standard deviation ($s$) represent unbiased estimates of the population parameters, $\mu$ and $\sigma$, respectively.

The mean $\bar{x}$ is the arithmetic average of values sampled from a population.[*] The standard deviation, $s$, describes how closely the individual values cluster around the mean value. For example, if all the measured values are within ± 0.1 g from the mean value in Test A and are within ±20.0 g from the mean value in Test B, the variability, or the scatter, of the data points in Test A is less than in Test B. The standard deviation is the value that portrays that range scatter, and does so in a very concise way. It just so happens that, in a large, normally-distributed data set, about 68% of the data are contained within ± one standard deviation ($s$) about the mean (Fig. 1.2).

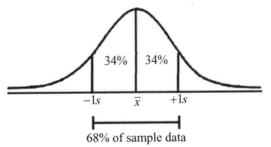

**Fig. 1.2** Normally distributed data, ± one standard deviation from the mean

So, for example, if the mean number ($\bar{x}$) of *Staphylococcus aureus* colonies on 100 tryptic soy agar plates is 100, and the standard deviation ($s$) is 20, then 68% of the plate counts are between 80 and 120 colonies per plate (Fig. 1.3).

---

[*] Also, note that, for a theoretical normal distribution, the mean will equal the median and the mode values. The median is the central value, and the mode is the most frequently occurring value.

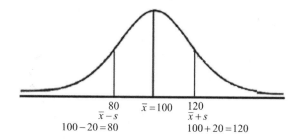

Fig. 1.3 Standard deviation of plate count values

If a second microbiologist counted colonies on the same 100 plates and also had an average plate count of $\bar{x} = 100$, but a standard deviation ($s$) of 10, then 68% of the count values would be between 90 and 110 (Fig. 1.4).

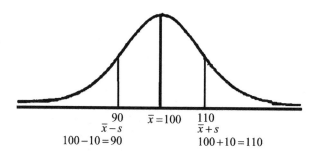

Fig. 1.4 Standard deviation of a second set of plate count values

The second microbiologist perhaps is more precise than the first in that the standard deviation, or scatter range of the data around the median, is smaller. On the other hand, he may consistently overcount/undercount. The only way to know is for both to count conjointly. Let's carry the discussion of standard deviations further.

± 1 standard deviation includes 68% of the data
± 2 standard deviations include 95% of the data
± 3 standard deviations include 99.7% of the data
Figure 1.5 provides a graphical representation.

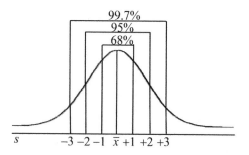

Fig. 1.5 Percentages of the area under the normal distribution covered by standard deviations

So, if one reads in a journal article that the $\bar{x}$ = 17.5 with an $s$ of 2.3, one would know that 68% of the data lie roughly between 17.5 ± 2.3, or data points 15.2 to 19.8, and 95% lie between 17.5 ±2(2.3), or 12.9 to 22.1. This gives a person a pretty good idea of how dispersed the data are about the mean. We know a mean of 15 with a standard deviation of 2 indicates the data are much more condensed around the mean than are those for a data set with a mean of 15 and a standard deviation of 10. This comparison is portrayed graphically in Fig. 1.6.

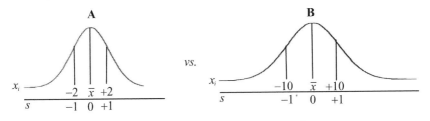

**Fig. 1.6 a** $\bar{x} = 15$ and $s = 2$, vs. **b** $\bar{x} = 15$ and $s = 10$

There is much more variability in the B group than in the A group. But then, what is variability? It is the scatter around the mean of the specific values.

Variability is measured by subtracting the mean of the data from each data point, i.e., $v = x_i - \bar{x}$. Take the data set [5, 6, 3, 5, 7, 4]. The mean is (5+6+3+5+7+4)/6 = 30/6 = 5. The variability of the data around the mean is $x_i - \bar{x}$.

$$
\begin{array}{rcr}
x_i - \bar{x} & = & v \\
5 - 5 & = & 0 \\
6 - 5 & = & 1 \\
3 - 5 & = & -2 \\
5 - 5 & = & 0 \\
7 - 5 & = & 2 \\
4 - 5 & = & -1 \\
\hline
\text{Sum} & = & 0
\end{array}
$$

The variability is merely a measure depicting how far a data point is from the mean. Unfortunately, if one adds the individual variability points, $v$, they sum to 0. This makes sense in that, if the data are distributed symmetrically about the mean, there should be the same value weights more than and less than the mean that cancel each other out. A correction factor will be introduced in the next section so that the variability points around the mean will not cancel each other out. Variability is often interchanged with the term *statistical error*. Statistical error does not mean a mistake, or that something is wrong; it means that a data point differs from the mean.

There are times when statistical variability can mean missing a target value. For example, suppose I cut three boards 36 inches long. Here, the variability of the board lengths is the difference between the actual and target value. Suppose the boards actually measure 35.25 in., 37.11 in., and 36.20 in..

$$
\begin{array}{rcrcl}
35.25 & - & 36.00 & = & -0.75 \\
37.11 & - & 36.00 & = & 1.11 \\
36.20 & - & 36.00 & = & 0.20
\end{array}
$$

I guess this is why I am not a carpenter. The biggest difference between the variability of data around a mean and variability of data around a target value is that the sum of the individual points of difference, $x - \overline{x}$, using the mean will add to 0, but using the $x$-target value usually will differ from 0.

$$
\begin{aligned}
x_i - \overline{x} &= -0.93 + 0.92 + 0.01 = 0 \\
x_i - \text{target} &= -0.75 + 1.11 + 0.20 = 0.56
\end{aligned}
$$

Let us now discuss the concepts of mean and standard deviation more formally.

## 1.2 Mean

The mean, or arithmetic average, plays a crucial role. We are interested in two classes of mean value – a population mean and a sample mean.

> $\mu$ = [mu] is the population mean. That is, it is the average of all the individual elements in an entire population – for example, the population mean number of bacteria found in the lakes of Wisconsin, or the population mean age of all the microbiologists in the world. Obviously, it is difficult, if not impossible to know the true population mean, so it is estimated by the sample mean, $\overline{x}$.

> $\overline{x}$ = [x bar, or overline x] is the sample mean, or the arithmetic average of a sample that represents the entire population. Given the data are normally distributed, the sample mean is taken to be the best point estimator of the population mean. The calculation of the sample mean is $\overline{x} = (x_1 + x_2 + \ldots + x_n)/n$, where $x_1$ = the first sample value, $x_2$ = the second sample value, and $\ldots x_n$ = the last sample value. The subscript designates the sample number. More technically, $\overline{x} = \sum_{i=1}^{n} x_i \Big/ n$

$\Sigma$ = [sigma] means "summation of." So anytime you see a $\Sigma$, it means add. The summation sign generally has a subscript, $i = 1$, and a superscript, $n$. That is, $\sum_{i=1}^{n} x_i$, where $i$, referring to the $x_i$, means "begin at $i = 1$" ($x_1$), and $n$ means end at $x_n$.

The sub- and superscripts can take on different meanings, as demonstrated in the following. For example, using the data set:

| $i$ | $x_i$ |
|---|---|
| 1 | 6 |
| 2 | 7 |
| 3 | 5 |
| 4 | 2 |
| 5 | 3 |

$i = 1$ through 5, and as the last $i$, 5 also represents $n$. $\sum_{i=1}^{n} x_i = 6+7+5+2+3$; however, if $i = 3$, then we sum the $x_i$ values from $x_3$ to $n = 5$. Hence,

$$\sum_{i=3}^{5} x_i = x_3 + x_4 + x_5 = 5+2+3$$

Likewise,

$$\sum_{i=1}^{n-1} x_i = x_1 + x_2 + x_3 + x_4 = 6+7+5+2$$

and

$$\sum_{i=1}^{3} x_i = x_1 + x_2 + x_3 = 6+7+5$$

Often, for shorthand, we will use an unadorned sigma, $\Sigma$, to represent $\sum_{i=1}^{n}$ with any changes in that generality clearly signaled to the reader.

## 1.3 Variance and Standard Deviation

As stated earlier, the variance and standard deviation are statistics representing the difference of the $x_i$ values from the $\bar{x}$ mean. For example, the mean of a data set, $\bar{x} = (118+111+126+137+148)/5 = 128$.

The individual data point variability around the mean of 128 is

$$
\begin{array}{rcr}
118 - 128 & = & -10 \\
111 - 128 & = & -17 \\
126 - 128 & = & -2 \\
137 - 128 & = & 9 \\
\underline{148 - 128} & = & \underline{20} \\
\text{Sum} & = & 0
\end{array}
$$

and the sum of the variability points $\Sigma(x_i - \bar{x})$ is zero. As previously noted, this makes sense – because the $\bar{x}$ is the central weighted value, this summation will never provide a value other than 0. So, we need to square each variability value $(x_i - \bar{x})^2$ and then add them to find their average. This average, $\dfrac{\Sigma(x_i - \bar{x})^2}{n}$, is referred to as the variance of the data.

$$\text{variance} = \frac{-10^2 + (-17)^2 + (-2)^2 + 9^2 + 20^2}{5} = \frac{874}{5} = 174.80$$

Likewise, the mean, the variance, and the standard deviation can be in terms of the entire population.

$\sigma^2 = $ [sigma squared] = the true population variance $= \dfrac{\sum(x-\mu)^2}{N}$

where $\mu = $ true population mean, and $N = $ true population size.

$\sigma = \sqrt{\sigma^2} = $ [sigma] = true population standard deviation

Again, because the entire population can rarely be known, the sample variance is computed as

$$s^2 = \frac{\sum(x-\bar{x})^2}{n-1}$$

Note that we divide by $n - 1$, not $n$. This is because we lose a degree of freedom when we estimate $\mu$ by $\bar{x}$.

$$s = \text{standard deviation} = \sqrt{s^2}$$

In hand-calculating $s$, a shortcut calculation is $s = \sqrt{\dfrac{\sum x^2 - n\bar{x}^2}{n-1}}$, which is generally faster to compute.

Two other important statistics of a data set are mode and median.

## 1.4 Mode of Sample Data

The mode is simply the most frequently-appearing numerical value in a set of data. In the set [4, 7, 9, 8, 8, 10], 8 is the mode.

## 1.5 Median of Sample Data

The median is the central numerical value in a set of data, essentially splitting the set in half. That is, there are as many individual values above it as below it. For data that are an odd number of values, it is the middle value of an ordered set of data. In the ordered data set [7, 8, 10], 8 is the median. For an ordered set of data even in the number of values, it is the sum of the two middle numbers, divided by 2. For the ordered data set [7, 8, 9, 10], the median is 8+9/2 = 8.5. This leads us directly into further discussion of normal distribution.

## 1.6 Using Normal Distribution Tables

As discussed earlier, the normal distribution is one of the most, if not the most, important distributions encountered in biostatistics. This is because it represents or models so many natural phenomena, such as height, weight, and IQs of individuals. Figure 1.7 portrays the normal distribution curve.

Because the normal distribution resembles a bell, it often is termed the "bell-shaped curve." The data have one central peak, that is, are "unimodal," and are symmetrical about the mean, median, and mode. Because of this symmetry, the mean = median = mode.

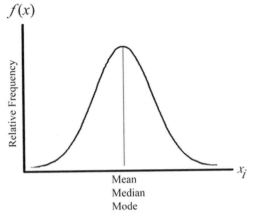

**Fig. 1.7** A normal distribution, the "bell-shaped curve"

Most statistical methods associated with normal distributions utilize the mean and the standard deviation in their calculations. We already know that 68% of the data lie between $\mu + \sigma$ and $\mu - \sigma$ standard deviations from the mean, that 95% of the data lie between the mean and two standard deviations ($\mu + 2\sigma$ and $\mu - 2\sigma$, or $\mu \pm 2\sigma$), and 99.7% of the data lie within three standard deviations of the mean ($\mu \pm 3\sigma$). These relationships describe a theoretical population, but not necessarily a small sample set using the same mean ($\bar{x}$) and standard deviation ($s$). However, they are usually very good estimators. So, whenever we discuss, say, the mean $\pm$ 2 standard deviations, we are referring to the degree to which individual data points are scattered around the mean (Fig. 1.8).

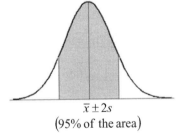

$$\bar{x} \pm 2s$$
$$\left(95\% \text{ of the area}\right)$$

**Fig. 1.8** The mean $\pm$ 2 standard deviations for a set of data

In any given sample, roughly ninety-five percent (95%) of the data points are contained within $\bar{x} \pm 2s$.

The $z$ Distribution is used to standardize a set of data points so that the mean will be zero, and the standard deviation, 1, or $\bar{x} = 0$, $s = 1$. The transformation is $z = (x_i - \bar{x})/s$. A $z = 2.13$ means the value is 2.13 standard deviations to the right of the mean. A $z = -3.17$ means the value is 3.17 standard deviations to the left of the mean. The normal $z$ Distribution table can be rather confusing to use, and we will use the Student's $t$ Distribution in its place, when possible. This will not always be the case, but for now, we will focus on the Student's $t$ Distribution. When the sample size is large, $n > 100$, the Student's $t$ Table and the normal $z$ Distribution are identical. The advantage of the Student's $t$ Distribution is that it compensates for small sample sizes. The normal curve is based on an infinite population. Because most statistical applications involve small sample sizes, certainly fewer than infinity, the normal distribution table is not appropriate, for it underestimates the random error effect. The Student's $t$ table (Table A.1) compensates for smaller samples by drawing the tails out farther and farther (Fig. 1.9).

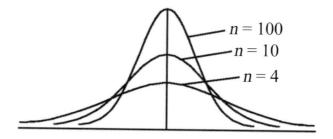

$n = 100$
$n = 10$
$n = 4$

**Fig. 1.9** Student's $t$ Distribution versus the normal distribution

To use the Student's $t$ Table (Table A.1), we need two values:
1. Sample size ($n$), and
2. $\alpha$ (alpha) level.

As you are now aware, the sample size is the number of experimental observations. The confidence level, $1 - \alpha$, is the amount of area under the distribution curve with which one is working. Generally, that value is 0.95; that is, it incorporates 95% of the area under the curve. The $\alpha$ level is the area outside the confidence area. If one uses two standard deviations, or a 95% confidence area, the $\alpha$ level is $1 - 0.95 = 0.05$. The $\alpha = 0.05$ means that 5% of the data are excluded. Note that there is nothing to figure out here. These are just statements.

In all the tests that we do using the Student's $t$ table, the sample size will always be provided, as will the degrees of freedom. The degrees of freedom will be designated as $df$. If $df = n - 1$, and $n = 20$, then $df = 20 - 1 = 19$. Or, if $df = n_1 + n_2 - 2$, and $n_1 = 10$ and $n_2 = 12$, then $df = 10 + 12 - 2 = 20$. The smaller the $df$ value, the greater the uncertainty, so the tails of the curve become stretched. In practice, this means the smaller the sample size, the more evidence one needs to detect differences between compared sets of data. We will discuss this in detail later.

The *df* value is used to find the Student's *t* tabled value at a corresponding $\alpha$ value. By convention, $\alpha$ is generally set at $\alpha = 0.05$. So, let us use Table A.1, when $\alpha = 0.05$ and for $n = 20$; hence, $df = n - 1 = 19$.

Using the *t* table:

*Step 1*. Go to Table A.1.

*Step 2*. Find *df* in the left column labeled *v*. Here, $v = 19$, so move down to $v = 19$.

*Step 3*. When you reach the 19, move right to the column corresponding with the $\alpha$ value of 0.05.

*Step 4*. Where the $v = 19$ row and the $\alpha = 0.05$ column meet, the tabled value = 1.729.

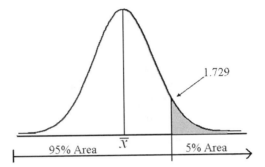

**Fig. 1.10** The *t* tabled value = 1.729, an upper-tail value

The table, being symmetrical, provides only the upper (positive), right-side *t* tabled value. The 1.729 means that a *t* test value greater than 1.729 is outside the 95% confidence area (Fig. 1.10). This is said to be an upper-tail value. A lower-tail value is exactly the same, except with a minus sign (Fig. 1.11).

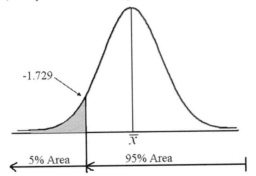

**Fig. 1.11** The *t* tabled value = –1.729, a lower-tail value

This means that any value less than –1.729 is outside the 95% region of the curve and is significant. Do not worry about the *t* test values yet. We will bring everything together in the six-step procedure.

Finally, a two-tail test takes into account error on the upper and lower confidence levels. Here, you use two values of $a$, but you divide it, $a/2$. $a = 0.05 = 0.05\ 2 = 0.25$. Using the $t$ table:

*Step 1.* Go to Table A.1.

*Step 2.* Find the $df$ in the left column labeled v. Here, $v = 19$, so move down to $v=19$.

*Step 3.* when you read the 19, move right to the column corresponding to $a/2 = 0.05\ 2 = 0.025$.

*Step 4.* Where the $v = 19$ row and the $a/2 = 0.025$ column meet, the tabled value is 2.093. This means $-0.2093$ and 2.093.

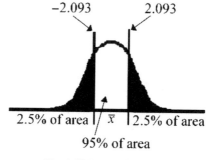

Fig. 1.12 Two-Tail Test

This table uses both values, $-2.093$ and 2.093. It is the upper and lower test jointly, with $a$ being divided by 2, $a/2$ .

## 1.7 Standard Error of the Mean

To this point, we have discussed the variability of the individual $x_i$ data around the mean, $\bar{x}$. Now, we will discuss the variability of the sample mean, $\bar{x}$ itself, as it relates to the theoretical ("true") population mean, $\mu$. The standard deviation of the mean, *not the data points*, is also termed the standard error of the mean. In most statistical tests, the means of samples are compared and contrasted, not the data points themselves. Error, in this sense, is variability. The computation for the standard error of the mean, $s_{\bar{x}}$ , is:

$$s_{\bar{x}} = \frac{s}{\sqrt{n}} \text{ or } \sqrt{\frac{s^2}{n}}$$

What does the standard deviation of the mean represent? Suppose ten sample sets are drawn randomly from a large population. One will notice that the sample mean is different from one sample set to another. This variability is of the mean, itself. Usually, one does not sample multiple sets of data to determine the variability of the mean. This calculation is done using only a single sample, because a unique relationship exists between the standard deviation of the mean and the standard deviation of the data that is proportional to data scatter. To determine the standard error of the mean, the standard deviation value of the sample data is simply divided by the square root of the sample size, $n$.

$$s_{\bar{x}} = \frac{s}{\sqrt{n}}$$

The standard deviation of the mean has the same relation to the mean as the standard deviation of data has to the mean, except it represents the variability of the mean, not the data (Fig. 1.13). That is, for the mean $\bar{x} \pm s_{\bar{x}}$, 68% of the time the true mean, $\mu$, is contained in that interval. About 95% of the time, the true mean value $\mu$ will lie within the interval, $\bar{x} \pm 2s_{\bar{x}}$. And about 99.7% of the time, the true mean value $\mu$ will lie within the interval, $\bar{x} \pm 3s_{\bar{x}}$.

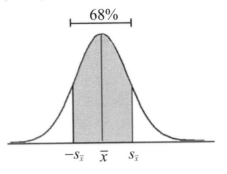

**Fig. 1.13** The standard error of the mean

Key Points

| | |
|---|---|
| $s$ = Standard deviation of the data set | $s_{\bar{x}} = s/\sqrt{n}$ = Standard deviation, not of the data set, but of the mean |

## 1.8 Confidence Intervals

Most of the work we will do with the normal distribution will focus on estimating the population mean, $\mu$. Two common approaches to doing this are: 1) a point estimate and 2) calculation of a confidence interval estimate. The point estimate of $\mu$ is simply the sample mean, $\bar{x}$. The interval estimate of $\mu$ is $\bar{x} \pm t_{(\alpha/2,\, n-1)} s/\sqrt{n}$, with a confidence level of $1 - \alpha$.

*Data Set.* Suppose you are to estimate the true mean weight of each of 10,000 containers in a single lot of bacterial growth medium. Each container is supposed to contain 1 kg of medium. Ten are randomly sampled and weighed. The weights, in grams, are

| $n$ | $x_i$ |
|---|---|
| 1 | 998 |
| 2 | 1003 |
| 3 | 1007 |
| 4 | 992 |
| 5 | 985 |
| 6 | 1018 |
| 7 | 1009 |
| 8 | 987 |
| 9 | 1017 |
| 10 | 1001 |

What are the point and interval mean estimates? Let $\alpha = 0.05$ and $df = v = n - 1$.

*Step 1.* Calculate the mean.

$$\frac{\sum x_i}{n} = \frac{998 + 1003 + \ldots + 1017 + 1001}{10} = \frac{10017}{10} = 1001.70$$

*Step 2.* Calculate the standard deviation.

$$s = \sqrt{\frac{\sum x^2 - n\bar{x}^2}{n-1}} = \sqrt{\frac{10035215 - 10(1001.7)^2}{9}} = 11.4843$$

*Step 3.* Go to Table A.1 (the Student's $t$ table).

Find $\alpha/2 = 0.05/2 = 0.025$, with $n - 1 = 10 - 1 = 9$ degrees of freedom $= v$. $t_{(0.025, 9)} = 2.262$. ($\alpha$ is divided by 2, because this is a two-tail test [to be fully discussed in detail later].)

*Step 4.* Compute the 95% confidence interval.

$$\mu = \bar{x} \pm t_{(\alpha/2, n-1)} s / \sqrt{n}$$

$$\mu = 1001.70 \pm 2.262 \left( 11.4843 / \sqrt{10} \right) = 1001.70 \pm 8.22$$

$$993.48 \leq \mu \leq 1009.92$$

This simply means that, if a large number of samples were collected, say 100, then 95 out of 100 times, the true weight mean will be contained in the interval:

$$993.48 \leq \mu \leq 1009.92$$

## 1.9 Hypothesis Testing

In statistics, one of the most, if not the most important goal is hypothesis testing. Hypothesis testing can be boiled down to three test questions. Suppose you are comparing two methods of product extraction, say Methods A and B. There are three questions we can test:

1. Does Method A produce greater extraction levels than Method B?
2. Does Method A produce lower extraction levels than Method B?
3. Do Methods A and B produce different extraction levels?

*Question 1.* Does Method A produce greater extraction levels than Method B? Notice that a connotation of "better" or "worse" is not a part of the hypothesis; that valuation is determined by the researcher. If greater extraction means better, fine, but it could also mean worse. The point is that, if this question is true, $A - B > 0$. An upper-tail test determines whether Method A produces greater extraction levels than does the second method, Method B, or simply if $A > B$.

*Question 2.* The second question, whether Method A produces lower extraction levels than does Method B, is in lower-tail form. If the answer is yes, then $A - B < 0$.

*Question 3.* The question as to whether Methods A and B are different from one another in extraction levels is a two-tail test. If the answer is "yes," then $A - B \neq 0$. Extractions by Method A do not have to be greater than those by Method B (an upper-tail test), nor do they have to be lower (a lower-tail test). They merely have to be different: $A > B$, or $A < B$.

Hypothesis testing is presented through two independent statements: the null hypothesis ($H_0$) and the alternative hypothesis ($H_A$, sometimes written as $H_1$). It is the $H_A$ statement that categorizes the tail of the test, and this should be articulated first. Do not let these concepts confuse you . . . we will make them crystal clear in coming chapters.

# Chapter 2

## One-Sample Tests

Often, a microbiologist needs to measure samples from only one population and 1) estimate its mean and standard deviation through confidence intervals, or 2) compare it to a standard. For example, through a series of dilutions, one can estimate the true population colony count, and then may want to compare that count measurement to a standard level for quality control purposes. In this chapter, we will be concerned only with normally distributed data.

## 2.1 Estimation of a One-Sample Mean

Recall from Chapter 1 that, in biostatistics, the true population mean, $\mu$, is estimated by the sample mean, $\bar{x}$, and the population standard deviation, $\sigma$, is estimated by the sample standard deviation, $s$. Recall also that the $\bar{x}$ is an unbiased estimate of $\mu$, given the sampling of the population provides a valid representation of the population.[*]

---

[*] In sampling a population, all elements of the population must be available to the sampling procedure. For example, if one wanted to identify the prevalence of Avian flu in India, the sampling must be throughout all of India, which is probably impossible from a practical viewpoint. If a microbiologist sampled from Calcutta, Delhi, and Mumbai (formerly Bombay) and stated that the sample "represented India," this statement would be erroneous. The most one could conclude would be the prevalence in these three cities. Even then, more than likely, certain destitute people would not be available to the sampling schema, so the study could not be generalized to "all individuals" in these three cities. These kinds of potential sampling bias and restriction must be evaluated before one can generalize sampled data to a larger population.

D.S. Paulson, *Biostatistics and Microbiology*, doi: 10.1007/978-0-387-77282-0_2,
© Springer Science + Business Media, LLC 2008

The variability in the data (variability of individual data points from the mean) measured by the standard deviation is really a measurement of uncertainty as to the true parameters of a population. Uncertainty is random variability inherent in any controlled experimentation grounded in the Second Law of Thermodynamics; that is, the forces at play in a controlled, nonrandom environment move it toward a non-controlled, random state. The larger the value of $s$, the greater the uncertainty.

Most of the time, one is more interested in the width of the confidence interval that contains the true mean than merely calculating the mean and standard deviation. The standard one-sample confidence interval formula is:

$$\mu = \bar{x} \pm t_{\alpha/2} \frac{s}{\sqrt{n}}$$

where $\mu$ = true, or population, mean, $\bar{x}$ = sample mean = $\sum (x)/n$, and $t_{\alpha/2}$ = Student's $t$ table value for a two-tail $\alpha$ value. If $\alpha = 0.05$, then $\alpha/2 = 0.05/2 = 0.025$. That is, 0.025 is used to identify the $t$-tabled value for a total significance level of $\alpha = 0.05$. The degrees of freedom used in the $t$ table is $n - 1$. $s$ = standard deviation = $\sqrt{\sum(x-\bar{x})^2/n-1}$ or, using the calculator formula, $s = \sqrt{\sum(x^2 - n\bar{x})^2/n-1}$, and $n$ = sample size.

A two-tail test includes both the lower and upper sides of the normal distribution (Fig. 2.1A), while a one-tail test refers either to the upper or the lower side of the normal distribution (Fig. 2.1B or Fig. 2.1C).

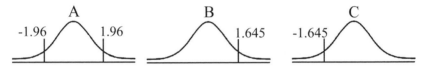

**Fig. 2.1** Normal distributions for two-tail (**A**), upper-tail (**B**), & lower-tail tests (**C**)

Let us look at the situation where $\alpha = 0.05$. For an upper-tail, as well as a lower-tail test, (Fig. 2.2A), the $\alpha$ region covers 5% of the total area of the curve. For a two-tail test, both the upper and lower regions are involved but still cover 5% of the curve, $\alpha/2 = 0.05/2 = 0.025$, or 2.5% of the normal curve in each tail (Fig. 2.2.B).

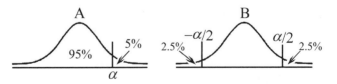

**Fig. 2.2** $\alpha$ Portion of normal distribution curve for one-tail (**a**) & two-tail (**b**) tests

*Example 2.1.* A microbiologist dispensed ten "10 mL" aliquot samples of tryptic soy broth using an automatic dispenser system. The microbiologist wants to provide a 99% confidence interval for the true dispensed volume. Hence, $\alpha = 0.01$. The volumes of the samples in milliliters are presented in Table 2.1.

**Table 2.1** Sample volumes in milliliters, Example 2.1

| $n$ | $x$ |
|---|---|
| 1 | 10.2 |
| 2 | 10.1 |
| 3 | 9.8 |
| 4 | 10.3 |
| 5 | 10.1 |
| 6 | 8.5 |
| 7 | 9.2 |
| 8 | 10.2 |
| 9 | 9.8 |
| 10 | 10.3 |

$$\Sigma x = 98.5$$

$$\bar{x} = \Sigma(x)/n = 98.5/10 = 9.85$$

$$s = \sqrt{\frac{(10.2-9.85)^2 + (10.1-9.85)^2 + \ldots + (9.8-9.85)^2 + (10.3-9.85)^2}{10-1}} = 0.58$$

$$t_{(\alpha/2,\, n-1)} = t_{(0.01/2,\, 10-1)} = t_{0.005,\, 9} = 3.250 \quad \text{(Table A.1, Student's } t \text{ Table).}$$

$$\mu = \bar{x} \pm t_{\alpha/2}\left(s/\sqrt{n}\right)$$

$$\mu = 9.85 \pm (3.250)0.58/\sqrt{10} \qquad\qquad 9.85 \pm 0.60$$

$$9.25 \leq \mu \leq 10.45 \text{, or rounded, } 9.2 \leq \mu \leq 10.4$$

This means that 99 out of 100 times, the true mean volume, $\mu$, will be contained within the interval 9.2–10.4 mL. Perhaps a new automatic dispenser system is in order.

## Key Point

In the field, most microbiologists will use a personal computer with statistical computation ability. This greatly speeds up the calculation processes and is valuable, so long as one understands the statistical reasoning behind the processes. Note that computer round-off procedures often create minor discrepancies, relative to the hand calculations.

Table 2.2 is a printout of a MiniTab® software routine. It provides the sample size, $n$, the mean, $\bar{x}$, the standard deviation, $s$ (St Dev), the standard error of the mean $s/\sqrt{n}$ (SE mean), and the 99% confidence interval (99% CI).

**Table 2.2** MiniTab® software printout of sample size, mean, standard deviation, standard error of mean, and 99% CI

| Variable | n | Mean ($\bar{x}$) | St Dev ($s$) | SE Mean $\left(s/\sqrt{n}\right)$ | 99% CI |
|---|---|---|---|---|---|
| x | 10 | 9.85000 | 0.57975 | 0.18333 | 9.25420, 10.44580 |

It is sometimes useful to plot the actual $x_i$ data points to visualize how they are spread in order to better understand the data set (Fig. 2.3).

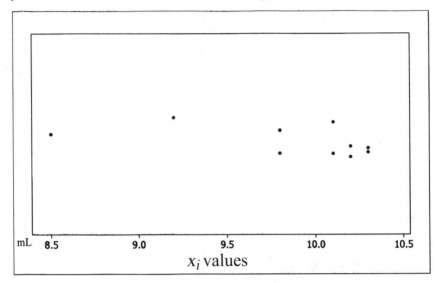

**Fig. 2.3** Individual $x_i$ plot, Example 2.1

Notice that the data are more "piled up" in the upper values. The mean value, 9.85, being the average volume, is not the center of the data set, as is the median value. Although this normal distribution discrepancy does not invalidate the confidence interval calculation, it is often worth assessing whether the values used are the correct ones and seeking to understand the microbiological reason, if any, for the results. Because there are only ten values, it is not to be expected that the mean would be located closer to the center of the data set.

*Example 2.2.* From a clinical specimen of sputum, a microbiologist wishes to estimate the number of microorganisms per milliliter of sputum. To do this, serial dilutions of the specimen were performed in triplicate, and dilutions were plated in duplicate. To replicate an experiment means to repeat the entire process (in triplicate, in this example). Performing duplicate plating is not duplicate sampling. Figure 2.4 presents a schema of the experiment, one replicate plated in duplicate, and Fig. 2.5 presents the data from three complete replicates at the $10^{-4}$ dilution.

Key Point

One of the requirements of the parametric statistics used in this text is that the data be linear. Microbial growth and the colony counts that represent it are exponential. Therefore, a $\log_{10}$ transformation of microbial colony counts must be made. For example, the $\log_{10}$ value of $10 \times 10^4 = 1.0 \times 10^5 = 5.0 \log_{10}$.

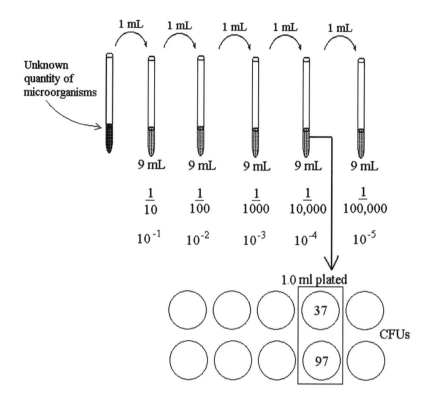

**Fig. 2.4** One replicate plated in duplicate, Example 2.2

Suppose the dilution selected is $10^{-4}$, and the duplicate plate count data from the $10^{-4}$ dilution replicates were as follows: *

| Replicate | 1 | 2 | 3 |
|---|---|---|---|
| Repeated measure (CFUs) | 37, 97 | 58, 79 | 102, 65 |
| Mean count (CFUs) | 67 | 68.5 | 83.5 |
| Next, convert to exponential | $67 \times 10^4$ | $68.5 \times 10^4$ | $83.5 \times 10^4$ |
| Multiply by exponent | 670,000 | 685,000 | 835,000 |
| $\log_{10}$ linearize the data | $\log_{10}(670,000)$ $= 5.8261$ | $\log_{10}(685,000)$ $= 5.8357$ | $\log_{10}(835,000)$ $= 5.9217$ |

The $\log_{10}$ populations from the triplicate samples are now averaged. $\bar{x} = (5.8261 + 5.8357 + 5.9217)/3 = 5.8612$ in $\log_{10}$ scale.

**Fig. 2.5** Experiment in triplicate, plated in duplicate, Example 2.2

The most probable number of organisms per milliliter $= 5.8612 \log_{10} \approx 7.25 \times 10^5$ per mL. The 95% confidence interval is $\mu = \bar{x} \pm t_{(\alpha/2;\, n-1)} s/\sqrt{n}$, where $\bar{x} = 5.8612$ and $n = 3$.

$$s = \sqrt{\frac{\sum(x - \bar{x})^2}{n-1}}$$

$$= \sqrt{\frac{(5.8261 - 5.8612)^2 + (5.8357 - 5.8612)^2 + (5.9217 - 5.8612)^2}{3-1}}$$

$$= 0.0526$$

$t_{(\alpha/2,\, n-1)} = t_{(0.05/2,\, 3-1)} = 4.303$, from Table A.1, the Student's $t$ table.

$$\bar{x} \pm t_{\alpha/2} s/\sqrt{n}$$

$$5.8612 \pm 4.303 \left(0.0526/\sqrt{3}\right)$$

$$5.8612 \pm 0.1307$$

$$5.7305 \leq \mu \leq 5.9919, \text{ which is the 95\% confidence interval.}$$

The same result is accomplished with the MiniTab® statistical software package (Table 2.3).

**Table 2.3** 95% Confidence interval, Example 2.2

| Variable | $n$ | Mean | St Dev | SE Mean | 95% CI |
|---|---|---|---|---|---|
| $x$ | 3 | 5.86117 | 0.05264 | 0.03039 | (5.73039, 5.99194) |

Please be mindful, it is important to understand the statistical process behind the software-generated data. Also, note that the 95% confidence interval represents the confidence interval for the true mean, not any particular value.

## 2.2 Comparing One Sample Group Mean to a Standard Value

Sometimes, one will want to determine if a particular sample set mean is significantly different, larger, or smaller than a standard value. This can occur, for example, in preparing a microbial aliquot or a reagent and comparing to a standard reference. The comparison can be done in two basic ways:

A. *Confidence intervals.* Confidence intervals provide an estimate of the $1 - \alpha$ range of values that may contain the true population mean, $\mu$, based on $\bar{x}$, the sample mean.

B. *t-test (use for sample vs. standard value testing).* This test compares a sample set to a standard value, but does not estimate the range within which the true mean can be found.

### 2.2.1 Confidence Interval Approach

Is there a significant difference between a test sample and a standard value in that the test sample is significantly more or less than the standard? The interval approach uses a $1 - \alpha$ confidence interval to determine if it differs from a sample interval, $\mu = \bar{x} \pm t_{(\alpha/2; n-1)} s/\sqrt{n}$ . If the standard value is contained within the sample confidence interval, they are considered not significantly different. If the standard value is outside the $1 - \alpha$ confidence interval, they are considered different at the $\alpha$ level. I will employ the six-step procedure to make certain we do not get lost.

---

**Confidence Interval Approach** (comparing a standard value to the sample mean)

*Step 1.* Formulate the hypothesis.
$H_0$: the sample mean is not different from the standard value (sample = standard)
$H_A$: the sample mean is different from the standard (sample $\neq$ standard)

*Step 2.* Determine $\alpha$, and select the sample size, $n$.
Often, these are already established from prior knowledge; if not, they need to be established now, prior to conducting the test or experiment.

*Step 3.* Select the statistical test to be used.
This is important, so as to plan the experiment/test, instead of running an experiment and then trying to determine what statistical test to use.

*Step 4.* Specify the acceptance/rejection of the alternative hypothesis, also known as the decision rule. At this time, one determines what constitutes the rejection of the null hypothesis at the specified $\alpha$ level. For this comparison of a sample and a standard, it will merely be that the sample confidence interval does not contain the standard value.

*Step 5.* Perform the experiment/test.
At this point, one actually conducts the experiment, not before, and computes the statistic based on the method selected in Step 3.

*Step 6.* Make the decision based on the computed confidence interval and the standard value. At this point, one evaluates the computed statistic on the basis of the decision rule and either accepts or rejects the $H_0$ (null) hypothesis.

---

*Example 2.3.* A microbiologist, in a time-kill kinetic study, needs to achieve a $10^8$ ($\approx 1.0 \times 10^8$) initial population of microorganisms per milliliter with a standard deviation of no more than 0.8 $\log_{10}$, prior to beginning the test. The problem, however, is that there is no way for the microbiologist to know if the initial population is $10^8$/mL prior to actually conducting the study.

The microbiologist would like to use the turbidity of a broth culture as an indicator for predicting the initial population. The microbiologist needs to do two things in this experiment. First, she/he must assure that the standard deviation of the data is within $0.80 \log_{10}$. If it is greater than $0.80 \log_{10}$, the microbiologist will not use the sample, because the standard cannot exceed $0.80 \log_{10}$, according to laboratory protocol. Second, the microbiologist wants to assure the mean $\log_{10}$ population is contained in a $1 - \alpha$ confidence interval.

*Step 1.* Specify the test hypothesis.

First, we must keep in mind that the standard deviation of the samples must not exceed $0.80 \log_{10}$. This will not be considered a hypothesis, but a restriction on the test. Let us, then, focus on the specific hypothesis.

$H_0$: mean initial population $= 10^8$ per mL, or $8.0 \log_{10}/\text{mL}$

$H_A$: $\log_{10}$ mean initial population $\neq 8.0 \log_{10} /\text{mL}$

*Step 2.* Set $\alpha$ and the sample size. We will set $\alpha = 0.05$, and the appropriate degree of broth turbidity will be established on the basis of nephelometer measurement and plating five samples of a suspension. Hence, $n = 5$. Note this is a two-tail test; therefore, $a = \alpha/2 = 0.05/2 = 0.025$

*Step 3.* Write the test statistic to be used. We will use $\mu = x \pm t_{(\alpha/2;\, n-1)} s / \sqrt{n}$ , where $t_{(\alpha/2;\, n-1)}$ is the tabled $t$ value in Table A.1.

*Step 4.* Formulate the decision rule. Because we must calculate $s$ to get the confidence interval of the sample, this is a good place to evaluate the restriction on $s$. So, if $s > 0.8$, do not continue. If $s$ is not greater than $0.8$, compute the confidence interval. If the $1 - \alpha$ interval, $\bar{x} \pm t_{(\alpha/2;\, n-1)} s / \sqrt{n}$ , does not contain $8.0 \log_{10}$ population, reject the $H_0$ hypothesis at $\alpha = 0.05$.

$$t_{(\alpha/2;\, n-1)} = t_{(0.05/2;\, 5-1)} = t_{0.025;\, 4} = 2.776, \text{ from Table A.1.}$$

*Step 5.* Perform the study.

The microbiologist plates three different turbidity levels, replicating each level five times. Based on the data in Table 2.4, Suspension 2 appears to be the optimal choice.

**Table 2.4** Microbial suspension sample counts $(\log_{10})$[a]

|           | Suspension |      |      |
|-----------|------------|------|------|
|           | 1          | 2    | 3    |
| Replicate |            |      |      |
| 1         | 7.21[a]    | 8.31 | 9.97 |
| 2         | 6.97       | 8.17 | 8.83 |
| 3         | 7.14       | 7.93 | 9.21 |
| 4         | 7.28       | 8.04 | 9.19 |
| 5         | 6.93       | 7.97 | 9.21 |

[a] Average of counts (each sample plated in duplicate)

Suspension 2 Data:

$\bar{x} = 8.084$

$s = 0.1558 \not> 0.8$; because the computed standard deviation is much less than the tolerance limit, the microbiologist continues.

$$\bar{x} \pm t_{(\alpha/2;\, n-1)} s / \sqrt{n} = 8.084 \pm 2.776 \left(0.1558 / \sqrt{5}\right) = 8.084 \pm 0.1934$$

$7.8906 \leq \mu \leq 8.2774$                    $7.89 \leq \mu \leq 8.28$

7.89 is the lower boundary, and 8.28 is the upper boundary of the 95% CI.

*Step 6.* Because 8.0 is within the 95% confidence interval (7.89–8.0–8.28), the initial population for Suspension 2 is considered not different from 8.00 at $\alpha = 0.05$. Also, the test standard deviation is well within the limit of 0.8 $\log_{10}$.

Some may argue with this conclusion because two samples were not at least 8.0 $\log_{10}$/mL. If the standard is based on the average value, then Suspension 2 should be used. If not, Suspension 3 provides values that each meet the minimum 8.0 $\log_{10}$ requirement, but we may need to assure that $s \ngtr 0.8 \log_{10}$.

## 2.2.1.1 Lower-Tail Test: Determining if a Sample Set Mean is Less than a Standard: Confidence Interval Approach

One can also determine if, on the average, a sample set of values is less than a standard value. The test range for this is $\bar{x}$, $\bar{x} - t_{(\alpha, n-1)} (s/\sqrt{n})$, with $df = n - 1$. For example, suppose that, from our Example 2.3, the goal was to assure that the sample population average was not less than the 8.0 $\log_{10}$/mL standard requirement.

*Step 1.* Specify the test hypothesis.

$H_0$: $\bar{x} \geq 8$ (the sample mean is at least 8.0 $\log_{10}$/mL)

$H_A$: $\bar{x} < 8$ (the sample mean is less than 8.0 $\log_{10}$/mL)

*Step 2.* Set $\alpha$ and specify $n$.

$\alpha = 0.05$

$n = 5$

*Step 3.* Write out the test statistic to be used.

$\bar{x} ; \bar{x} - t_{(\alpha, n-1)} (s/\sqrt{n})$

In other words, the interval from the mean value to the lower bound.

*Step 4.* State the decision rule.

If the test statistic interval calculated is, on the average, less than 8.0 $\log_{10}$ (lower bound of confidence interval), the mean $\bar{x}$ and $\bar{x} - t_{(\alpha, n-1)} (s/\sqrt{n})$ interval will not include 8.0, and $H_0$ is rejected at $\alpha = 0.05$.

*Step 5.* Perform the experiment.

At $\alpha = 0.05$, for a one-tail confidence interval with $n - 1$, or $5 - 1 = 4$ degrees of freedom, $t_\alpha = 2.132$ (from Table A.1). We already have the data necessary to our calculations: $\bar{x} = 8.084$, $s = 0.1558$, and $n = 5$.

$8.084; 8.084 - 2.132 (0.1558/\sqrt{5}) = 8.084 - 0.149 = 7.935$

The interval ranges from 8.084 to 7.935 $\log_{10}$/mL.

*Step 6.* Make decision.

Because 8.0 is included in the sample interval, one cannot reject $H_0$ at $\alpha = 0.05$.[*]

---

[*] This is the average sample value. However, if one wants to assure that the average can be no less than 8.0 $\log_{10}$/mL, the $H_0$ hypothesis is rejected, because 7.935 is lower than 8.0. It is important to identify what one means in statistical testing.

## 2.2.1.2 Upper-Tail Test: Determining if a Sample Set Mean is Greater than a Standard: Confidence Interval Approach

The same approach used in the lower-tail test can be used in upper-tail tests, except the test statistic is $\bar{x}, \bar{x} + t_{(\alpha, n-1)} (s/\sqrt{n})$. Suppose the microbiologist could have an initial population average no greater than 8.0 $\log_{10}/mL$.

*Step 1.* State the hypothesis.

$H_0$: $\bar{x} \leq 8.0$ $\log_{10}/mL$ (sample mean is less than or equal to the standard)

$H_A$: $\bar{x} > 8.0$ $\log_{10}/mL$ (sample mean is greater than the standard)

*Step 2.* Set $\alpha$, and select $n$.

$\alpha = 0.05$

$n = 5$

*Step 3.* Write out the test statistic.

$$\bar{x}; \ \bar{x} + t_{(\alpha, n-1)} (s/\sqrt{n})$$

*Step 4.* Make the decision rule.

If the computed test interval contains 8.0, one cannot reject $H_0$ at $\alpha$.

*Step 5.* Perform the experiment.

$t_{(0.05; 4)} = 2.132$ (as before), a one-sided tabled value from Table A.1.

$\bar{x} = 8.084$

$s = 0.1558$

$8.084; 8.084 + 2.132 (0.1558/\sqrt{s}) = 8.233$

The interval ranges from 8.084 to 8.233 $\log_{10}/mL$.

*Step 6.* Make the decision.

Because the upper confidence interval does not contain 8.0, reject $H_0$ at $\alpha = 0.05$. There is good reason to believe the sample set mean is greater than 8.0 $\log_{10}/mL$ of the initial population. Here, the microbiologist would be better advised to choose turbidity sample 1.

## 2.2.2 Use of the Student's $t$ Test to Make the Determination of a Sample Mean Different, Less than, or Greater than a Standard Value

The basic formula for making this determination is:

$$t_c = \frac{\bar{x} - c}{\dfrac{s}{\sqrt{n}}}$$

where $t_c$ = calculated $t$ value, $\bar{x}$ = sample mean, $c$ = standard value, $s$ = standard deviation of the sample, and $n$ = sample size.

### 2.2.2.1 Two-Tail Test

Let us continue with the data from Example 2.3. Here, recall that 8.0 $\log_{10}$ microorganisms per milliliter was the standard for the initial population. Now, for the two-tail $t$ test, we want to know if the sample mean population and the standard differ significantly. We can use the six-step procedure to determine this.

*Step 1.* State the hypothesis.

$H_0$: $\bar{x}$ = the standard of 8.0 $\log_{10}$/mL

$H_A$: $\bar{x} \neq$ the standard of 8.0 $\log_{10}$/mL

*Step 2.* Set $\alpha$, and specify $n$.

As before, we will use $\alpha = 0.05$ and $n = 5$.

*Step 3.* Write out the test statistic calculated ($t_c$).

$$t_c = \frac{\bar{x} - c}{\dfrac{s}{\sqrt{n}}}$$

*Step 4.* Write out the decision rule, based on the $t$-tabled value ($t_t$).

$t_t = t_{tabled} = t_{(\alpha/2;\ n-1)}$

$t_t = t_{(0.025,\ 4)} = 2.776$ from Table A.1.

Because this is a two-tail test, we have both lower and upper tabled values to consider, or $t_t = -2.776$ ($t$-tabled lower value) and 2.776 ($t$-tabled upper value). So, if $t_c > 2.776$ or $t_c < -2.776$, reject $H_0$ at $\alpha = 0.05$ (Fig. 2.6).

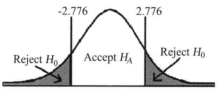

**Fig. 2.6** Example 2.3, Step 4, two-tail test diagram

*Step 5.* Conduct the experiment.

We already know that:

$\bar{x} = 8.084$ $\log_{10}$/mL

$s = 0.1558$ $\log_{10}$/mL

$n = 5$

$c = 8.0$ $\log_{10}$/mL

$$\frac{8.084 - 8.0}{\dfrac{0.1558}{\sqrt{5}}} = \frac{0.084}{0.0697} = 1.2052$$

*Step 6.* Make the decision.

Because $t_c = 1.2052 \ngtr 2.776$, one cannot reject $H_0$ at $\alpha = 0.05$. There is no strong evidence that the sample and control values differ.

## 2.2.2.2 Lower-Tail Test

For this test, we want to be sure that the sample mean $\bar{x}$ is not less than the standard, 8.0 $\log_{10}$/mL.

*Step 1.* State the hypothesis.

$H_0$: $\bar{x} \geq$ standard (true population mean is at least 8.0 $\log_{10}$/mL)

$H_A$: $\bar{x} <$ standard (true population mean is less than 8.0 $\log_{10}$/mL)

*Step 2.* Specify $\alpha$, and set $n$.

$\alpha = 0.05$

$n = 5$

*Step 3.* Write out the test statistic.

$$t_c = \frac{\bar{x} - c}{\frac{s}{\sqrt{n}}}$$

*Step 4.* State the decision rule.

Because this is a lower-tail test, $t_t$ will be a negative tabled value from Table A.1.

$t_t = t_{(\alpha;\, n-1)} = t_{(0.05;\, 4)} = 2.132$, or $-2.132$ (Fig. 2.7).

So, if $t_c < -2.132$, reject $H_0$ at $\alpha = 0.05$.

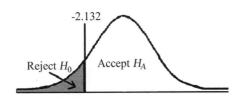

**Fig. 2.7** Example 2.3, Step 4, lower-tail test diagram

*Step 5.* Conduct the experiment.

$\bar{x} = 8.084\ \log_{10}$/mL

$s = 0.1558\ \log_{10}$/mL

$n = 5$

$c = 8.0\ \log_{10}$/mL

$$t_c = \frac{8.084 - 8.0}{\frac{0.1558}{\sqrt{5}}} = 1.2052$$

*Step 6.* Make the decision rule.

Because $t_c = 1.2052 \not< -2.132$, one cannot reject $H_0$ at $\alpha = 0.05$. There is no evidence that the sample mean is less than the standard at $\alpha = 0.05$.

### 2.2.2.3 Upper-Tail Test

*Step 1.* State the hypothesis.

$H_0$: $\bar{x} \leq$ standard of 8.0 $\log_{10}$/mL

$H_A$: $\bar{x} >$ standard of 8.0 $\log_{10}$/mL

*Step 2.* Set $\alpha$ and $n$.

As before, let us set $\alpha = 0.05$ and $n = 5$.

*Step 3.* Write out the test statistic calculated.

$$t_c = \frac{\bar{x} - c}{\frac{s}{\sqrt{n}}}$$

*Step 4.* State decision rule.

If $t_c > t_t$, reject $H_0$ at $\alpha$.

$t_t = t_{(0.05, 4)} = 2.132$ from Table A.1 (Fig. 2.8).

So, if $t_c > 2.132$, reject $H_0$ at $\alpha = 0.05$.

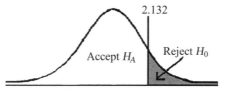

**Fig. 2.8** Example 2.3, Step 4, upper-tail test diagram

*Step 5.* Conduct the experiment.

$\bar{x} = 8.084 \ \log_{10}/mL$

$s = 0.1558$

$n = 5$

$c = 8.0 \ \log_{10}/mL$

$$t_c = \frac{8.084 - 8.0}{\frac{0.1558}{\sqrt{5}}} = 1.2052$$

*Step 6.* Make the decision.

Because $t_c = 1.2052 \not> 2.132$, one cannot reject $H_0$ at $\alpha = 0.05$. Although there is not enough evidence to claim that $8.084 > 8.0$, it can be. This is not the conclusion reached using the confidence interval approach, which ranged from 8.084 to 8.204 $\log_{10}/mL$.

Key Point

REMEMBER: CONFIDENCE INTERVALS ESTIMATE PARAMETERS,

AND STATISTICAL TESTS TEST PARAMETERS

The reason for this is that the confidence interval approach provides an estimate of the $1 - \alpha$ range of the true mean in terms of an interval. In this case, it is the upper one-half of the interval that did not contain the standard. Yet, when the standard is compared or "tested" to the sample, the sample is not significantly greater than the standard at $\alpha$. This subtlety must be recognized. Tests test if differences exist;

confidence intervals predict the region where one expects the true population mean to be found at $1 - \alpha$ confidence.

If 8.0 is a control value, it would be wiser to use the confidence interval approach. If one wants to determine if 8.084 is greater than 8.0, and it turns out to be, there is no penalty; then, the significance test can be used.

## 2.3 Determining Adequate Sample Sizes for One-Sample Statistical Tests

A fundamental consideration of statistics is determination of an adequate sample size. This is exceedingly important in order to detect differences that are significant. So the question of interest, when using a $t$ test or other statistic, is whether the sample size is large enough to detect a significant difference if one is present. The smaller the sample size, the greater the difficulty in detecting differences between a sample and a standard value, because the confidence interval of the sample increases in range due to increasing uncertainty. The $t$ test and other hypothesis-testing statistics are weighted heavily to guard against Type I, or alpha ($\alpha$) error (stating a difference exists when one does not) at the expense of committing a Type II, or beta ($\beta$) error (stating no significant difference exists, when it actually does).

So, at the beginning of any experiment, one must define microbiologically just what constitutes "a significant difference" between a sample set average value and a control value. For example, if the sample population differs from a standard by 1.0 $\log_{10}$, that may be important, but what about 0.5, 0.1, or 0.001 $\log_{10}$? The value for the $\alpha$ error often is set at $\alpha = 0.05$ in microbiology. An $\alpha$ of 0.05 (a Type I error 5 times out of 100) is good for most general work, but what about flying on a plane? No one would fly at that risk level; it is too high. The $\alpha$ value would have to be more like 0.0000001, at least for me.

Also, note that what microbiologists call significant difference (technically termed detection level) is inversely related to sample size. It is the value chosen as being important for the statistician to detect. As the sample size increases, the detectable difference between a standard and a sample gets smaller, because the test becomes more sensitive. That is, one is able to detect smaller and smaller differences. Of course, economics comes into play, in that one is limited by what sample size can actually be accommodated, due to cost, supplies, or logistics, so balance is necessary.

The variability, or standard deviation of the sample set is also important. The more variable the data – that is, the greater the spread between the individual data points and the sample mean – the greater is the standard deviation, and the greater the sample size must be in order to detect important differences. The same goes for the alpha level. An $\alpha$ of 0.05 requires a larger sample size than does an $\alpha$ of 0.10 to conclude significant difference.

## 2.3.1 Quick Sample Size Formula: Sample Set Mean Versus a Standard Value

A quick formula for determining sample size is

$$n \geq \frac{z_{\alpha/2}^2 s^2}{d^2}$$

where $n$ = sample size, $z$ = normal distribution tabled value[*], $d$ = detection level considered to be important, and $s$ = standard deviation of the sample data, based on previous knowledge.[**]

Key Point

---

Statistical tests are but an extension of your thinking. Ground results of analyses in your knowledge of microbiology.

---

**Table 2.5** Brief z value table for sample size determination

| $\alpha$ | $z_{\alpha/2}$ |
|---|---|
| 0.001 | 3.291 |
| 0.0025 | 3.090 |
| 0.005 | 2.807 |
| 0.01 | 2.576 |
| 0.025 | 2.326 |
| 0.05 | 1.960 |
| 0.10 | 1.645 |
| 0.25 | 1.282 |

*Example 2.4.* A microbiologist wants to be able to detect a 0.1 $\log_{10}$ difference from a standard value. Anything less than 0.1 $\log_{10}$, in her opinion, will not be of value. So, she sets $\alpha = 0.01$, finds that $z_{\alpha/2} = z_{0.005} = 2.576$ (Table 2.5), and lets $s = 0.1558$, based on previous work. Therefore, $n \geq (2.576^2(0.1558)^2)/0.1^2 = 16.1074$. Hence, $n \geq 17$ (whenever there is a remainder, round up). A sample size of 17 should enable the microbiologist to detect 0.1 $\log_{10}$ differences between the standard value and the sample mean.

---

[*] To be completely correct in sample size determination, the $t$ distribution can be used, but it requires that one perform a series of iterations. A more straight-forward approach based on the $z$ distribution is presented here that, in practice, works satisfactorily for the vast majority of microbiological applications. For the moment, we will use Table 2.5 to represent the $z$ normal distribution table. We will look at the use of the $t$ distribution later in this chapter.

[**] Here, a practical problem exists in determining the sample size prior to running the experiment. Before the sample size can be estimated, one must know the variability (variance $[s^2]$) of the data. But one most often cannot know that until the experiment is run ☹. The variance must be estimated based on other similar experiments, if possible. If none exist, guess. But be conservative, and use your experience in microbiology. After the experiment has been completed, recalculate the sample size based on the known variance ($s^2$) for future reference.

## 2.4 Detection Level

Also of importance is determining a relevant detection level, $d$. A good microbiologist cannot be satisfied with a sample size that provides an arbitrary value, so s/he will want to know what the limit of detection is for a given sample size. The basic sample size formula can easily be restated to determine the detection level.

$$n = \frac{z_{\alpha/2}^2 s^2}{d^2}$$

$$d^2 = \frac{z_{\alpha/2}^2 s^2}{n}$$

$$d = \sqrt{\frac{z_{\alpha/2}^2 s^2}{n}}$$

Suppose $n = 10$, $\alpha = 0.01$, and $s = 0.1558$.

$$d = \sqrt{\frac{2.576^2 (0.1558)^2}{10}} = 0.1269$$

The detection level is 0.13 $\log_{10}$. So, if a difference between a sample set and a standard value is 0.13 $\log_{10}$ or greater, it will be detected.

## 2.5 A More Accurate Method of Sample Size Determination

The use of the Student's $t$ distribution (Table A.1) provides a more accurate estimation of the sample size, $n$, but also is a more difficult computation, requiring an iterative process. The basic sample size formula, adapted to the Student's $t$ distribution, is:

$$n \geq \frac{t_{(\alpha/2; n-1)}^2 s^2}{d^2}$$

The problem is the $n - 1$ degrees of freedom value that is required for determining the $t$ value, when one does not know what the sample size, $n$, should be. To solve this problem, one uses a "seed" value of $n$ that is excessively large, not smaller than what one suspects will be the actual $n$, because the algorithm will solve the problem faster. The value of $n$ predicted from solution of that first iteration becomes the new seed value for the next, and so on. This process continues until a $n$ seed value that is calculated equals the $n$ value of the previous iteration.

### Key Point

For preliminary work:
– Use the $z$ calculation method
– Estimates as "worst case"
Once one has a reliable value of $s$ that can be applied, one can switch to the $t$ table method. This, of course, presumes the experiment in question will be repeated often enough to be worth the extra effort.

The $z$ computation is easy, but if one does not have a good estimate of a standard deviation before the experiment is conducted, neither computation will be of much value. However, once the study has been conducted, recompute the sample size calculation, using the actual $s$ value. For example, suppose on completion of the above study, $s = 0.317$. The sample size chosen was 10, but with the knowledge of $s$ being greater than the original estimate, one would like to know how this increase affected the experiment. Computing the actual detection level is of great value here.

$$d = \sqrt{\frac{t^2_{(\alpha/2;\, n-1)}s^2}{n}} = \sqrt{\frac{3.250^2(0.317)^2}{10}} = 0.3258,$$

which is unacceptably larger than the expected level of detection. This has greatly reduced the statistic's ability to detect significant differences.

The problem is that, with this experiment, the actual standard deviation is $0.317/0.1 = 3.17$ times larger than the desired detection level. To have a detection level of $0.1 \log_{10}$ will require the sample size to be larger, probably much larger. In this type of situation, it may be better to retrain personnel or implement automation such that the experiments can be performed with a much lower standard deviation, instead of ratcheting up the sample size. Finally, one can always increase the value of $d$ (minimum detectable difference), but accepting that this will make the experiment less accurate.

Let us now, using the iterative process and the Student's $t$ distribution (Table A.1), compute the sample size required. What should the sample size for $\alpha = 0.01$, $s = 0.317$, and $d = 0.10 \log_{10}$ be? First, we overestimate $n$, such as $n = 150$. We stop the iterative process once the current sample size calculated equals that previously calculated.

Sample size formula: $n \geq \dfrac{t^2_{(\alpha/2;\, n-1)}s^2}{d^2}$

*Iteration 1.*    $t_{(0.01/2;\, 150-1\,=\,149)} = 2.576$ (Table A.1), so
$n_1 = [2.576^2(0.317)^2]/0.1^2 = 66.68 \approx 67$

*Iteration 2.*    $t_{(0.01/2;\, 67-1\,=\,66)} = 2.660$, so
$n_2 = [2.660^2(0.317)^2]/0.1^2 = 71.10 \approx 72$

Next, the calculated $n$ value for Iteration 2 is used as the tabled $n$ for Iteration 3.

*Iteration 3.*    $t_{(0.01/2;\, 71-1\,=\,70)} = 2.660$, so
$n_3 = [2.660^2(0.317)^2]/0.1^2 = 71.10 \approx 72$

Because $n_3 = 72$ and $n_2 = 72$, the iterative process ceases. The experiment needs about 72 people to have a resolution of $0.1 \log_{10}$.

Let us compare the $t$-tabled sample size calculations with those calculated earlier via the $z$ distribution using the data from Example 2.3.

$s = 0.1558$ $\qquad\qquad\qquad$ $d = 0.10 \log_{10}$ $\qquad\qquad\qquad$ $\alpha = 0.01$

Let us first estimate $n = 25$.

*Iteration 1.*    $t_{(0.01/2,\ 25-1=24)} = 2.797$ (Table A.1)
$n_1 = [2.797^2(0.1558)^2]/0.10^2 = 18.9898 \approx 19$

*Iteration 2.*    $t_{(0.01/2,\ 19-1=18)} = 2.878$
$n_2 = [2.878^2(0.1558)^2]/0.10^2 = 20.1056 \approx 21$

*Iteration 3.*    $t_{(0.01/2,\ 21-1=20)} = 2.845$
$n_3 = [2.845^2(0.1558)^2]/0.10^2 = 19.6471 \approx 20$

*Iteration 4.*    $t_{(0.01/2,\ 20-1=19)} = 2.861$
$n_4 = [2.861^2(0.1558)^2]/0.10^2 = 19.8688 \approx 20$

The correct sample size is 20, using the $t$ table approach. As you will recall, the $z$ table estimates $n = 17$.

The $t$ table method requires three more replicate calculations than did the $z$ computation, because the $z$ table is a large-sample table, where $n > 120$. The $t$ table is more accurate and reliable when dealing with smaller samples.

## 2.6 (Optional) Equivalency Testing

All the statistical work that we have done so far is for determining if a significant difference exists between a sample and a standard; that is, if a sample and a standard are significantly different at the preset $\alpha$ level. If one cannot detect a statistical difference using the test, this does not necessarily translate as "equivalent," because testing is against the alternative hypothesis that a difference exists (Fig. 2.9).

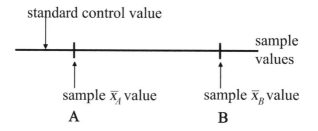

**Fig. 2.9** Difference testing

In the case of **A**, where the standard and sample values are close, the test would probably conclude no difference existed between the standard and sample mean value **A**. That is, the $H_0$ hypothesis of no difference would not be rejected. Although the test concludes they are not statistically different, they may actually be significantly different. The sample size may have been too small to detect a significant difference at a given $\alpha$ level. When the $\bar{x}$ value is **B**, the $H_0$ hypothesis of no difference would probably be rejected, because the sample mean **B** is relatively far from the standard. From a normal curve perspective, the relationship between standard and means **A** and **B** can be portrayed, as in Fig. 2.10.

The statistical testing we have discussed thus far is designed to evaluate differences and is conservative in making difference claims. If **A** is not within the detection limit of the standard, it is simply considered not different from the standard, as in Fig. 2.10. However, if one wants to test for equivalence, a different statistical approach must be employed.

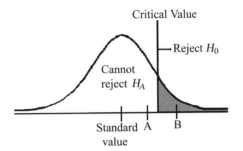

**Fig. 2.10** Normal curve: Standard value versus means **A** and **B**

Equivalence testing, a relatively new family of methods in biostatistics, is designed, for example, to determine if the effect of a generic drug, $X$, is equivalent to that of drug $Y$. That is, are they interchangeable? Unlike difference testing, where accepting $H_0$ is considered "not different," in equivalence testing, $H_0$ is reversed with $H_A$, so $H_0$ is the difference hypothesis, and $H_A$ is the equivalence one. For example, using a two-tail test for difference testing, as we have been doing, the hypothesis is:

$H_0$: $\bar{x}$ = standard

$H_A$: $\bar{x} \neq$ standard

But for equivalence testing, there are two $H_0$ hypothesis tests that are compared to a preset standard value, $\Delta$. The standard may be $\bar{x} \pm 0.05$, or some other range value.

This provides an upper and lower limit for $\Delta$. The hypotheses are:

$H_{0_L}$ : $\bar{x} < \Delta_L$, where $\Delta_L$ = lower limit

$H_{0_U}$ : $\bar{x} > \Delta_U$, where $\Delta_U$ = upper limit

$H_A$: $\Delta_L \leq \bar{x} \leq \Delta_U$

In equivalence testing, we are particularly interested in two parameters:

$\Delta$ = set limit of equivalence (any value more extreme is considered different), and often, this is considered to be ± 20% of the standard target value of interest, termed $L$.

$d$ = difference between the $\bar{x}$ value and the standard set value, $L$;

that is, $\bar{x} - L = d$

There are three different tests in equivalence testing: nonsuperiority, noninferiority, and equivalence, which incorporates both the nonsuperiority and the noninferiority tests. We will discuss this in greater detail later.

The basic test formulas are:

| Nonsuperiority | Noninferiority | Equivalence |
|---|---|---|
| $t_{cs} = \dfrac{\Delta - d}{s_x}$ | $t_{ci} = \dfrac{d - (-\Delta)}{s_x}$ | $t_{cs} = \dfrac{\Delta - d}{s_x}$ and $t_{ci} = \dfrac{d - (-\Delta)}{s_x}$ |

where $L$ is the target value and $\Delta$ is the target value tolerance.
$d = (\sum x_i / n) - L$, and $s_x = s/\sqrt{n}$ is the standard error of the $\bar{x}$. Tabled values are, as before, drawn from the Student's $t$ table (Table A.1).

We will outline this analysis, as always, using the six-step procedure.

*Step 1.* Formulate the hypothesis.

| Nonsuperiority | Noninferiority | Equivalence (This requires 2 one-sided tests.) |
|---|---|---|
| $H_0$: $d > \Delta$ <br> $H_A$: $d \leq \Delta$ <br> If $H_0$ is rejected, conclude that the sample group is nonsuperior to the standard (not larger) at $\alpha$. | $H_0$: $d < \Delta$ <br> $H_A$: $d \geq \Delta$ <br> If $H_0$ is rejected, conclude that the sample group is noninferior to the standard (larger) at $\alpha$. | $H_0$: $d > \Delta_U$; $H_0$: $d < \Delta_L$ <br> $H_A$: $\Delta_L \leq d \leq \Delta_U$ <br> If $H_0$ is rejected, the sample group is equivalent to the standard. |

*Step 2.* Set $\alpha$ and specify $n$.
*Step 3.* Write test statistic.

| Nonsuperiority | Noninferiority | Equivalence |
|---|---|---|
| $t_{cs} = \dfrac{\Delta - d}{s_x}$ | $t_{ci} = \dfrac{d - (-\Delta)}{s_x}$ | $t_{cs} = \dfrac{\Delta - d}{s_x}$ and <br><br> $t_{ci} = \dfrac{d - (-\Delta)}{s_x}$ |
| $t_{t(\alpha, n-1)}$ <br> where <br> $d = \bar{x} - L$ <br> and <br> $s_x = s/\sqrt{n}$, standard error of the $\bar{x}$ value <br> $n$ = number of $x_i$s | $t_{t(\alpha, n-1)}$ <br> where <br> $d = \bar{x} - L$ <br> and <br> $s_x = s/\sqrt{n}$, standard error of the $\bar{x}$ value <br> $n$ = number of $x_i$s | $t_{t(\alpha, n-1)}$ <br> where <br> $d = \bar{x} - L$ <br> and <br> $s_x = s/\sqrt{n}$, standard error of the $\bar{x}$ value <br> $n$ = number of $x_i$s |

*Step 4.* Write out the decision rule.

| Nonsuperiority | Noninferiority | Equivalence |
|---|---|---|
| If $t_{c_s} > t_t$, reject $H_0$ and conclude nonsuperiority at $\alpha$. | If $t_{c_i} > t_t$, reject $H_0$ and conclude noninferiority at $\alpha$. | If $t_{c_s} > t_t$ and $t_{c_i} > t_t$, reject $H_0$ at $\alpha$ and accept equivalence. |

*Step 5.* Perform the experiment.
*Step 6.* Make the decision to accept or reject $H_0$.

## 2.6.1 Nonsuperiority Test

*Example 2.5 – Testing for Nonsuperiority.* Nonsuperiority tests are less common than the other two tests (noninferiority and equivalence), but still are used. In a microbiological food laboratory, a food mixer is cleaned with an antimicrobial dishwashing detergent containing the antibacterial agent, Triclosan. It was found that, after several cleanings with the detergent, a Triclosan residual layer had accumulated on the stainless steel surfaces and may have inhibited bacterial growth, biasing the experiments, as well as contaminating the food. Unfortunately, the cleaning standards procedure was written using the Triclosan product, and many earlier experiments had been conducted evaluating the procedure. The goal of the microbiologist was to design a post-wash rinsing procedure that would assure that build-up of Triclosan did not exceed a set limit, $L$, the standard, plus a 20% of $L$ tolerance, $\Delta$. The microbiologist proposed to perform a one-sided nonsuperiority test, comparing samples post-wash with the Triclosan dishwashing detergent with samples following a wash and rinse with an assortment of surfactants, including Triton X-100, and a high pressure dishwasher water spray cycle. To prevent excess round-off error, all Triclosan removal values and the $\Delta$ value were multiplied by 100.
 *Step 1.* State hypothesis.
 This is a nonsuperiority test. The Triclosan residual was not expected to diminish with repeated washing.

> $H_0$: $d > \Delta$, the mean residual of Triclosan exceeds the standard allowable level
> $H_A$: $d \leq \Delta$, the mean residual of Triclosan does not exceed the set limit (nonsuperior)

*Step 2.* Set $\alpha$ and $n$.
   $\alpha = 0.05$ and $n = 10$ replicates
*Step 3.* State the test statistic.

$$t_{c_s} = \frac{\Delta - d}{s_x}$$

*Step 4.* State the acceptance/rejection criterion.

If $t_{c_s} \geq t_t$, reject $H_0$ at $\alpha = 0.05$. The Triclosan residual does not exceed (is nonsuperior to) the baseline standard.

$t_t = t_{t(\alpha, n-1)} = t_{t(0.05, 10-1=9)} = 1.833$ (Table A.1).

*Step 5.* Perform experiment.

The residual amounts of Triclosan contained in the ten samples were each multiplied by 100, because the amounts were very small (Table 2.6). The baseline standard was determined to be $4.8 = L$. The limit was set at 20% of the standard. So 20% would be considered the limit, $\Delta$, $(0.20 \times 4.8) = 0.96$.

**Table 2.6** Experimental results, Example 2.5

| $n$ | $x_i$ |
|---|---|
| 1 | 5.7 |
| 2 | 7.1 |
| 3 | 6.8 |
| 4 | 3.9 |
| 5 | 5.5 |
| 6 | 7.0 |
| 7 | 8.1 |
| 8 | 3.5 |
| 9 | 8.2 |
| 10 | 7.3 |

$\sum x_i = 63.1$ and $\bar{x} = 6.31$

$d = \bar{x} - L$

$d = 1.51$

$d = 6.31 - 4.8$

$$s^2 = \frac{\sum(x_i - \bar{x})^2}{n-1} = \frac{(5.7-6.31)^2 + (7.1-6.31)^2 + \ldots + (7.3-6.31)^2}{10-1} = 2.6477$$

$s = 1.627$

$$s_x = \frac{s}{\sqrt{n}} = \frac{1.627}{\sqrt{10}} = 0.515 \qquad t_{cs} = \frac{\Delta - d}{s_x} = \frac{0.96 - 1.51}{0.515} = -1.068$$

*Step 6.* Decision.

Because $t_{c_s} = -1.068 \ngeq t_t = 1.833$, one cannot reject $H_0$ at $\alpha = 0.05$. There is evidence to conclude the post-wash rinsing method does not prevent the Triclosan from building up (standard). Hence, the residual is greater than the baseline. The baseline standard and post-wash samples are not equivalent at $\alpha = 0.05$.

## 2.6.2 Confidence Interval Approach to Superiority/Inferiority Testing

*Nonsuperiority Test.* A $1 - \alpha$, one-sided confidence interval can be constructed to determine equivalence – nonsuperiority, in this case – using the formula, $d + t_{(\alpha/2)}s_x$. The $H_0$ hypothesis (that the standard and the sample are not equivalent) is rejected if $\Delta > d + t_{\alpha}s_x$.

For the data on triclosan residuals in Example 2.5, a $1 - \alpha$ (one-sided confidence interval upper limit) $= d + t_{\alpha}s_x$. At $1 - 0.05 = 0.95$, the nonsuperiority confidence interval extends to $1.51 + 1.833 (0.515)$, or $2.4540$. If $\Delta \geq 2.4540$, reject $H_0$ at $\alpha = 0.05$; the standard and the sample are equivalent. Because $\Delta = 0.96$, and it does not exceed the upper portion of the 95% confidence interval, $H_0$ cannot be rejected at $\alpha = 0.05$. $d$ is greater than the set tolerance level, $\Delta$.

Many find the confidence interval procedure more user-friendly than the significance test we performed first.

*Example 2.6 – Testing for Noninferiority.* Testing for noninferiority is similar to the nonsuperiority test just performed, except the lower tail is used with the $H_0$ being the test hypothesis.

A microbiology laboratory wishes to prepare antigen samples for use over the course of a month. Some of the antigen will be used right away, and some will be used later, but all will be used within a month – 30 days. The goal is to determine if the antigenicity is stable over the course of a month – that is, does it degrade from the initial antigenicity level? Usually, a target value, $L$, is provided, and here, it is 0.97, or $L = 0.97$, or 97% of the theoretical 100% yield of antigen. That is, antigenicity of a sample needs to be about 97% of the theoretical yield to be consistent. The tolerance value, $\Delta$, is 10%, or 0.10. The actual difference between the mean and the set limit, $L$, for antigenicity cannot exceed $-10\%$. We are only interested in the $-10\%$ in this test. We can work this validation experiment with the six-step procedure.

*Step 1.* State the hypothesis for this noninferiority test.

$H_0$: $d < \Delta$ (that is, $d$ is the difference between the mean of individual values and the set value, $L$, and $\Delta$ is the maximum allowable difference between the mean and the target value – in this case, $-0.10$.

$H_A$: $d \geq \Delta$

*Step 2.* Set $\alpha$ and $n$.

The $\alpha$ level was set at $\alpha = 0.01$, and $n = 15$.

*Step 3.* State the test statistic.

$$t_{c_i} = \frac{d - (-\Delta)}{s_x}$$

*Step 4.* Specify the rejection value of $H_0$.

$t_{t(\alpha, n-1)} = t_{t(0.01, 14)} = 2.624$ (from Table A.1)

Note: Although this is a lower-tail test, the test is constructed to be an upper-tail one.

If $t_{c_i} > 2.624$, reject $H_0$ at $\alpha = 0.01$. The sample is not inferior to the set standard.

*Step 5.* Perform the experiment (Table 2.7)

**Table 2.7** Experimental results, Example 2.6

| $n$ | $x_i$ | | $n$ | $x_i$ |
|---|---|---|---|---|
| 1 | 0.72 | | 9 | 0.78 |
| 2 | 0.80 | | 10 | 0.72 |
| 3 | 0.79 | | 11 | 0.61 |
| 4 | 0.63 | | 12 | 0.59 |
| 5 | 0.76 | | 13 | 0.68 |
| 6 | 0.70 | | 14 | 0.91 |
| 7 | 0.62 | | 15 | 0.83 |
| 8 | 0.61 | | | |

$$\sum x = 10.75$$
$$\bar{x} = \sum x / n = 0.7167$$
$$L = 0.97$$
$$d = \bar{x} - L = 0.7167 - 0.97 = -0.2533$$
$$s^2 = \frac{\sum(x_i - \bar{x})^2}{n-1} = \frac{(0.72 - 0.7167)^2 + (0.80 - 0.7167)^2 + \ldots + (0.83 - 0.7167)^2}{14} = 0.0090,$$

and $s = 0.0948$ .

$$s_x = \frac{s}{\sqrt{n}} = \frac{0.0948}{\sqrt{15}} = 0.0245$$
$$t_c = \frac{d - (-\Delta)}{s_x} = \frac{-0.2533 - (-0.10)}{0.0245} = -6.2571$$

*Step 6.* Because $t_c = -6.62571 \not> t_t = 2.624$, one cannot reject $H_0$ at $\alpha = 0.01$.

There is significant evidence to suggest the antigen did degrade over the course of 30 days.

As stated before, some find the confidence interval approach more user-friendly. The $1 - \alpha$ confidence interval lower level is $d - t_t s_x$. The 99% CI lower level is $-0.2533 - 2.624 (0.0245) = -0.3176$.

The confidence interval test for noninferiority is, if $-\Delta \le d - t_t s_x$, reject $H_0$ at $\alpha$. $-\Delta = -0.10 \not< d - t_t s_x = -0.3176$, so one cannot reject $H_0$ at $\alpha = 0.01$. Notice that in the data, there was significant variability that inflated the $s_x$ value. The range of the $x_i$ values was $0.91 - 0.59 = 0.32$, which the microbiologist should investigate further for solutions to maintenance of stability.

*Example 2.7 – Testing for Equivalency (Two One-Sided Tests).* Equivalence testing, or the two one-sided test schema is, by far, the most common. Here, we are interested in whether $d$ is more extreme than each tabled test value, such that $H_0$ is rejected and equivalence stated.

Unlike the test hypothesis for differences in which a two-tail test is $\alpha/2$, here, we perform two (2) one-tail tests (the nonsuperiority and noninferiority calculations) at $\alpha$. The overall $\alpha$ level is $\alpha$, not $2\alpha$. But when we use the confidence interval approach, $\alpha$ is used for each side, for a total of $2\alpha$, as in significant-difference testing. In other words, at an $\alpha = 0.025$, both the one-sided tests combined provided an $\alpha = 0.025$, not $\alpha = 0.05$. However, using the confidence interval approach, the same strategy as in difference testing is used. If the end $\alpha$ is to be $\alpha = 0.05$, the tabled value to use in the confidence interval is $\alpha/2$.

In a diagnostic microbiology laboratory, blood cell counts are calibrated against a standard that provides volumetric cell count in a two-dimensional computer array. The tolerance limit, $\Delta$, is $\pm 0.50 \log_{10}$ cells on a standard of $10^6$ cells/mL. The microbiologist wants to know if the calibration results at the end of the day are equivalent to the standard, $L$ (set level), at the beginning of the day, or $\log_{10} 10^6 = 6$.

*Step 1.* State the hypotheses; there are three: two for $H_0$ and one for $H_A$.

$H_0$: $d > \Delta_U$ (difference between the sample mean and set level exceeds the upper tolerance limit)

$H_0$: $d < \Delta_L$ (difference between the sample mean and set level is less than the lower tolerance limit)

$H_A$: $\Delta_L \leq d \leq \Delta_U$ (the difference between the sample mean and set level does not exceed the lower or upper tolerance limits)

*Step 2.* Set $\alpha$ and $n$.

$\alpha = 0.05$, and $n = 10$ samples

*Step 3.* State the test statistic.

There are two (2) one-tail tests.

$$t_{cs} = \frac{\Delta - d}{s_x} \quad \text{and} \quad t_{ci} = \frac{d - (-\Delta)}{s_x}$$

*Step 4.* Write the decision rule.

If $t_{cs} \geq t_{t(\alpha; n-1)} = t_{(0.05; 9)} = 1.833$ (Table A.1), and

If $t_{ci} \geq t_{t(\alpha; n-1)} = t_{(0.05; 9)} = 1.833$, reject $H_0$ at $\alpha = 0.05$.

*Step 5.* Perform the experiment (Table 2.8).

**Table 2.8** Experimental results, Example 2.7

| $n$ | $x_i$ |
|---|---|
| 1 | 5.8 |
| 2 | 6.2 |
| 3 | 7.6 |
| 4 | 6.5 |
| 5 | 6.8 |
| 6 | 7.3 |
| 7 | 6.9 |
| 8 | 6.7 |
| 9 | 6.3 |
| 10 | 7.1 |

$$\sum x = 67.2$$
$$\bar{x} = \sum x / n = 6.72$$
$$L = 6.0 \log_{10} \text{ target value}$$
$$d = \bar{x} - L = 6.72 = 6.0 = 0.72$$
$$d = 0.72$$

$$s^2 = \frac{\sum(x_i - \bar{x})^2}{n-1} = \frac{(5.8 - 6.72)^2 + (6.2 - 6.72)^2 + \ldots + (7.1 - 6.72)^2}{9} = 0.2929$$

$$s = 0.5412$$

$$s_x = \frac{s}{\sqrt{n}} = 0.1711$$

$$t_{cs} = \frac{\Delta - d}{s_x} = \frac{0.5 - 0.72}{0.1711} = -1.2858$$

$$t_{ci} = \frac{d - (-\Delta)}{s_x} = \frac{0.72 - (-0.5)}{0.1711} = 7.1303$$

*Step 6.* Make the decision.

Although $t_{ci} = 7.1303 > 1.833$, $t_{cs} = -1.2858 \not> 1.833$, so one cannot reject $H_0$ at $\alpha = 0.05$. The standard, $L$, and the mean of samples are not the same (equivalent) at the end of the day.

The confidence interval approach uses the interval $d \pm t_{(\alpha/2; \, n-1)}s_x$, where $\alpha = 0.10$ and $\alpha/2 = 0.05$.

  Upper: 0.72 + 1.833 (0.1711) = 1.0336
  Lower: 0.72 – 1.833 (0.1711) = 0.4064

If $\Delta$ falls outside the $d$ confidence interval, $H_0$ is rejected at $\alpha = 0.10$. $d = 0.4064$ to 1.0336. Because $\Delta = 0.5$ is contained in the interval, one cannot reject $H_0$ at $\alpha = 0.10$ (Fig. 2.11).

You now have a strong background in one-sample testing to a standard value. In Chapter 3, we will carry this process to comparing two samples to one another.

$$( 0.4064 \xrightarrow[0.5]{\qquad} 1.0336 )$$

**Fig. 2.11** 90% Confidence interval on $d = 0.72$

# Chapter 3

## Two-Sample Statistical Tests, Normal Distribution

The two-sample comparative test can be considered the "workhorse" of microbial biostatistics. Here, one is evaluating two different sample groups such as treatments, methods, procedures, or product formulations, for example, to determine if the means of data differ, or if the mean for one is smaller or larger than that for the other. No matter what the test hypothesis, two sample sets are compared.

### Key Point

There are three basic categories of test designs for the two-sample tests:

*1. Two-Sample Independent t-Test.* This test assumes the variances of the data from the two sample groups are independent and, therefore, one does not confirm that the two samples have the same variances, $s^2$. Hence, they are considered different, $\sigma_1^2 \neq \sigma_2^2$. So what? When the variances are unequal, greater sample sizes are required than if they were equal, to detect significant statistical differences. However, I am not talking about a small difference, but one where one variance is two or three times larger than the other. If one wants to be extremely conservative in rejecting $H_0$ (saying the two samples differ when they do not), and the sample variances are unequal $\left( s_1^2 / s_2^2 \neq 1.0 \right)$, use this independent test.

*2. Two-Sample Pooled Variance t-Test (Pooled t-Test).* The variances of the two sample groups are considered equal $\left( s_1^2 / s_2^2 = 1.0 \right)$. This is a very reasonable assumption when, for example, microbiologists are exposing the same microbial species to different antimicrobials, or for compound comparisons, such as media types and different use methods. Fortunately, the pooled *t*-test is also robust to variance differences $\left( s_1^2 / s_2^2 \neq 1.0 \right)$. This test is more powerful and can detect true differences between sample groups with smaller sample sizes than can the independent *t*-test, because the variances are combined.

*3. Matched Pair t-Test.* This test is extremely valuable and powerful. It is based on blocking (matching) sample pairs that are very similar. That is, the members of Group A and Group B are assigned according to common or similar attributes. For example, Groups A and B may represent a pre- and post-treatment reading on the same experimental unit. Hence, each pre-treatment sample will be compared to itself at the post-treatment.

D.S. Paulson,*Biostatistics and Microbiology*, doi: 10.1007/978-0-387-77282-0_3,
© Springer Science + Business Media, LLC 2008

41

## 3.1 Requirements of all *t* Tests

The requirements of all three versions of the two-sample *t*-test are:

1. Sample collection is by random selection of both treatments; that is, if one has two treatments, A and B, then from a common pool, the assignments are randomly made. For example, suppose lawns of *Escherichia coli* on agar plates are to be exposed to two drugs, A and B, impregnated on discs, and the zones of inhibition are to be measured and compared. Here, the inoculated plates would be randomized to one of two treatments, A or B (if, for some reason, both treatment discs A and B could not be placed on the same agar plate).

2. Samples from both groups are tested simultaneously. That is, Group A is not run this week and Group B next week. Depending on the randomization schema, runs of A and B are mixed in order to avoid any "time" effect.

3. The sampled populations are normally distributed.

Now, let us investigate each version of the Student's *t* test in some detail.

### 3.1.1 Two-Sample Independent *t* Test: Variances are not Assumed Equivalent, $\sigma_1^2 \neq \sigma_2^2$

In this statistic, the samples acquired from each group are assumed independent of each other. That is, what affects one group does not affect the other. The variances, $s^2$, as well as the standard deviations, $s$, are assumed different, or at least one does not claim they are the same. The test statistic is:

$$t_c = \frac{\bar{x}_A - \bar{x}_B}{\sqrt{\dfrac{s_A^2}{n_A} + \dfrac{s_B^2}{n_B}}}, \text{ with } n_1 + n_2 - 2 \text{ degrees of freedom.}$$

where $\bar{x}_A, \bar{x}_B$ are the means of sample groups A and B, $s_A^2, s_B^2$ are the variances of samples groups A and B, and $n_A$, $n_B$ are the sample sizes of sample groups A and B.

Three tests for significance can be conducted: lower-tail, upper-tail, and two-tail.

---

### Key Point

**The six-step procedure makes the test strategy logical to follow.**

---

*Step 1.* Construct the hypothesis test.

| Lower-Tail Test[*] | Upper-Tail Test | Two-Tail Test |
|---|---|---|
| $H_0: A \geq B$ | $H_0: A \leq B$ | $H_0: A = B$ |
| $H_A: A < B$ | $H_A: A > B$ | $H_A: A \neq B$ |

---

[*] Remember, the terms, lower-, upper-, and two-tail refer to the test hypothesis, $H_A$.

*Step 2.* Set $\alpha$ and specify sample sizes. Try to keep the sample sizes equal, because these values play a role in determining the critical acceptance/rejection values in the $t$ tables.

*Step 3.* State the test statistic. $t_c = \dfrac{\bar{x}_A - \bar{x}_B}{\sqrt{\dfrac{s_A^2}{n_A} + \dfrac{s_B^2}{n_B}}}$

where $\bar{x}_A$ and $\bar{x}_B = \dfrac{\sum x_i}{n}$ for each group and $s^2 = \dfrac{\sum (x - \bar{x})^2}{n-1}$ for each group.

*Step 4.* Write the decision rule.

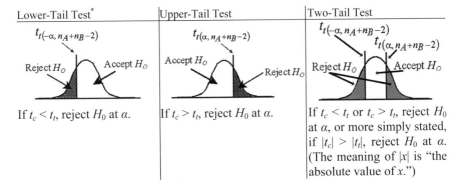

| Lower-Tail Test* | Upper-Tail Test | Two-Tail Test |
|---|---|---|
| $t_{t(-\alpha,\, n_A+n_B-2)}$ | $t_{t(\alpha,\, n_A+n_B-2)}$ | $t_{t(-\alpha,\, n_A+n_B-2)}$   $t_{t(\alpha,\, n_A+n_B-2)}$ |
| Reject $H_O$   Accept $H_O$ | Accept $H_O$   Reject $H_O$ | Reject $H_O$   Accept $H_O$ |
| If $t_c < t_t$, reject $H_0$ at $\alpha$. | If $t_c > t_t$, reject $H_0$ at $\alpha$. | If $t_c < t_t$ or $t_c > t_t$, reject $H_0$ at $\alpha$, or more simply stated, if $|t_c| > |t_t|$, reject $H_0$ at $\alpha$. (The meaning of $|x|$ is "the absolute value of $x$.") |

*Step 5.* Perform the experiment and collect and analyze the data.
*Step 6.* Make the decision based on the calculated test statistic.

*Example 3.1.* A microbiologist wants to compare Drugs A and B in terms of their antimicrobial properties versus a clinical isolate (i.e., isolated from a medical patient) of *Escherichia coli*. The *E. coli* was subcultured and then inoculated as a lawn onto Mueller-Hinton agar plates. A disc impregnated with either Drug A or Drug B was then placed onto the agar in the middle of the plate, and plates were incubated at 35 $\pm$ 2°C for 24 h. The width of the zone (at its widest) in which growth of *E. coli* was prevented was measured on ten plates exposed to each drug. Let's work this through, using all three possible hypotheses.

*Step 1.* Specify the test hypothesis.

| Lower-Tail Test* | Upper-Tail Test | Two-Tail Test |
|---|---|---|
| $H_0$: $A \geq B$ $H_A$: $A < B$ | $H_0$: $A \leq B$ $H_A$: $A > B$ | $H_0$: $A = B$ $H_A$: $A \neq B$ |
| Drug A produces a smaller zone of inhibition than does Drug B. | Drug A produces a larger zone of inhibition than does Drug B. | The zones of inhibition produced by Drugs A and B are not the same. |

---

* Remember, the terms, lower-, upper-, and two-tail refer to the test hypothesis, $H_A$.

*Step 2.* Set $\alpha$ and $n_A$ and $n_B$. Let us use $\alpha = 0.05$, and $n_A = n_B = 10$.
*Step 3.* State the decision rule.

| Lower-Tail Test | Upper-Tail Test | Two-Tail Test |
|---|---|---|
| $t_{(-\alpha;\, n_A+n_B-2)}$ | $t_{(\alpha;\, n_A+n_B-2)}$ | $t_{(\alpha/2;\, n_A+n_B-2)}$ |
| $t_{(-0.05;\, 18)}$ | $t_{(0.05;\, 18)}$ | $t_{(0.05/2;\, 18)}$ |
| Using the Student's $t$ table (Table A.1), | Using the Student's $t$ table (Table A.1), | Using the Student's $t$ table (Table A.1), |
| $t_t = -1.734$ | $t_t = 1.734$ | $t_{t(0.025,\,18)} = \pm 2.101$ |

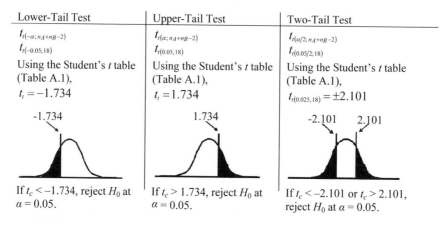

| | | |
|---|---|---|
| If $t_c < -1.734$, reject $H_0$ at $\alpha = 0.05$. | If $t_c > 1.734$, reject $H_0$ at $\alpha = 0.05$. | If $t_c < -2.101$ or $t_c > 2.101$, reject $H_0$ at $\alpha = 0.05$. |

*Step 4.* Choose the test statistic.

$$t_c = \frac{\bar{x}_A - \bar{x}_B}{\sqrt{\dfrac{s_A^2}{n_A} + \dfrac{s_B^2}{n_B}}} \quad \text{(The same formula is used for each of the hypothesis tests.)}$$

*Step 5.* Perform the experiment. Results are presented in Table 3.1.

**Table 3.1** Zones of inhibition in millimeters, Example 3.1

| $n$ | Drug A | Drug B |
|---|---|---|
| 1 | 10.5 | 11.2 |
| 2 | 7.8 | 10.3 |
| 3 | 9.3 | 9.2 |
| 4 | 8.7 | 9.7 |
| 5 | 10.2 | 8.9 |
| 6 | 8.9 | 10.7 |
| 7 | 7.4 | 9.9 |
| 8 | 9.3 | 10.1 |
| 9 | 8.7 | 7.8 |
| 10 | 7.9 | 8.9 |

$$\bar{x}_A = \frac{\sum x_i}{n} = \frac{10.5+7.8+\cdots+8.7+7.9}{10} = 8.87$$

$$s_A^2 = \frac{\sum(x_i-\bar{x}_A)^2}{n-1} = \frac{(10.5-8.87)^2+(7.8-8.87)^2+\cdots+(8.7-8.87)^2+(7.9-8.87)^2}{10-1} = 1.01$$

$$\bar{x}_B = \frac{\sum x_i}{n} = \frac{11.2+10.3+\cdots+7.8+8.9}{10} = 9.67$$

$$s_B^2 = \frac{\sum(x_i - \overline{x}_B)^2}{n-1} = \frac{(11.2-9.67)^2 + (10.3-9.67)^2 + \cdots + (7.8-9.67)^2 + (8.9-9.67)^2}{10-1} = 1.00$$

$$t_c = \frac{8.87-9.67}{\sqrt{\dfrac{1.01}{10} + \dfrac{1.00}{10}}} = -1.7844 \approx -1.78$$

*Step 6.* Make the decision.

| Lower-Tail | Upper-Tail | Two-Tail |
|---|---|---|
| Because $-1.78 < -1.734$, reject $H_0$ at $\alpha = 0.05$; Drug A has smaller zones of inhibition than does Drug B. | Because $-1.78 \not> 1.734$, one cannot reject $H_0$ at $\alpha = 0.05$; Drug A does not have larger zones of inhibition than does Drug B. | Because $-1.78 \not< -2.101$, and $-1.78 \not> 2.101$, one cannot reject $H_0$ at $\alpha = 0.05$; Drug A and Drug B are not significantly different in zones of inhibition. |

Note that the lower-tail test rejected $H_0$, but the two-tail test did not. This is because the two-tail test is less sensitive to directions of difference. It must account for both a lower and upper critical value. If you can, use a single-tail test for greater power in your analysis.

### 3.1.2 Two-Sample Pooled $t$ Test: Variances are Equivalent, $\sigma_1^2 = \sigma_2^2$

Most times, the variances of two sampled populations will be equivalent, or close enough to use a pooled variance statistic; that is, the two variances are combined into a common one. A two-sample pooled $t$-test is statistically more powerful than is an independent variance, two-sample test, given all else is the same.

$$t_c = \frac{\overline{x}_A - \overline{x}_B}{S_{pooled}}$$

where $\overline{x}_A$ and $\overline{x}_B = \dfrac{\sum x_i}{n}$, and $S_{pooled} = \sqrt{\dfrac{(n_A-1)s_A^2 + (n_B-1)s_B^2}{n_A + n_B - 2}} \sqrt{\dfrac{1}{n_A} + \dfrac{1}{n_B}}$,

where $s_A^2$ and $s_B^2 = \dfrac{\sum(x - \overline{x})^2}{n-1}$ for each sample group.

The same three hypotheses can be evaluated, using the six-step procedure.

*Step 1.* Formulate the hypothesis.

| Lower-Tail | Upper-Tail | Two-Tail |
|---|---|---|
| $H_0$: $A \geq B$ | $H_0$: $A \leq B$ | $H_0$: $A = B$ |
| $H_A$: $A < B$ | $H_A$: $A > B$ | $H_A$: $A \neq B$ |

*Step 2.* Set $\alpha$ and the sample size for $n_A$ and $n_B$.

| Lower-Tail Test | Upper-Tail Test | Two-Tail Test |
|---|---|---|
| $-\alpha$ | $\alpha$ | $-\alpha/2$ and $\alpha/2$ |

*Step 3.* Select the test statistic.

$$t_C = \frac{\bar{x}_A - \bar{x}_B}{s_{pooled}}$$

*Step 4.* State the acceptance/rejection criteria.

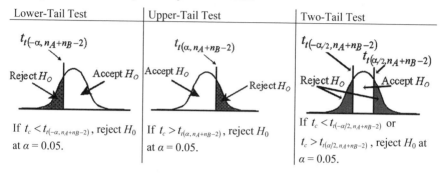

| Lower-Tail Test | Upper-Tail Test | Two-Tail Test |
|---|---|---|
| $t_{t(-\alpha, n_A+n_B-2)}$ | $t_{t(\alpha, n_A+n_B-2)}$ | $t_{t(-\alpha/2, n_A+n_B-2)}$  $t_{t(\alpha/2, n_A+n_B-2)}$ |
| Reject $H_O$   Accept $H_O$ | Accept $H_O$   Reject $H_O$ | Reject $H_O$   Accept $H_O$ |
| If $t_c < t_{t(-\alpha, n_A+n_B-2)}$, reject $H_0$ at $\alpha = 0.05$. | If $t_c > t_{t(\alpha, n_A+n_B-2)}$, reject $H_0$ at $\alpha = 0.05$. | If $t_c < t_{t(-\alpha/2, n_A+n_B-2)}$ or $t_c > t_{t(\alpha/2, n_A+n_B-2)}$, reject $H_0$ at $\alpha = 0.05$. |

*Step 5.* Perform experiment.
*Step 6.* Make the decision.

Let's analyze the data from Example 3.1, to see if anything changes from what we found when assuming unequal sample variances.

*Step 1.* Formulate the hypothesis.

| Lower-Tail | Upper-Tail | Two-Tail |
|---|---|---|
| $H_0: A \geq B$<br>$H_A: A < B$ | $H_0: A \leq B$<br>$H_A: A > B$ | $H_0: A = B$<br>$H_A: A \neq B$ |
| Zones of inhibition are smaller for Drug A than for Drug B. | Zones of inhibition are larger for Drug A than for Drug B. | Zones of inhibition for Drug A and Drug B are different. |

*Step 2.* Set $\alpha$ and sample size, $n$. $\alpha = 0.05$ and $n_1 = n_2 = 10$
*Step 3.* Write the test statistic: $t_c = \frac{\bar{x}_A - \bar{x}_B}{s_{pooled}}$.
*Step 4.* State the decision rule.

| Lower-Tail Test | Upper-Tail Test | Two-Tail Test |
|---|---|---|
| $t_{t(-\alpha; n_A+n_B-2)}$<br>$t_{t(-0.05; 18)}$<br>Using the Student's $t$ table (Table A.1)<br>$t_t = -1.734$ | $t_{t(\alpha; n_A+n_B-2)}$<br>$t_{t(0.05; 18)}$<br>Using the Student's $t$ table (Table A.1)<br>$t_t = 1.734$ | $t_{t(\alpha/2; n_A+n_B-2)}$  $t_{t(0.05/2; 18)}$<br>Using the Student's $t$ table (Table A.1)<br>$t_{t(0.025; 18)} = \pm 2.101$ |

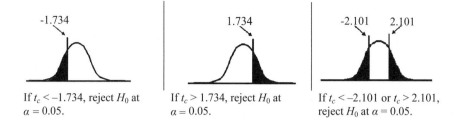

If $t_c < -1.734$, reject $H_0$ at $\alpha = 0.05$.

If $t_c > 1.734$, reject $H_0$ at $\alpha = 0.05$.

If $t_c < -2.101$ or $t_c > 2.101$, reject $H_0$ at $\alpha = 0.05$.

*Step 5.* Perform the experiment and calculate the test statistic. Results have been presented already in Table 3.1 (reproduced here for convenience).

**Table 3.1** Zones of inhibition in millimeters, Example 3.1

| $n$ | Drug A | Drug B |
|---|---|---|
| 1 | 10.5 | 11.2 |
| 2 | 7.8 | 10.3 |
| 3 | 9.3 | 9.2 |
| 4 | 8.7 | 9.7 |
| 5 | 10.2 | 8.9 |
| 6 | 8.9 | 10.7 |
| 7 | 7.4 | 9.9 |
| 8 | 9.3 | 10.1 |
| 9 | 8.7 | 7.8 |
| 10 | 7.9 | 8.9 |

$$\overline{x}_A = \frac{\sum x_i}{n} = \frac{10.5 + 7.8 + \cdots + 8.7 + 7.9}{10} = 8.87$$

$$s_A^2 = \frac{\sum (x_i - \overline{x}_A)^2}{n-1} = \frac{(10.5 - 8.87)^2 + (7.8 - 8.87)^2 + \cdots + (8.7 - 8.87)^2 + (7.9 - 8.87)^2}{10-1} = 1.01$$

$$\overline{x}_B = \frac{\sum x_i}{n} = \frac{11.2 + 10.3 + \cdots + 7.8 + 8.9}{10} = 9.67$$

$$s_B^2 = \frac{\sum (x_i - \overline{x}_B)^2}{n-1} = \frac{(11.2 - 9.67)^2 + (10.3 - 9.67)^2 + \cdots + (7.8 - 9.67)^2 + (8.9 - 9.67)^2}{10-1} = 1.00$$

$$s_{pooled} = \sqrt{\frac{(n_A - 1)s_A^2 + (n_B - 1)s_B^2}{n_A + n_B - 2}} \sqrt{\frac{1}{n_A} + \frac{1}{n_B}}$$

$$s_{pooled} = \sqrt{\frac{9(1.01) + 9(1.00)}{10 + 10 - 2}} \sqrt{\frac{1}{10} + \frac{1}{10}} = 0.4483$$

$$t_c = \frac{\overline{x}_A - \overline{x}_B}{s_{pooled}} = \frac{8.87 - 9.67}{0.4483} = -1.78$$

*Step 6.* Make the decision.

| Lower-Tail | Upper-Tail | Two-Tail |
|---|---|---|
| $t_c = -1.78 < t_t = -1.74$; reject $H_0$ at $\alpha = 0.05$. | $t_c = -1.78 \not> t_t = 1.74$; one cannot reject $H_0$ at $\alpha = 0.05$. | $|t_c| = 1.78 \not> |t_t| = 2.01|$; one cannot reject $H_0$ at $\alpha = 0.05$. |

### 3.1.3 Paired *t* Test

A number of experiments lend themselves to pairing, where the variability between the two groups contrasted can be minimized. Consider an experimental design to measure the microbial colony-forming counts at the human abdominal site after prepping the skin with two antiseptic formulations. The standard two-sample *t*-test would require sampling subjects from a population size, $N$, and from that, sample the $n_A$ and $n_B$ subgroups, which are randomly assigned to one of the two formulations. Suppose 20 subjects were selected, and 10 were randomly assigned to Antiseptic A and 10 to Antiseptic B. Both males and females, having different ages, weights, heights, health status, and other components, would be allowed to participate. The statistical variances, more than likely, would be relatively large, because of how such inter-subject differences contribute to differences in individual skin microorganisms. The greater the statistical variability (the variance), the greater the number of subjects necessary to detect any significant difference between the products. Figure 3.1 provides an example of a completely randomized design in which the average microbial reductions from Group 1 are compared to those from Group 2.

$Log_{10}$ Microbial
Counts (CFU)

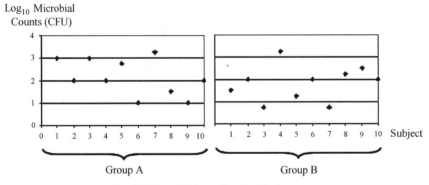

**Fig. 3.1** Completely randomized design

In a matched-pair design, each subject would receive both treatments; that is, statistically speaking, each subject would constitute a block; only the data points in each block are compared. In the abdominal evaluation, each subject receives both treatments. The areas of skin to each side of the umbilicus are randomly assigned for treatment, one with each of the two test products. The differences between treatments are then measured within each subject, not between subjects. Figure 3.2 portrays the kind of results one would see from this study design. The many sources of variability that exist between each subject can be ignored, in that only the differences within the block (an individual subject's abdomen) are compared, making the design more powerful than a completely randomized one.

Other examples of a similar kind would be experiments using the same culture plate in a Kirby-Bauer test to compare two antibiotics – each plate is challenged with both antibiotics – or using subjects paired based on similarities in weight, sex, blood pressure, liver functions, etc. Whenever possible, a matched-pair test is desirable.

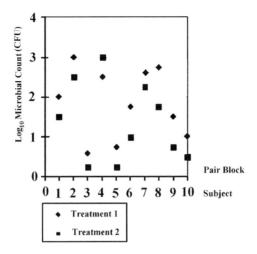

**Fig. 3.2** Paired design (the variance in this case is not between subjects but within each subject)

The formula used for statistical analysis is $t_c = \dfrac{\overline{d}}{\frac{s_{paired}}{\sqrt{n}}}$, where $\overline{d} = \dfrac{\sum(x_A - x_B)}{n} = \dfrac{\sum d}{n}$,

or the average difference between pairs within a block, and $s_{paired} = \sqrt{\dfrac{\sum(d_i - \overline{d})^2}{n-1}}$, or the standard deviation of the differences.

*Six-Step Procedure.* The paired test can be configured as a lower-, upper-, or two-tail test, using the six-step procedure.

*Step 1.* State the hypothesis.

| Lower-Tail Test | Upper-Tail Test | Two-Tail Test |
|---|---|---|
| $H_0: A \geq B$ | $H_0: A \leq B$ | $H_0: A = B$ |
| $H_A: A < B$ | $H_A: A > B$ | $H_A: A \neq B$ |

*Step 2.* Set $\alpha$ level and $n_A$ and $n_B$. In matched-pair designs, the samples sizes, $n_A$ and $n_B$ must be equal, $n_A = n_B$.

*Step 3.* Write out the test statistic.

$$t_c = \dfrac{\overline{d}}{\frac{s_d}{\sqrt{n}}}$$

*Step 4.* Write out the acceptance/rejection decision rule.

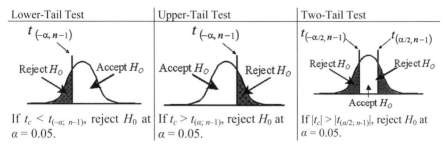

| Lower-Tail Test | Upper-Tail Test | Two-Tail Test |
|---|---|---|
| $t_{(-\alpha, n-1)}$ Reject $H_o$  Accept $H_o$ | $t_{(-\alpha, n-1)}$ Accept $H_o$  Reject $H_o$ | $t_{(-\alpha/2, n-1)}$  $t_{(\alpha/2, n-1)}$ Reject $H_o$  Reject $H_o$  Accept $H_o$ |
| If $t_c < t_{(-\alpha; n-1)}$, reject $H_0$ at $\alpha = 0.05$. | If $t_c > t_{(\alpha; n-1)}$, reject $H_0$ at $\alpha = 0.05$. | If $|t_c| > |t_{(\alpha/2; n-1)}|$, reject $H_0$ at $\alpha = 0.05$. |

**Note**: The tabled value is dependent on $n$ being the number of pairs (blocks), not test sites. We lose degrees of freedom in the paired test, so if the pairing does not reduce variability, one should not use the paired approach.

*Step 5.* Perform the experiment.

*Step 6.* Make the decision based on the outcome of the experiment and the decision rule in Step 4.

*Example 3.2.* A microbiologist wants to compare two media preparations for their effects on determining the resistance of *Bacillus subtilis* (*B. atrophaeus*) spores to sodium hypochlorite. The microbiologist wants the experimental conditions to be equivalent in all areas, except the growth medium. The two media to be used are soil extract nutrient agar made using reagent grade water (control) and the same medium made using an on-site source of deionized water. The objective is to determine if the survival of the bacteria following exposure to the sodium hypochlorite is affected by the medium used to germinate and grow vegetative bacteria from the spores, thereby indicating an inflated sporicidal efficacy. Variables that must be controlled include *Bacillus subtilis* spores being taken from a single inoculum suspension and use of the same commercial powdered nutrient agar, the same lot of Petri dishes, the same incubator, the same microbiologist, and the same methods of treatment. Because sodium hypochlorite (bleach) is an effective sporicidal agent, a diluted form will be used (300 ppm) for a 5-min exposure of the spores. The initial spore population will be at least $1.0 \times 10^5$ colony-forming units per milliliter (CFU/mL), and all computations will be performed on data linearized by $\log_{10}$ transformation. The microbiologist wants to know if the test medium (test variable – deionized water vs. on-site water) affects assessment of spore resistance to sodium hypochlorite. The analysis of the data from this experiment can be easily set into the six-step procedure.

*Step 1.* First, write out the test (alternative) hypothesis, then the $H_0$, or null hypothesis.

The concern is that medium prepared using the on-site water may cause spores to appear more sensitive to bleach than does the control medium made using reagent grade deionized water. If this is so, following the 300 ppm sodium hypochlorite exposure, counts of bacterial colonies on the test medium will be lower than counts from the standard medium. Let $A$ = test medium (on-site water) and $B$ = control medium (reagent grade water). The test hypothesis, then, is $H_A$: $A < B$. Hence, a lower-tail test is used.

$H_0$ : $A \geq B$

$H_A$ : $A < B$

*Step 2.* Set $\alpha$ and $n$. The microbiologist wants to have at least 90% confidence that a type I error is not committed; that is, stating that a true difference exists between the two post-treatment colony counts when one does not. Hence, $1 - 0.90 = 0.10$, and $\alpha = 0.10$. The sample replicates are nine ($n = 9$), plated in duplicate. The duplicate plate counts of colonies from each medium formulation will be averaged as a single $x_i$ value.

*Step 3.* Select the test statistic. Here, the statistic is $t_c = (\bar{d}/s_d)/\sqrt{n}$ .

*Step 4.* Write out the decision rule. Because this is a lower-tail test, a negative *t*-tabled value will be used (Fig. 3.3). $t_{(-\alpha;\, n-1)} = t_{(-0.10;\, 8)} = -1.397$ (Table A.1).

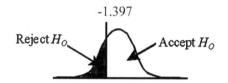

Fig. 3.3 Step 4, Example 3.2

If *t*-calculated < *t*-tabled ($-1.397$), reject the $H_0$ hypothesis at $\alpha = 0.10$. Medium prepared with the on-site water (A) indicates that spores are more sensitive to sodium hypochlorite than does medium made with the reagent-grade water (B).

*Step 5.* Perform the experiment. The microbiologist dispensed 1.0 mL aliquots of a $10^7$–$10^8$ spore/mL suspension into a pair of tubes (1 test, 1 control), vortexed for 15 s, and then added the sodium hypochlorite for a 5-min exposure, followed by neutralization. Serial dilutions of the spore suspension were then transferred to media preparations $A$ and B for incubation at $35 \pm 2°C$ for 24–48 h. The bacterial colonies were counted, and the raw count data were $\log_{10}$ linearized. The collected data are presented in Table 3.2.

**Table 3.2** Collected data for Example 3.3

| Block pair ($n$) | Medium $A$ | Medium $B$ | $d = x_A - x_B$ |
|---|---|---|---|
| 1 | 3.8 | 4.1 | −0.3 |
| 2 | 4.7 | 4.3 | 0.4 |
| 3 | 5.2 | 4.9 | 0.3 |
| 4 | 4.7 | 4.7 | 0.0 |
| 5 | 3.9 | 3.8 | 0.1 |
| 6 | 3.3 | 3.2 | 0.1 |
| 7 | 3.6 | 3.4 | 0.2 |
| 8 | 3.2 | 3.0 | 0.2 |
| 9 | 3.3 | 3.1 | 0.2 |
| | | | $\Sigma d = 1.2$ |

$$\bar{d} = \frac{\sum d_i}{n} = \frac{1.2}{9} = 0.1333$$

$$s_d^2 = \frac{\sum(d_i - \bar{d})^2}{n-1} = \frac{(-0.3 - 0.1333)^2 + \ldots + (0.2 - 0.1333)^2}{8} = 0.04$$

$$t_c = \frac{\bar{d}}{\frac{s_d}{\sqrt{n}}} = \frac{0.1333}{\frac{\sqrt{0.04}}{\sqrt{9}}} = 0.222$$

*Step 6.* Make the decision. Because $t_c = 0.222 \not< t_t = -1.397$, one cannot reject $H_0$ at $\alpha = 0.10$. There is no substantial evidence to support the hypothesis that using the on-site water (A group) in the growth medium results in the appearance of significantly more sensitive spores than does the control medium.

### 3.1.4 Sample Size Determination

In microbiological experiments, it is very easy to bias the statistical results. Although lack of randomization is probably the greatest contributor to biased studies, so is an inadequate sample size. In standard hypothesis testing, what primarily is evaluated is the validity of the alternative hypothesis. That is, is there sufficient evidence to claim that a significant difference between the two groups exists? The rejection of an alternative hypothesis ($H_A$) does not translate to acceptance of the null hypothesis ($H_0$). Yet, in many research papers, this is exactly what is claimed, at least by inference. The more appropriate statement is that one *cannot* reject $H_0$ at the set $\alpha$ level, a subtle difference in meaning.

For example, in comparing 5-min exposures of bacterial spores to a chemical disinfectant at full strength and at $x$ strength using a sample size of four in each group, the microbiologist may claim no difference exists in the sporicidal activity of the two concentrations. But unknown to the microbiologist was that, given the 1.0 $\log_{10}$ variability in the study ($s = 1.0$), it was under-powered to detect a true and significant difference. So, before conducting a study, it is necessary to select a sample size that, considering the variability of the data, will actually support a statistical conclusion.

Selection of an appropriate statistical sample size depends upon four indices: 1) $\sigma$, 2) $\alpha$, 3) $\beta$, and 4) $\delta$. $\delta$ is estimated by $s$, the standard deviation of the data and is computed differently for independent, pooled, and paired tests, as we have already seen. The standard deviation is a measure of precision. The larger the sample size, the greater the precision, i.e., the smaller the $s$ value. The alpha, or $\alpha$ value is the acceptance level for Type I error – stating a difference is significant, when it is not – and is set by the researcher. The level of $\alpha$ is often, by convention, specified as 0.05. This means that, over the long haul, a Type I error will be committed 5% of the time, or 5 out of 100 repetitions of the experiment. The beta, or $\beta$ value is the acceptance level for Type II error – stating that data sets are not significantly different, when they actually are. The value of $\beta$ also is set by the researcher. It is important to note that $1 - \beta$ is the "Power of the Statistic." Beta error is often set at 0.20. Hence, 20% of the time, a true difference between data sets will not be detected. The power of the statistic is 0.80. Delta ($\delta$) is the specified detection level of the statistic used; that is, the ability of the statistical model to detect differences between data sets. $\delta$ is estimated by the minimum critical value needed to detect the magnitude of difference between two experimental groups that will be considered significant. It is set by the researcher.

A microbiologist may need to detect 0.5 $\log_{10}$ differences between treatments to call them significantly different from one another. In other cases, $d$ may be set lower or higher, as required. If it is set at 0.5 $\log_{10}$, then, given adequate sample sizes, the statistic will detect differences between data sets that are greater than or equal to 0.5 $\log_{10}$. Generally, before the experiment is conducted, these four indices are used to compute a sample size value.

### 3.1.4.1 Sample Size Determination, Independent Two-Sample $t$ Test, $\sigma_1^2 \neq \sigma_2^2$

The equation is $n \geq \dfrac{(s_1^2 + s_2^2)(z_{\alpha/2} + z_\beta)^2}{d^2}$, where $n$ = minimum sample size of each group compared, $s_1^2 = \sum(x - \bar{x}_1)/(n_1 - 1)$, $s_2^2 = \sum(x - \bar{x}_2)/(n_2 - 1)$, $z_{\alpha/2}$ = normal tabled value (Table A.2) of $\alpha/2$, $z_\beta$ = normal tabled value (Table A.2) of $\beta$, and $d$ = the specified detection level (not to be confused with "$d$," as used to represent the difference between pairs in a matched-pair $t$-test).

Suppose $s_1^2 = 4.3$, $s_2^2 = 7.1$, $d = 2.0$, $\alpha = 0.05$, and $\beta = 0.20$. Then $z_{\alpha/2} = 0.5 - 0.025 = 0.4750$ (see page 54 for the normal $z$ table computations), which covers the area $1 - \alpha/2$, when $\alpha = 0.05$. The value, 0.4750, in Table A.2 corresponds to 1.96, so $z_{\alpha/2} = 1.96$. $z_\beta = 0.5 - 0.20 = 0.3$, which corresponds to 0.842, for $\beta = 0.20$. $d = 2.0$ specifies a detection level of 2.0 $\log_{10}$.

$$n \geq \frac{(4.3 + 7.1)(1.96 + 0.842)^2}{2^2} = 22.38$$

Hence, a sample size of at least 23 in each test group is required, a total of 46 subjects at $\alpha = 0.05$, $\beta = 0.20$, and $d = 2.0$.

Once the experiment has been completed with the projected sample, it is a good idea to recheck the sample size, once one knows what $s_1^2$ and $s_2^2$ actually are. It is the variance component of the predicted sample size equation that generally is found to differ, post hoc. To solve for the detection level, one calculates:

$$d = \sqrt{\frac{(s_1^2 + s_2^2)(z_{\alpha/2} + z_\beta)^2}{n}}$$

It is very valuable to do this check, particularly if no significant statistical difference is reported between the data sets. One wants to be assured that, if a significant difference actually existed, the statistic has the power to catch it. For example, suppose $s_1^2 + s_2^2$ were found to actually be 6.2 and 7.9, respectively, after the experiment was concluded.

## Key Point
## Using the Normal $z$ Distribution Table

The $z$ Table (Table A.2) is a little more difficult to use than the Student's $t$ Table (Table A.1). But by following these instructions, the process can be easy.

*1.* The $z$ distribution is symmetric with an area of 1.0. Hence, by convention, only one-half of it is used, which covers 0.5 area. The values in the $z$ table are normalized values expressing $\bar{x} = 0$, $s = 1$. The normalizing calculation is $z = (\bar{x} - \mu / \sigma)$.

*2.* Upper-tail and lower-tail critical values are found the same way, except the upper-tail carries a positive sign and the lower-tail, a negative one. To determine the $z$ tabled value at $\alpha$, subtract $\alpha$ from 0.5 and find that value in the $z$ table. The first two significant digits are found on the left-most column, and the last or third significant digit in the upper-most row.

Example: Find the $z$ tabled value for $\alpha = 0.05$.

*Upper-Tail Test:*
   *Step 1.* $0.5 - 0.05 = 0.450$
   *Step 2.* Go into the $z$ table matrix and find the value 0.450, or the one closest to it. That value is 0.4495.
   *Step 3.* Move your finger horizontally to the left-most $z$ value (1.6) and vertically from 0.4495 to the upper-most row to find 0.04. This critical value is 1.64, the $z$-tabled value, or $z_t$.

*Lower-Tail Test:*
If a lower-tail test is used, the same procedure (upper-tail test) is done but a negative sign is placed in front of the $z$ value, in this case, –1.64.

*Two-Tail Test:*
Instead of subtracting $\alpha$ from 0.5, subtract $\alpha/2$ from 0.5. This is because both the upper- and lower-table regions need to be considered. Then, the same steps are repeated.
   *Step 1.* $0.05 \div 2 = 0.025$
   *Step 2.* $0.5 - 0.025 = 0.475$
   *Step 3.* Find the 0.475 value in the $z$ table matrix, which is exactly 0.475.
   *Step 4.* Move your finger horizontally from 0.475 to the left-most column and find the value, which is 1.9. Next, move your finger vertically up from 0.475 to the top-most row and find that value, which is 0.06. Combining those values, 1.9 + 0.06, gives the value 1.96. Because this is a two-tail test, there are two critical $z_t$ values, –1.96 and 1.96.

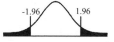

$d = \sqrt{(6.2 + 7.9)(1.96 + 0.842)^2/23} = 2.19$ detection level, which is pretty close to our set level of 2.0 $\log_{10}$. To get a detection level of 2.0 with $s_1^2 = 6.2$ and $s_2^2 = 7.9$ would require 28 in each test group. $n \geq \dfrac{(6.2 + 7.9)(1.96 + 0.842)^2}{2^2} = 27.68 \approx 28.$

Now, the big question is whether "0.19 $\log_{10}$" is worth the extra subjects. One could have concluded that no difference between groups existed, when really, the sample size of 23 was inadequate to detecting a significant difference such as 2.10 $\log_{10}$.

### 3.1.4.2 Sample Size Determination: Pooled Variance $t$ Test, $\sigma_1^2 = \sigma_2^2$

$n \geq \dfrac{s_{pooled}^2 \left(z_{\alpha/2} + z_\beta\right)^2}{d^2}$ . Use the same formula, except the pooled variance, $s_{pooled}^2$, is used in place of $s_1^2 + s_2^2$.

### 3.1.4.3 Sample Size Determination, Two-Sample Matched-Pair Test

$n \geq \dfrac{s_d^2 \left(Z_{\alpha/2} + Z_\beta\right)^2}{d^2}$ . Use the same formula, but use the paired variance ( $s_d^2$ ) instead of the independent or pooled variance.

## 3.2 Other Topics

### 3.2.1 Proportions

Some times, microbiologists deal with qualitative data that take the form of one of two values, 1 or 0. They are used to represent binary outcomes and are termed nominal data. The numbers themselves have no arithmetic meaning. The binary values may represent success/failure, growth/no growth, sterile/nonsterile, or positive/negative, for example. These qualitative values can be quantitatively represented as "proportions." Say there are 100 trials, of which five are successes and 95 are failures. Let $s$ = proportion of successes = number of successes/total trials = 5/100 = 0.05, and the failure proportion equals 1 − 0.05 = 0.95. These converted proportion values may then be used in two-sample statistical tests.

Testing proportions requires a larger sample size than do quantitative studies, all other things held constant, so the sample size should be as large as possible, $\geq 30$. Because proportions provide for less power, statistically, than do actual measurements, we will apply only a single kind of two-sample test, and not have options for nonequal variances, equal variances, or matched pairs. The population proportion parameter is $\pi$, which is estimated by $P$. The population standard deviation is $\sigma_\Delta = \sqrt{\dfrac{\pi_A(1 - \pi_A)}{n_A} + \dfrac{\pi_B(1 - \pi_B)}{n_B}}$ . It is estimated by the sample standard deviation $s = \sqrt{\dfrac{P_c(1-P_c)}{n_A + n_B}}$ or $s = \sqrt{P_c(1-P_c)\left(\dfrac{1}{n_A} + \dfrac{1}{n_B}\right)}$, where $P_c$ is the pooled

(combined) proportions of Group A and Group B and is calculated as $((n_A P_A + n_B P_B))/(n_A + n_B)$. The test statistic to use is $t_c = \dfrac{P_A - P_B}{\sqrt{\dfrac{P_c(1-P_c)}{n_A + n_B}}}$.

Let us perform the two-sample comparison using the six-step procedure.

*Example 3.3.* A microbiologist wants to compare two types of fluid medium (egg/meat medium and garden soil extract medium) to see if they produce different growth results for *Bacillus subtilus* survivors post-exposure of the spores to a 121°C steam-sterilization cycle – that is, are the surviving populations significantly different. This is a two-tail test. For instructive purposes, I will also compare the results using upper- and lower-tail tests. Thirty (30) test tubes of each medium will be used. *A* is the egg/meat medium, and *B* is the garden soil extract.

*Step 1.* Formulate the hypothesis. Let *P* be the proportion of successes. Here, tubes with no growth are designed as 0. Note that, often, success is defined as growth, or a 1. It does not matter from a statistical perspective, as long as one is clear about the assignment value of success and failure.

| Lower-Tail | Upper-Tail | Two-Tail |
|---|---|---|
| $H_0: P_A \geq P_B$ | $H_0: P_A \leq P_B$ | $H_0: P_A = P_B$ |
| $H_A: P_A < P_B$ | $H_A: P_A > P_B$ | $H_A: P_A \neq P_B$ |

*Step 2.* Set $\alpha$ and $n_A$ and $n_B$. In this pilot, we will set $\alpha = 0.10$ and $n_A = n_B = 30$.

*Step 3.* Choose the test statistic. $t_c = (P_A - P_B)/\sqrt{(P_c(1-P_c))/n_A + n_B}$.

*Step 4.* State the decision rule.

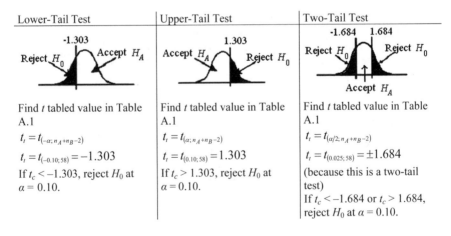

| Lower-Tail Test | Upper-Tail Test | Two-Tail Test |
|---|---|---|
| Find $t$ tabled value in Table A.1 | Find $t$ tabled value in Table A.1 | Find $t$ tabled value in Table A.1 |
| $t_t = t_{(-\alpha; n_A + n_B - 2)}$ | $t_t = t_{(\alpha; n_A + n_B - 2)}$ | $t_t = t_{(\alpha/2; n_A + n_B - 2)}$ |
| $t_t = t_{(-0.10; 58)} = -1.303$ | $t_t = t_{(0.10; 58)} = 1.303$ | $t_t = t_{(0.025; 58)} = \pm 1.684$ |
| If $t_c < -1.303$, reject $H_0$ at $\alpha = 0.10$. | If $t_c > 1.303$, reject $H_0$ at $\alpha = 0.10$. | (because this is a two-tail test) |
| | | If $t_c < -1.684$ or $t_c > 1.684$, reject $H_0$ at $\alpha = 0.10$. |

*Step 5.* Perform the experiment. The researcher randomly selected 30 Group A and 30 Group B tubes, inoculated them with *Bacillus subtilis* spores, and placed them in the autoclave for 15 min at 121°C. The tubes were then removed and read after 24 h of incubation for growth (1) or no growth (0). Those tubes that read no growth were re-incubated and re-read everyday for seven consecutive days. The results are shown in Table 3.3.

**Table 3.3** Growth results, Example 3.3

| Group A | Group B |
|---|---|
| 0 0 0 1 1 | 0 0 1 0 0 |
| 0 1 0 1 0 | 1 0 0 1 0 |
| 0 1 1 0 1 | 0 0 1 0 1 |
| 1 1 0 0 1 | 0 0 0 1 0 |
| 0 0 1 0 1 | 0 1 0 1 0 |
| 0 1 1 0 1 | 0 0 0 1 0 |
| $n_A = 30$ | $n_B = 30$ |

$$P_A = \frac{\text{number of 0s}}{30} = \frac{15}{30} = 0.50 \qquad P_B = \frac{\text{number of 0s}}{30} = \frac{21}{30} = 0.70$$

$$P_c = \frac{n_A P_A + n_B P_B}{n_A + n_B} = \frac{30(0.50) + 30(0.70)}{30 + 30} = \frac{36}{60} = 0.60$$

$$t_c = \frac{P_A - P_B}{\sqrt{\dfrac{P_c(1 - P_c)}{n_A + n_B}}} = \frac{0.5 - 0.7}{\sqrt{\dfrac{0.6(1 - 0.6)}{30 + 30}}} = -3.16$$

*Step 6.* Make the decision concerning the test.

| Lower-Tail Test | Upper-Tail Test | Two-Tail Test |
|---|---|---|
| Because $t_c = -3.16 < t_t = -1.303$, one rejects $H_0$ at $\alpha = 0.10$. Configuration $A$ produces fewer "no growth" tubes than does Configuration $B$. | Because $t_c = -3.16 \ngtr t = 1.303$, one cannot reject $H_0$ at $\alpha = 0.10$. Configuration $A$ does not produce more "no growth" tubes than does Configuration $B$. | Because $t_c = -3.16 < t_t = -1.684$, one must reject $H_0$ at $\alpha = 0.10$. Configurations $A$ and $B$ differ in that Configuration $A$ produces significantly fewer "no growth" tubes than does Configuration $B$. |

## 3.2.2 Optional Two-Sample Bioequivalency Testing

Recall that bioequivalency testing – including noninferiority and nonsuperiority testing – is different from hypothesis testing for significance. In general significance testing, one is comparing two treatments to determine if they are significantly different – larger, smaller, or either. The null hypothesis ($H_0$) is rejected if outcomes are different enough to be determined significant at the $\alpha$ level specified. However, as discussed in Chapter 2, a rejection of the $H_A$ hypothesis does *not* mean there is no difference between samples groups, but only that a difference could *not* be detected by the statistic used.

Because many occasions occur when equivalence between data sets from two samples is the important question, an actual equivalency test can be used to assess this. For example, a microbiologist may want to replace one nutrient medium with another due to cost, substitute one supplier for another, use an alternate method, instead of the standard one, or even replace one stain with another – but only if these prove to be equivalent – that is, they can be substituted for one another in use. Recall from Chapter 2 that the alternative hypothesis ($H_A$) and the null hypothesis ($H_0$) are reversed in equivalence testing. That is, if two methods, samples, or items produce measurable data that are statistically equivalent – interchangeable – the $H_0$ hypothesis, now the hypothesis of difference, is rejected at $\alpha$. Conversely, if the two data sets cannot be demonstrated equivalent, the null hypothesis of nonequivalence is accepted.

General Hypothesis Construction for Equivalence Testing

| Lower-Tail | Upper-Tail | Two-Tail |
|---|---|---|
| $H_0$: $A < B$ <br> $H_A$: $A \geq B$ | $H_0$: $A > B$ <br> $H_A$: $A \leq B$ | $H_0$: $A \neq B$ <br> $H_A$: $A = B$ |
| Noninferiority | Nonsuperiority | Equivalence |
| If the test is significant, $H_0$ is rejected. | If the test is significant, $H_0$ is rejected. | If the test is significant, $H_0$ is rejected. |

In equivalence testing, one specifies what degree of difference constitutes a significant difference, and it is symbolized as $\Delta$. It may be a one $\log_{10}$ difference between methods, or 1/10 of one $\log_{10}$. $\Delta$ is a set value that sets a difference that is worth detecting. One concludes that at a specific difference, $\Delta$, the two samples are different if $A - B < -\Delta$, or $A - B > \Delta$. Noninferiority and nonsuperiority are one-sided tests of bioequivalence.[*]

### 3.2.3 Two Independent Samples: Sample Variances Assumed Equal

Let us look at the test structures carefully.

| Noninferiority Test: <br> $H_A$: $A \geq B$ | Nonsuperiority Test: <br> $H_A$: $A \leq B$ | Equivalence <br> (Both Nonsuperiority and Noninferiority Tests are used.) |
|---|---|---|
| Two-sample equivalence tests follow the same logic as testing one sample compared to a standard, as discussed in Chapter 2. In this case, one sample group is compared to another sample group arbitrarily labeled A and B. | | Two one-sided tests for equivalence rule out superiority and inferiority. |

---

[*] More precisely, a bioequivalence test is composed of two one-sided tests; note that "one-sided," the term used in bioequivalence testing, means the same as "one-tail."

Test Statistic Used:

| Noninferiority | Nonsuperiority | Equivalence |
|---|---|---|
| $t_{c_i} = \dfrac{d - (-\Delta)}{s_d}$ | $t_{c_s} = \dfrac{\Delta - d}{s_d}$ | $t_{c_i} = \dfrac{d - (-\Delta)}{s_d}$ and $t_{c_s} = \dfrac{\Delta - d}{s_d}$ |

where $d$ is the difference between sample group means. $\Delta$ is the set level of what constitutes a significant difference.

The six-step procedure makes the process easy.

*Step 1.* State the hypothesis.

| Noninferiority | Nonsuperiority | Equivalence |
|---|---|---|
| $H_0$: Group $A$ < Group $B$ <br> $H_A$: Group $A \geq$ Group $B$ <br> If $H_0$ is rejected, conclude Group A is noninferior to Group B at $\alpha$. | $H_0$: Group $A$ > Group $B$ <br> $H_A$: Group $A \leq$ Group $B$ <br> If $H_0$ is rejected, conclude Group A is nonsuperior to Group B at $\alpha$. | $H_0$: Group $A$ > Group $B$ <br> and <br> $H_0$: Group $A$ < Group $B$ <br> $H_A$: Group $A$ = Group $B$ <br> If both one-sided tests reject $H_0$, Group A = Group B at $\alpha$. |

*Step 2.* Set $\alpha$ and sample sizes, $n_A$ and $n_B$.
*Step 3.* Write out the test statistic used.
*Step 4.* Write the decision rule.

| Noninferiority Test | Nonsuperiority Test | Equivalence Test (Both noninferiority and nonsuperiority tests are run.) |
|---|---|---|
| If $t_{c_i} > t_t$, reject $H_0$ <br> $t_t = t_{(\alpha, n_A + n_B - 2)}$ | If $t_{c_s} > t_t$, reject $H_0$ <br> $t_t = t_{(\alpha, n_A + n_B - 2)}$ | Using 2 one-sided tests at $\alpha$: <br> If $t_{c_s} > t_t = t_{t(\alpha, n_A + n_B - 2)}$, and if <br> $t_{c_i} > t_t = t_{t(\alpha, n_A + n_B - 2)}$, accept equivalence at $\alpha$. |

*Step 5.* Perform the experiment.
*Step 6.* Make the decision.

*Example 3.4.* For many years, silver halide preparations have been used to treat dicubitus ulcers. A microbiologist working with a new formula that purports to make silver more readily absorbed into infected tissues wants to be sure that plasma blood levels of silver in test animals have not changed from those associated with the standard, less available silver compound. The following levels of Ag were detected in blood sera after 2 weeks of continuous use of the product formulae in laboratory test animals, 30 samples ($n$) per formula. $A$ is the new compound, and $B$ is the standard compound. The researcher can accept equivalence, if the two compounds do not differ more than 20%; $\Delta = \pm 20\%$.

*Step 1.* State the hypothesis.

| Noninferiority Test | Nonsuperiority Test | Equivalence Test |
|---|---|---|
| $H_0: A < B$ | $H_0: A > B$ | $H_0: A > B$ and $H_0: A < B$ |
| $H_A: A \geq B$ | $H_A: A \leq B$ | $H_A: A = B$ |
| $H_0$: The new treatment outcome is less than $-\Delta$. | $H_0$: The new treatment outcome is more than $\Delta$. | $H_0$: The new treatment outcome is different than that of the standard (larger or smaller). |

*Step 2.* Set $\alpha$, $n$, and $\Delta$.

| Noninferiority | Nonsuperiority | Equivalence |
|---|---|---|
| Let us set $\alpha$ at 0.10, $n_A = n_B = 30$, and $\Delta =$ 20% of standard results. | Let us set $\alpha$ at 0.10, $n_A = n_B = 30$, and $\Delta = 20\%$ of standard results. | Because we use 2 one-way tests, let us set $\alpha$ at 0.10 for each, $n_A = n_B = 30$, and $\Delta = 20\%$ of standard results. |

*Step 3.* Write out the test statistic.

| Noninferiority | Nonsuperiority | Equivalence |
|---|---|---|
| $t_{c_i} = \dfrac{d - (-\Delta)}{s_d}$ | $t_{c_s} = \dfrac{\Delta - d}{s_d}$ | $t_{c_i} = \dfrac{d - (-\Delta)}{s_d}$ and $t_{c_s} = \dfrac{\Delta - d}{s_d}$ |

*Step 4.* Write out the decision rule.

| Noninferiority | Nonsuperiority | Equivalence |
|---|---|---|
| If $t_{c_i} \geq t_t$, reject $H_0$; non-inferior at $\alpha = 0.10$. $t_t = t_{(0.10,\ 30 + 30 - 2)} = 1.303$. | If $t_{c_s} \geq t_t$, reject $H_0$; non-superior at $\alpha = 0.10$. $t_t = t_{(0.10,\ 30 + 30 - 2)} = 1.303$ | If $t_{c_i} > t_t$ and $t_{c_s} > t_t$, reject $H_0$; equivalent at $\alpha = 0.10$. $t_t = t_{(0.10,\ 30 + 30 - 2)} = 1.303$ |

*Step 5.* Perform the experiment. The collected data are presented in Table 3.4.

**Table 3.4** Collected data for Example 3.4

| n | A | B | n | A | B | n | A | B |
|---|---|---|---|---|---|---|---|---|
| 1 | 0.17 | 0.23 | 11 | 0.12 | 0.24 | 21 | 0.26 | 0.33 |
| 2 | 0.15 | 0.26 | 12 | 0.13 | 0.26 | 22 | 0.50 | 0.22 |
| 3 | 0.41 | 0.12 | 13 | 0.21 | 0.47 | 23 | 0.29 | 0.26 |
| 4 | 0.22 | 0.14 | 14 | 0.17 | 0.40 | 24 | 0.24 | 0.20 |
| 5 | 0.41 | 0.30 | 15 | 0.29 | 0.25 | 25 | 0.20 | 0.21 |
| 6 | 0.44 | 0.27 | 16 | 0.30 | 0.24 | 26 | 0.21 | 0.17 |
| 7 | 0.26 | 0.24 | 17 | 0.13 | 0.20 | 27 | 0.34 | 0.28 |
| 8 | 0.20 | 0.23 | 18 | 0.13 | 0.17 | 28 | 0.33 | 0.32 |
| 9 | 0.28 | 0.24 | 19 | 0.20 | 0.06 | 29 | 0.21 | 0.24 |
| 10 | 0.14 | 0.25 | 20 | 0.20 | 0.16 | 30 | 0.23 | 0.26 |

$$n_A = n_B = 30$$
$$\bar{x}_A = 0.2457 \qquad s_A = 0.0986$$
$$\bar{x}_B = 0.2407 \qquad s_B = 0.0785$$

We have set $\Delta$ as a 20% difference from $\bar{x}_B$; $\Delta = 0.20 \times 0.2407 = 0.0481$, which takes on two values, $\pm 0.0481$, corresponding to $\pm \Delta$. The nonsuperiority and noninferiority tests will be in terms of Group A to Group B. This is arbitrary, for we could also set the statistic in terms of Group B to Group A.

$d = \bar{x}_A - \bar{x}_B = 0.2457 - 0.2407 = 0.005$. This calculation provides the difference, $d$, between Groups A and B mean values. $\Delta$ provides the limit that these values can differ. A – B cannot be greater than $\Delta$ in order to reject $H_0$.

$$s_d = \sqrt{\frac{(n_A - 1)s_A^2 (n_B - 1)s_B^2}{n_A + n_B - 2}\left(\frac{1}{n_A} + \frac{1}{n_B}\right)}$$

$$s_d = \sqrt{\frac{29(0.0986)^2 + 29(0.0785)^2}{30 + 30 - 2}\left(\frac{1}{30} + \frac{1}{30}\right)} = 0.0230$$

We will compute both $t_{c_s}$ and $t_{c_i}$, because we need both in this example. If you are performing a nonsuperiority test, you need only compute $t_{c_s}$, and if you are only doing a noninferiority test, you need only compute $t_{c_i}$. If you are doing an equivalence test, you compute both.

$$t_{c_i} = \frac{d - (-\Delta)}{s_d} = \frac{0.005 - (-0.0481)}{0.023} = 2.3087$$

$$t_{c_s} = \frac{\Delta - d}{s_d} = \frac{0.0481 - 0.005}{0.023} = 1.8739$$

*Step 6.* Make the decision.

| Noninferiority | Nonsuperiority | Equivalence |
|---|---|---|
| As $t_{c_i} = 2.3087 > t_t = 1.303$, the $H_0$ hypothesis is rejected at $\alpha = 0.10$. Group $A$ is non-inferior to Group $B$. | As $t_{c_s} = 1.8739 > t_t = 1.303$, the $H_0$ hypothesis is rejected at $\alpha = 0.10$. Group $A$ is non-superior to Group $B$. | For the two one-tail tests, if each $t_c > t_t$, reject the $H_0$ hypothesis at $\alpha = 0.10$. We consider the groups equivalent. As $t_{c_s} = 1.8339 > 1.303$ and $t_{c_i} = 2.3087 \geq 1.303$, reject $H_0$ at $\alpha$. The groups are equivalent. |

## 3.2.4 Confidence Interval Approach

The procedures used in Chapter 2 also are used for two-sample tests.

*Nonsuperiority:* If $\Delta > d + t_\alpha s_d$, reject $H_0$ at $\alpha$.
*Noninferiority*: If $\Delta < d - t_\alpha s_d$, reject $H_0$ at $\alpha$.
*Equivalence*: If $\Delta$ falls outside of $d \pm t_{\alpha/2} s_d$, reject $H_0$ at $\alpha$.

# Chapter 4

## Analysis of Variance

Key Point

Analysis of Variance is used in
comparing more than two sample groups.

One very important statistical method for microbiologists to understand is Analysis of Variance (ANOVA). In its most basic form, it is but an extension of two-sample testing that includes testing of more than two sample groups. For example, there may be four sample groups in a study to compare or three media types to contrast. The main difference from our previous work is that ANOVA uses variances, $s^2$, as its prime measurement. The $F$ distribution used in ANOVA is just the $t$-tabled value squared. A $t_t$ tabled value of 1.96, when computing two samples in the $F$ distribution is $F = 3.84$, which is $t^2$.

The general statistical rationale behind ANOVA lies in the comparison of two variances: the variance between the treatments and the variance within the treatments. If the between- and within-variances are equal, then no treatment effect is present. ANOVA variances are calculated by computing the sum of squares, e.g., $\sum (x - \bar{x})^2$, and dividing by the set degrees of freedom, which converts the sum of squares calculation into a variance estimate. The mean square error ($MS_E$) term in ANOVA is the variance estimate, $\sigma_E^2$, within the treatments, that is, the random variability. The mean square treatment ($MS_T$) term consists of the variance estimate, $\sigma_T^2$, between the treatments; that is, the treatment effect. If the $H_0$ hypothesis is true, there is no treatment effect – then both terms ($MS_T$ and $MS_E$) are unbiased estimates of the random variability, and no treatment effect is present. That is, $MS_T \approx MS_E$.

If a treatment effect is present, $MS_T$ is composed of both random error and the treatment effect, so $MS_{Treatment} = \sigma_E^2 + \sigma_T^2$. If the treatment effect is significant, $MS_{Error}$ remains the measurement of random error, or $\sigma_E^2$. If $\sigma_T^2 = 0$, or no treatment

D.S. Paulson, *Biostatistics and Microbiology*, doi: 10.1007/978-0-387-77282-0_4,
© Springer Science + Business Media, LLC 2008

64        4. Analysis of Variance

effects are present, then the treatment effect drops out of the equation; hence, $MS_{Treatment} = MS_{Error}$. There are no upper- or lower-tail tests in ANOVA, only a two-tail test, or a measure of differences.

Let us now look at one-factor Analysis of Variance models – one factor in that only one variable is evaluated between and within sample groups. We will discuss two forms of this ANOVA.

1. The Completely Randomized Design, and
2. The Randomized Complete Block Design.

Key Point

The completely randomized design is analogous to the two sample independent *t*-test, and the randomized complete block design is analogous to the two-sample matched-pair test.

## 4.1 The Completely Randomized One-Factor ANOVA

Suppose a microbiologist decides to compare four spore recovery methods A, B, C, and D (the treatment factor) measured by the colony counts recovered of *Bacillus subtilis* spores. The easiest way to depict this is presented in Table 4.1.

**Table 4.1** One factor, four treatments, and three replicates

|  | Suspension (treatment factor) | | | |
| --- | --- | --- | --- | --- |
|  | A | B | C | D |
| Replicates | 1 | 1 | 1 | 1 |
|  | 2 | 2 | 2 | 2 |
|  | 3 | 3 | 3 | 3 |

A completely randomized design is distinguished in that the order of the experimental units tested is selected at random across all treatments. That is, one draws the run order from the complete data set of 12 points (4 treatment suspensions × 3 replicates) in completely randomized sampling. Hence, each of the 12 data points has a 1/12 probability of being selected over the 12 sample run times. In order to do this, the cells can be recoded sequentially 1 through 12, as in Table 4.2.

**Table 4.2** Coded cells for population recovery of *Bacillus subtilis* spores

|  | Suspension (treatment factor) | | | |
| --- | --- | --- | --- | --- |
|  | A | B | C | D |
| Replicates | 1 | 4 | 7 | 10 |
|  | 2 | 5 | 8 | 11 |
|  | 3 | 6 | 9 | 12 |

Then each of the 12 data points are selected at random, say, with an outcome as depicted in Table 4.3.

**Table 4.3** Randomization schematic for population recovery of *Bacillus subtilis* spores

|  | Suspension (treatment factor) | | | |
|---|---|---|---|---|
|  | A | B | C | D |
| Replicates | 1 (4) | 4 (9) | 7 (2) | 10 (5) |
|  | 2 (10) | 5 (7) | 8 (8) | 11 (12) |
|  | 3 (3) | 6 (11) | 9 (1) | 12 (6) |

Look at suspension C and the third value point, 9. To the immediate right, in parenthesis, is (1). The "1" designates that this is the first sample run through the experimental process. This is followed by suspension C's seventh value point, (2). This, in turn, is followed by suspension A's third value point, (3), and is continued until the 12th replicate of suspension D, 11, is run.

The linear statistical model for this design is:

$$y_{ij} = \mu + T_i + e_{ij} \qquad \begin{cases} i = 1, 2, \ldots k \text{ treatments} \\ j = 1, 2, \ldots n \text{ replicates} \end{cases}$$

where $y_{ij}$ is the quantitative value of the *jth* treatment (colony-forming units, in this case), $\mu$ is the common mean of all treatment factors, $T_i$ is the *ith* treatment effect, where $i$ = A, B, C, and D treatments, and $e_{ij}$ is the random error, or noise component that cannot be accounted for via the treatments.

To keep things as simple and straight-forward as possible, let us work an example to learn this statistical method.

*Example 4.1.* A microbiologist is working with a new antibiotic drug, which has proved very effective on bacteria producing $\beta$-lactamase, specifically *Staphylococcus aureus*, Methicillin-resistant strains. Five replicate drug samples will be used with four different preservatives. The drug samples are calculated to undergo an aging process of 90 days at 45°C prior to the study's commencement. The study will expose each drug sample to a 10 mL suspension of $1.0 \times 10^5$ organisms per mL for twenty (20) min. The bacterial cells will then be cleansed of any extracellular drugs, plated in duplicate, and incubated at $35 \pm 2$°C for 24–48 h. The microbial reductions from baseline will then be calculated and compared, using a completely randomized, one-factor analysis of variance method.

### Key Point

Analysis of variance tests are for detecting differences between treatments.

The six-step procedure simplifies the ANOVA calculation process.

*Step 1.* State the test hypotheses.

$H_0$: Treatment A = Treatment B = Treatment C = Treatment D in microbial reductions.

$H_A$: At least one treatment differs from the others in terms of microbial reductions.

*Step 2.* Specify $\alpha$ and $n$. Because this is a preliminary research and development evaluation, $n = 5$ replicates of each treatment, for a total of $5 \times 4 = 20$ samples, and we will arbitrarily set $\alpha$ at 0.10.

*Step 3.* Select the test statistic.

We will use the completely randomized analysis of variance design to compute an $F_c$ ($F_{calculated}$) value.

*Step 4.* Present the acceptance/rejection criteria.

If $F_c > F_t$, reject $H_0$ at $\alpha$.

$$F_t = F_{t(\alpha; \text{ numerator, denominator})} = F_{t(\alpha;\ [a-1],\ a[n-1])}$$

where $a$ = number of treatment factors compared (here, 4), $n$ = number of replicates within each treatment factor (here, 5), and $\alpha$ = set level (here, $\alpha = 0.10$).

$F_{t(0.10);(4-1,\ 4(5-1))} = F_{t(0.10;\ 3,\ 16)}$ from Table A.4, the $F$ Distribution Table. 3 is the numerator, 16 is the denominator, and $\alpha = 0.10$.

## The Key to Using the $F$ Table

First, find in Table A.4 the appropriate $\alpha$ level to be used. Find the degrees of freedom value (here, it is 3) for the numerator, $v_1$. This is presented in the first row. Then, find the degrees of freedom for the denominator, $v_2$ (here, it is 16). Where 3 and 16 meet at $\alpha = 0.10$, the $F$ value resides. It is 2.46, or $F_{t(0.10;\ 3,\ 16)} = 2.46$.

So, if $F_c > 2.46$, reject $H_0$ at $\alpha = 0.10$.

*Step 5.* Perform the experiment.

The microbiologist researcher randomized the run order of the experiment across treatments and replicates ($4 \times 5 = 20$ samples). Twenty samples were selected at random from the stability chamber. Next, twenty (20) labels, numbered 1 through 20, were placed in a box and shaken for 2 min. The researcher pulled one label out with the value, "3." The sample labeled "3" would be the first sampled. Then, the label "3" was put back in the box, which was again shaken for 2 min. The next drawing was "17." The 17th sample would be the second one tested. This process continued until all the treatments were assigned a run order, which is listed in Table 4.4.

Table 4.4 Randomly-selected drug sample labels

| Treatment A | | Treatment B | | Treatment C | | Treatment D | |
|---|---|---|---|---|---|---|---|
| Sample | Run order | Sample | Run order | Sample | Run order | Sample | Run order |
| 1 | (3) | 6 | (5) | 11 | (12) | 16 | (14) |
| 2 | (17) | 7 | (10) | 12 | (16) | 17 | (7) |
| 3 | (9) | 8 | (4) | 13 | (18) | 18 | (20) |
| 4 | (11) | 9 | (15) | 14 | (8) | 19 | (13) |
| 5 | (1) | 10 | (19) | 15 | (2) | 20 | (6) |

Undoubtedly, after a number of drawings, a sample previously drawn is redrawn. It is then merely placed back into the box, reshuffled, and a new value drawn. This process is termed sampling with replacement, and it assures that each value has

$1/N = 1/20$ probability of being selected. $N$, here, represents the total number of samples (number of treatments $\times$ number of replicates). If the drawn labels are not placed back into the box, the sample schema becomes biased. In this example, the first drawn is $1/N$, or $1/20$ probability, the second drawn would be $1/N - 1 = 1/19$ probability, and then $1/N - 2$, and all the way to $1/20 - 19$, or $1/1$.

An easier way to do this is with a computer-generated random number program. After all 20 samples have been acquired, the completely randomized run order is presented (Table 4.5). This schema is easier to follow.

**Table 4.5** Completely randomized run order for Example 4.1

| Treatment number | Run order |
|---|---|
| 1 | 3 |
| 2 | 17 |
| 3 | 9 |
| 4 | 11 |
| 5 | 1 |
| 6 | 5 |
| 7 | 10 |
| 8 | 4 |
| 9 | 15 |
| 10 | 19 |
| 11 | 12 |
| 12 | 16 |
| 13 | 18 |
| 14 | 8 |
| 15 | 2 |
| 16 | 14 |
| 17 | 7 |
| 18 | 20 |
| 19 | 13 |
| 20 | 6 |

After the study was conducted, according to the run order, the colony count values were recorded in $\log_{10}$ scale (Table 4.6).

**Table 4.6** Treatment values for Example 4.1

|  |  | Treatments | | | |
|---|---|---|---|---|---|
|  |  | A | B | C | D |
|  | 1 | 3.17 | 2.06 | 2.27 | 4.17 |
|  | 2 | 2.91 | 3.21 | 3.78 | 4.01 |
| Replicates | 3 | 4.11 | 2.57 | 3.59 | 3.92 |
|  | 4 | 3.82 | 2.31 | 4.01 | 4.29 |
|  | 5 | 4.02 | 2.71 | 3.15 | 3.72 |

In the completely randomized analysis of variance design, one is interested in comparing the variances among the treatments (treatment effects) and the variance within the treatments (random error effect). Both these components are what make up the total variability.

| Total Variability | = | Variability due to Treatments | + | Variability due to Random Error |
|---|---|---|---|---|
| $SS_{Total}$ | = | $SS_{Treatment}$ | + | $SS_{Error}$ |
| $\sum\sum(x_{ij}-\overline{\overline{x}})$ | = | $a\sum(\overline{x}._{.j}-\overline{\overline{x}})^2$ | + | $\sum\sum(x_{ij}-\overline{x}._{.j})^2$ |

which is the difference of each value from the grand average = difference of each treatment mean from grand mean + difference of each replicate from its own mean.

$SS_{Total}$ (sum of squares total) consists of the variability between each $x_{ij}$ observation and the mean of all observations, the grand mean, $\overline{\overline{x}}$. The total sum of squares in ANOVA is compartmentalized into variability among the groups (treatment effect) and within the groups (random error effect). The treatment effect is the difference between the treatment means, $\overline{x}._{.j}$, and the grand or overall mean, $\overline{\overline{x}}$. Variability within each treatment group (the random error component) is the difference between individual $x_{ij}$ observations within a specific treatment group, $\overline{x}._{.j}$.

All of these differences are squared; if they are not, their sums will equal zero.

Because the formulas look complex, we will use an ANOVA table and simplify the work (Table 4.7).

**Table 4.7** ANOVA table for Example 4.1

| Treatment A $(x_{ij}-\overline{x}._{.1})^2$ | Treatment B $(x_{ij}-\overline{x}._{.2})^2$ | Treatment C $(x_{ij}-\overline{x}._{.3})^2$ | Treatment D $(x_{ij}-\overline{x}._{.4})^2$ |
|---|---|---|---|
| $(3.17-3.61)^2 = 0.1936$ | $(2.06-2.57)^2 = 0.2601$ | $(2.27-3.36)^2 = 1.1881$ | $(4.17-4.02)^2 = 0.0225$ |
| $(2.91-3.61)^2 = 0.4900$ | $(3.21-2.57)^2 = 0.4096$ | $(3.78-3.36)^2 = 0.1764$ | $(4.01-4.02)^2 = 0.0001$ |
| $(4.11-3.61)^2 = 0.2500$ | $(2.57-2.57)^2 = 0.0000$ | $(3.59-3.36)^2 = 0.0529$ | $(3.92-4.02)^2 = 0.0100$ |
| $(3.82-3.61)^2 = 0.0441$ | $(2.31-2.57)^2 = 0.0676$ | $(4.01-3.36)^2 = 0.4225$ | $(4.29-4.02)^2 = 0.0729$ |
| $(4.02-3.61)^2 = 0.1681$ | $(2.71-2.57)^2 = 0.0196$ | $(3.15-3.36)^2 = 0.0441$ | $(3.72-4.02)^2 = 0.0900$ |
| $\sum(x_{ij}-\overline{x}._{.1})^2 = 1.1458$ | $\sum(x_{ij}-\overline{x}._{.2})^2 = 0.7569$ | $\sum(x_{ij}-\overline{x}._{.3})^2 = 1.8840$ | $\sum(x_{ij}-\overline{x}._{.4})^2 = 0.1955$ |
| $\overline{x}_A = \dfrac{18.03}{5} = 3.61$ | $\overline{x}_B = \dfrac{12.86}{5} = 2.57$ | $\overline{x}_C = \dfrac{16.80}{5} = 3.36$ | $\overline{x}_D = \dfrac{20.11}{5} = 4.02$ |

*Step 1.* Find the mean values $(\overline{x}._{.j})$ of each treatment. This is done using the information computed in Table 4.7. In this example, there are 4: $\overline{x}_A$, $\overline{x}_B$, $\overline{x}_C$, and $\overline{x}_D$. Next, find the grand mean, or the mean of the means ($\overline{\overline{x}}$).

$$\overline{\overline{x}} = \frac{\overline{x}_A+\overline{x}_B+\overline{x}_C+\overline{x}_D}{4} = \frac{3.61+2.57+3.36+4.02}{4} = 3.39$$

*Step 2.* Find $(x_{ij}-\overline{x}._{.j})^2$, or the difference within each treatment, the error estimate. This is the individual treatment values minus the means of the individual treatments for that group. These individual values were determined in Table 4.7.

$$SS_{Error} = \sum\sum(x_{ij}-\overline{x}._{.j})^2 = (1.1458+0.7569+1.8840+0.1955) = 3.982$$

*Step 3.* Find the $SS_{Total}$. This is simply the grand mean subtracted from each $x_{ij}$ value squared, then summed.

$$SS_T = \sum\sum (x_{ij} - \overline{\overline{x}})^2$$
$$= (3.17 - 3.39)^2 + (2.91 - 3.39)^2 + \ldots + (4.29 - 3.39)^2 + (3.72 - 3.39)^2$$
$$= 9.563$$

*Step 4.* Find the $SS_{Treatment}$ by subtraction.

$$SS_{Total} - SS_{Error} = SS_{Treatment}$$
$$9.563 - 3.982 = 5.581 = SS_{Treatment}$$

*Step 5.* Create the ANOVA procedure table (Table 4.8). Based on the formulas listed in Table 4.8, create Table 4.9. Enter the computed values $SS_{Total}$, $SS_{Error}$, and $SS_{Treatment}$ in Table 4.9. Then, determine the degrees of freedom for the treatment and error terms. Enter those values into Table 4.9. Next, divide the mean square treatment value ($MS_{Treatment}$) by the mean square error ($MS_{Error}$) to get the $F$ calculated value ($F_c$).

Key Point

| |
|---|
| Remember this table, and use it! |

**Table 4.8** Completely randomized design ANOVA procedure table, Example 4.1

| Source | $DF$ | Sum of squares | Mean square | $F_c$ |
|---|---|---|---|---|
| Treatment | $c - 1$ | $SS_{Treatment}$ | $\dfrac{SS_{Treatment}}{c-1} = MS_{Treatment}$ | $\dfrac{MS_{Treatment}}{MS_{Error}} = F_c$ |
| Error | $c(r-1)$ | $SS_{Error}$ | $\dfrac{SS_{Error}}{c(r-1)} = MS_{Error}$ | |
| Total | $rc - 1$ | $SS_{Total}$ | | |

$DF$ = degrees of freedom. The sum of squares must be averaged by the appropriate degrees of freedom; $c$ = number of columns or treatments; $r$ = number of rows or replicates; $c = 4$; $c - 1 = 4 - 1 = 3$ degrees of freedom; $c(r-1) = 4(5-1) = 4 \times 4 = 16$ degrees of freedom; $rc - 1 = 5 \times 4 - 1 = 20 - 1 = 19$, or just add degrees of freedom for treatments and error, $3 + 16 = 19$.

**Table 4.9** Values from ANOVA table for Example 4.1

| Source | DF | Sum of squares | Mean square | $F_c$ |
|---|---|---|---|---|
| Treatment | 3 | 5.581 | 5.581/3 = 1.860 | 1.860/0.249 = 7.47 |
| Error | 16 | 3.982 | 3.982/16 = 0.249 | |
| Total | 19 | 9.563 | | |

The calculated $F$ value is $F_c = 7.47$.

*Step 6.* Make the decision. If $F_c > F_t$, reject $H_0$ at the $\alpha$ level.

Because $F_c = 7.47 > F_t = 2.46$, $F_{t(\alpha;\, c-1,\, c(r-1))} = F_{t(0.10;\, 3,\, 16)} = 2.46$ (from Table A.4). Reject $H_0$ at $\alpha = 0.10$. In fact, looking at the $F$ table (Table A.4) at 3, 16 degrees of freedom, $F_c$ is larger than $F_t$ at $\alpha = 0.01 = 5.29$. Hence, $p < 0.01$, which is highly significant.

## Key Point

Remember $p < 0.01$ means the probability of computing an $F$ value as large or larger than 7.47, given the null hypothesis is true, is less than 0.01, or in statistical terms, $p(F_c \geq 7.47 | H_0 \text{ true}) < 0.01$.

The same results can be achieved using a statistical software package. We will use MiniTab®. For MiniTab®, the data are input into two columns. Here, column 1, or C1, is the treatment response, and Column 2, or C2, is the $j^{th}$ treatment value, coded 1 through 4, where 1 = Treatment A, 2 = Treatment B, 3 = Treatment C, and 4 = Treatment D.

Table 4.10 represents the actual MiniTab® input values. The $\log_{10}$ treatment values are in Column 1 (C1) and the treatments, 1 – 4, in Column 2. Table 4.11 is the actual MiniTab® ANOVA output. Note also that the exact $p$ value, $p = 0.002$, is computed, as well as a 95% confidence interval for each treatment.

**Table 4.10** Actual MiniTab® input table, Example 4.1

| $n$ (Row) | C1 (Treatment value) | C2 (Treatment code) | $n$ (Row) | C1 (Treatment value) | C2 (Treatment code) |
|---|---|---|---|---|---|
| 1 | 3.17 | 1 | 11 | 2.27 | 3 |
| 2 | 2.91 | 1 | 12 | 3.78 | 3 |
| 3 | 4.11 | 1 | 13 | 3.59 | 3 |
| 4 | 3.82 | 1 | 14 | 4.01 | 3 |
| 5 | 4.02 | 1 | 15 | 3.15 | 3 |
| 6 | 2.06 | 2 | 16 | 4.17 | 4 |
| 7 | 3.21 | 2 | 17 | 4.01 | 4 |
| 8 | 2.57 | 2 | 18 | 3.92 | 4 |
| 9 | 2.31 | 2 | 19 | 4.29 | 4 |
| 10 | 2.71 | 2 | 20 | 3.72 | 4 |

MiniTab® also provides a table of $1 - \alpha$ confidence intervals. The sample groups that overlap in their confidence intervals are not different from each other at the $\alpha$ level of significance. In this example, Level 1 (Treatment A) and Level 2 (Treatment B) confidence intervals do not overlap, so they are different from one another at $\alpha = 0.05$. However, Level 1 (Treatment A) and Level 3 (Treatment C) confidence intervals overlap, so they are not significantly different at $\alpha = 0.05$.

**Table 4.11** Actual MiniTab® ANOVA output

One-way ANOVA with 95% confidence intervals for each treatment

| Source | DF | SS | MS | F | P |
|---|---|---|---|---|---|
| C2 (Treatment) | 3 | 5.581 | 1.860 | 7.47 | 0.002 |
| Error | 16 | 3.982 | 0.249 | | |
| Total | 19 | 9.563 | | | |

Individual 95% CIs for Mean Based on Pooled St Dev

| Level | N | Mean | StDev | |
|---|---|---|---|---|
| 1 | 5 | 3.6060 | 0.5352 | ( -------+------- ) |
| 2 | 5 | 2.5720 | 0.4350 | ( -------+------- ) |
| 3 | 5 | 3.3600 | 0.6863 | ( -------+------- ) |
| 4 | 5 | 4.0220 | 0.2211 | ( -------+------- ) |

Pooled StDev = 0.4989

```
    -----+----------+----------+----------+-----
        2.40       3.00       3.60       4.20
```

## 4.2 Contrasts

The ANOVA table (Table 4.11) has clearly demonstrated that significant differences exist among treatments, but which one or ones? The graph in Table 4.11 is a valuable aid in determining which treatments differ but is not actually a contrast procedure. Instead, it is a set of individual confidence intervals with a common variance. In practice, one will want to actually perform a series of contrast tests. There are a number of methods that can be used to determine where differences exist. We will use the Tukey method, for it is neither too liberal, nor too conservative.

The Tukey method requires the microbiologist to use a Studentized range value ($q_{\alpha;\,a,\,f}$) (Table A.3), where $\alpha$ is the Type I error level, $a$ is the number of treatments or groups, and $f = N - a$, the total number of observations minus the number of treatment groups, $a$. The $q_\alpha$ value provides a critical Studentized value for all pairwise contrasts, much like a $t_{\alpha/2}$ value in the two sample $t$-test. All possible treatment pairs are compared to one another. If $|\bar{x}_i - \bar{x}_j| > q_{(\alpha;\,a,\,f)}s_x$, reject $H_0$. The two treatments ($\bar{x}_i$ and $\bar{x}_j$) differ significantly at $\alpha$.

The $s_x$ value is computed as:

$$s_x = \sqrt{s^2/n} = \sqrt{MS_E/n}$$

where $n$ = number of replicates per sample set.

There are $a(a-1)/2$ pair-wise contrasts possible. In this four-treatment group example, there are $a(a-1)/2 = 4(4-1)/2 = 12/2 = 6$ pair-wise contrasts. They are 1 vs. 2, or A vs. B; 1 vs. 3, or A vs. C; 1 vs. 4, or A vs. D; 2 vs. 3, or B vs. C; 2 vs. 4, or B vs. D; and 3 vs. 4, or C vs. D. $n = 5$ replicates.

$MS_E = 0.249$ (from Table 4.11).

$$s_x = \sqrt{\frac{MS_E}{n}} = \sqrt{\frac{0.249}{5}} = 0.2232$$

Let us use $\alpha = 0.05$, $a = 4$, $f = N - a = 20 - 4 = 16$, and $q_{(0.05;\ 4,\ 16)} = 4.05$ (from Table A.3). $q\,s_x = 4.05(0.2232) = 0.9040 =$ the critical value.
If $|\bar{x}_i - \bar{x}_j| > q\,s_x$, the pairs differ at $\alpha$.

| Mean Difference | $q\,s_x$ | Significant/ Not Significant at $\alpha$ |
|---|---|---|
| $\|\bar{x}_A - \bar{x}_B\| = \|3.606 - 2.572\| = 1.034$ | $> 0.9040$ | Significant |
| $\|\bar{x}_A - \bar{x}_C\| = \|3.606 - 3.360\| = 0.300$ | $\ngtr 0.9040$ | Not Significant |
| $\|\bar{x}_A - \bar{x}_D\| = \|3.606 - 4.022\| = 0.416$ | $\ngtr 0.9040$ | Not Significant |
| $\|\bar{x}_B - \bar{x}_C\| = \|2.572 - 3.360\| = 0.788$ | $\ngtr 0.9040$ | Not Significant |
| $\|\bar{x}_B - \bar{x}_D\| = \|2.572 - 4.022\| = 1.450$ | $> 0.9040$ | Significant |
| $\|\bar{x}_C - \bar{x}_D\| = \|3.606 - 4.022\| = 0.660$ | $\ngtr 0.9040$ | Not Significant |

A plot of the treatments shows where these differences are (Fig. 4.1). Table 4.11, the MiniTab® plot of 95% confidence intervals, also did this in this example.

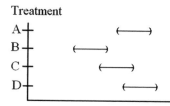

**Fig. 4.1** Differences among treatments, Example 4.1

At the 95% confidence level, Treatment A is different from Treatment B but not Treatments C or D. Treatment B is different from Treatments A and D but not Treatment C. Treatment C is not different from the others. Treatment D is different from Treatment B but not Treatments A or C.

## 4.3 Confidence Intervals

A $100(1 - \alpha)$ confidence interval for each treatment group can be calculated.

$$\mu_i = \bar{x}_i \pm t_{(\alpha/2;\ N-a)}\sqrt{\frac{MS_E}{n}}$$

For example, an $\alpha = 0.01$ confidence interval for $\mu_1$ and $\mu_3$ will be presented.

$a = $ number of treatments $= 4$

$n_1 = n_2 = n_3 = n_4 = 5$

$t_{(0.01/2;16)} = 2.921$ (from Table A.1)

$MS_E = 0.249$

$$\mu_1 = \bar{x}_1 \pm t_{(\alpha/2;\ N-a)}\sqrt{\frac{MS_E}{n}} = 3.606 \pm 2.921\sqrt{\frac{0.249}{5}}$$

$$\mu_1 = 3.606 \pm 0.6518$$

$$2.9542 \le \mu_1 \le 4.2578$$

$$\mu_3 = 3.36 \pm 0.6518$$

$$2.7082 \le \mu_3 \le 4.0118$$

## 4.4 Sample Size Calculation

The appropriate sample size should always be determined when beginning the six-step procedure, specifically at Step 2. Also, if the alternative hypothesis ($H_A$) is rejected at the end of the experiment (Step 6), it is prudent to check the sample size to assure that the statistic was powered adequately to detect significant differences.

We will employ a general formula for determining the sample size of each treatment.

$$n \ge \frac{as^2(z_{\alpha/2} + z_\beta)^2}{\delta^2}$$

where $n$ = individual treatment replicate size, $a$ = number of treatment groups to compare, $s^2 = MS_E$, $z_{\alpha/2}$ = normal table value (Table A.2) for $\alpha/2$, $z_\beta$ = normal table value (Table A.2) for $\beta$, and $\delta$ = detection level required by the statistic.

In example 4.1, $a = 4$ and $s^2 = MS_E = 0.249$. Let us use $\alpha = 0.05$, $\beta = 0.20$, and an $\delta$ value of 1.5. So,

$$z_{\alpha/2} = 1.96 \ \text{(Table A.2)}$$

$$z_\beta = 0.842 \ \text{(Table A.2)}$$

$$\delta = 1.5 \ \log_{10}$$

$$n \ge \frac{as^2(z_{\alpha/2} + z_\beta)^2}{\delta^2}$$

$$n \ge \frac{4(0.249)(1.96 + 0.842)^2}{1.5^2} = 3.48 \approx 4$$

Because we used five replicates, the detection level, $\delta$, will be smaller than 1.5 $\log_{10}$ microorganisms. What is the detection level? Simply rearrange algebraically the terms of the sample size determination formula, solving for $\delta$.

$$\delta = \sqrt{\frac{as^2(z_{\alpha/2} + z_\beta)^2}{n}}$$

$$\delta = \sqrt{\frac{4(0.249)(1.96 + 0.842)^2}{5}}$$

$$\delta = 1.25 \ \log_{10}$$

So, if one required a 1.5 $\log_{10}$ detection level, the sample size would be more than adequate. If a detection level of one $\log_{10}$ was needed, the study could fail to detect true differences. That is, it would be underpowered. The correct sample size minimum would be per treatment group.

$$n \geq \frac{4(0.249)(1.96+0.842)^2}{1.0^2} = 7.82 \approx 8 \text{ replicates per group}$$

## 4.5 Randomized Block Design

Like the paired two-sample $t$-test, the randomized block design attempts to add power to the statistical method by blocking or comparing treatments within the homogeneous groups. For example, to evaluate the incidence of post-surgical infections in patients who have been presurgically prepped with one of three preoperative compounds – a 4% chlorhexidine gluconate (CHG), a 10% povidone iodine (PVP-I), and a parachlorametaxylenol (PCMX) – four hospitals in different regions of the United States were evaluated. Each hospital (block) applied the three topical antimicrobials, which were randomized to the patients in each hospital (Fig. 4.2).

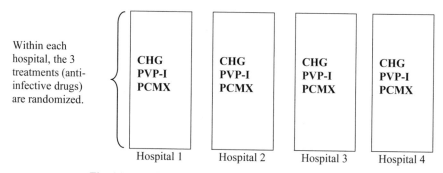

**Fig. 4.2** Randomization is restricted to within each block

Each hospital may vary in average post-surgical infection rates (Fig. 4.3), but blocking will remove the hospital effect, by comparing the three products' actual effects in each hospital. Hence, in Hospital 1, Block 1 randomizes who gets each of the three drugs. The same process is followed by Hospitals 2, 3, and 4. Hospitals are not compared to one another and are known to be different in infection rates. If they were not, blocking would be of no use. The treatments are confined to their evaluation within each hospital, not among them.

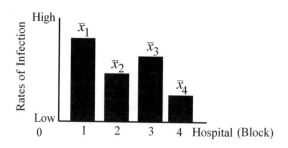

**Fig. 4.3** Infection rates per hospital depicted by a bar chart

The critical difference between the completely randomized design previously discussed and the randomized block design is that, in the latter design, the randomization is within each block, not among blocks.

In the completely randomized design (Table 4.12), each $N$ observation is equally likely to be sampled, because the randomization is over all $N$ values ($N = 15$).

**Table 4.12** Completely randomized design

|  | | Treatments | | |
|---|---|---|---|---|
|  | $n$ | 1 | 2 | 3 |
|  | 1 | 7 | 3 | 8 |
|  | 2 | 12 | 13 | 4 |
| Replicates | 3 | 9 | 1 | 11 |
|  | 4 | 14 | 2 | 15 |
|  | 5 | 10 | 5 | 6 | $N=15$ |

In the randomized block design, randomization occurs only over the three treatments within in each block (Table 4.13).

**Table 4.13** Randomized block design

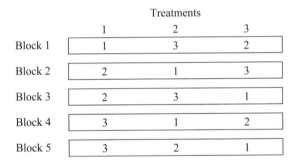

|  | Treatments | | |
|---|---|---|---|
|  | 1 | 2 | 3 |
| Block 1 | 1 | 3 | 2 |
| Block 2 | 2 | 1 | 3 |
| Block 3 | 2 | 3 | 1 |
| Block 4 | 3 | 1 | 2 |
| Block 5 | 3 | 2 | 1 |

The randomized block design is usually the more powerful design, for it isolates differences in blocks (hospitals) to get a more accurate computation of the treatment effect (surgical preparation product). So, the randomized block model requires the computation of the sum of squares for the blocks, or $SS_B$.

The sum of squares for treatment effect is $SS_T = b\sum(\bar{x}_{.j} - \bar{\bar{x}})^2$, and the mean square treatment = $MS_T = \dfrac{SS_T}{a-1}$, where $b$ = number of blocks and $a$ = number of treatments. The sum of squares for blocks is $SS_B = a\sum(\bar{x}_{i.} - \bar{\bar{x}})^2$, and the mean square blocks = $MS_B = \dfrac{SS_B}{b-1}$. The sum of squares total is $\sum\sum(x_{ij} - \bar{\bar{x}})^2$. The mean square total is not computed. The sum of squares error is $\sum\sum(x_{ij} - \bar{x}_{.j})^2$, and the mean square error = $MS_E = \dfrac{SS_E}{(a-1)(b-1)}$. The complete model for the randomized block design is:

$$\sum\sum\left(x_{ij}-\bar{\bar{x}}\right)^2 \quad = \quad b\sum\left(\bar{x}_{.j}-\bar{\bar{x}}\right)^2 \quad + \quad a\sum\left(\bar{x}_{i.}-\bar{\bar{x}}\right) \quad + \quad \sum\sum\left(x_{ij}-\bar{x}_{.j}\right)^2$$
$$SS_{Total} \qquad = \qquad SS_{Treatment} \qquad + \qquad SS_{Block} \qquad + \qquad SS_{Error}$$

Combining these in an ANOVA table, we have Table 4.14.

**Table 4.14** ANOVA table

| Variation | Degrees of freedom | Sum of squares | Mean square | $F_c$ | $F_t$ |
|---|---|---|---|---|---|
| Treatments (Columns) | $a-1$ | $SS_T$ | $\dfrac{SS_T}{a-1}=MS_T$ | $\dfrac{MS_T}{MS_E}=F_{cT}$ | $F_{t(\alpha;\,a-1,\,(a-1)(b-1))}$ (Table A.4) |
| Blocks (Rows) | $b-1$ | $SS_B$ | $\dfrac{SS_B}{b-1}=MS_B$ | $\dfrac{MS_B}{MS_E}=F_{cB}$ | $F_{t(\alpha;\,b-1,\,(a-1)(b-1))}$ (Table A.4) |
| Error | $(a-1)(b-1)$ | $SS_E$ | $\dfrac{SS_E}{(a-1)(b-1)}=MS_E$ | | |
| Total | $N-1$ | Total $SS$ | | | |

*Example 4.2.* A microbiologist was asked to evaluate the effectiveness of three antimicrobials in hospitals throughout the continental United States. Four large teaching hospitals were chosen: A = Southeast, B = Northeast, C = Northwest, and D = Southwest. Three generally supplied bulk antimicrobial preoperative skin preparation formulations were randomly assigned to surgical patients during a six-month period. Out of 1000 procedures, the numbers of post-surgical infections were tallied. We will use the six-step procedure to perform this example.

*Step 1.* Write out the test hypothesis.
   $H_0$: 4% CHG = Alcohol + CHG = PVP-I in infection rates per 1,000 procedures
   $H_A$: At least one antimicrobial formulation is different from the others in infection rates per 1,000 procedures
Also, we want to assure the blocking was useful; that is, was the blocking significant, although we have no need to distinguish blocks?
   $H_0$: The blocks (hospitals) are not significantly different
   $H_A$: At least one hospital is different from the others
   (If the block effect is not significant, there is no need to use the block design. It is better to use the completely randomized design.)
*Step 2.* Select $\alpha$ and the number of treatments.
   $\alpha = 0.05$
   $a = 3$ products (treatments)
   $b = 4$ hospital groups (blocks)
*Step 3.* Write out the test statistic to use.
   This will simply be the randomized block design.
*Step 4.* Make the decision rule. First, we need some intermediate calculations.
   $a$ = number of antimicrobial product formulations = 3
   $b$ = number of hospitals = 4
   $(a-1)(b-1)$ = degrees of freedom for error term = $(3-1)(4-1) = 2 \times 3 = 6$

$(a-1)$ = degrees of freedom for treatments = $3-1=2$
$(b-1)$ = degrees of freedom for blocks = $4-1=3$

*Treatments*

$F_{t(\alpha;\ a-1;\ (a-1)(b-1))} = F_{t(0.05;\ 2,\ 6)} = 5.14$ (Table A.4)
If $F_c > 5.14$, reject $H_0$ at $\alpha = 0.05$. The treatments are significantly different from one another at $\alpha = 0.05$.

*Blocks*

$F_{t(\alpha;\ b-1,\ (a-1)(b-1))} = F_{t(0.05;\ 3,\ 6)} = 4.76$ (Table A.4)
If $F_c > 4.76$, reject $H_0$ at $\alpha = 0.05$. The blocks are significantly different from one another at $\alpha = 0.05$, so the block design is applicable.

*Step 5.* Conduct the experiment. The sample design looked like this:

| Antimicrobial Formulation | 1 4% CHG | | 2 Alcohol + CHG | | 3 Povidone Iodine | |
|---|---|---|---|---|---|---|
| Block | | | | | | |
| Hospital A | 12 | (1) | 19 | (3) | 31 | (2) |
| Hospital B | 11 | (2) | 9 | (3) | 17 | (1) |
| Hospital C | 17 | (1) | 12 | (2) | 25 | (3) |
| Hospital D | 22 | (2) | 15 | (1) | 35 | (3) |

$(x)$ = experimental run order by block (hospital)

Notice that the randomization was restricted to within each block. After the study had been completed, the actual rate per 1000 infections were put into the table for computational ease.

| Blocks $(i)$ | Antimicrobial Formulation $(j)$ | | | |
|---|---|---|---|---|
| | $1._j$ | $2._j$ | $3._j$ | |
| Hospital $A_{j.}$ | 21 | 19 | 31 | $\bar{x}_{A.} = 23.67$ |
| Hospital $B_{j.}$ | 11 | 9 | 17 | $\bar{x}_{B.} = 12.33$ |
| Hospital $C_{j.}$ | 17 | 12 | 25 | $\bar{x}_{C.} = 18.00$ |
| Hospital $D_{j.}$ | 22 | 15 | 35 | $\bar{x}_{D.} = 24.00$ |
| | $\bar{x}_{.1} = 17.75$ | $\bar{x}_{.2} = 13.75$ | $\bar{x}_{.3} = 27.00$ | $19.50 = \bar{\bar{x}}$ |

$$SS_T = SS_{Treatment} = b\sum\left(\bar{x}_{.j} - \bar{\bar{x}}\right)^2$$
$$= 4\left[(17.75 - 19.5)^2 + (13.75 - 19.5)^2 + (27.00 - 19.5)^2\right] = 369.50$$
$$SS_B = SS_{Blocks} = a\sum\left(\bar{x}_{i.} - \bar{\bar{x}}\right)^2$$
$$= 3\left[(23.65 - 19.5)^2 + (12.33 - 19.5)^2 + (18.00 - 19.5)^2 + (24.00 - 19.5)^2\right] = 273.39$$
$$SS_T = SS_{Total} = \sum\sum\left(x_{ij} - \bar{\bar{x}}\right)^2$$
$$= (21.00 - 19.5)^2 + (11.00 - 19.5)^2 + \ldots + (25.00 - 19.5)^2 + (35.00 - 19.5)^2 = 683.00$$
$$SS_E = SS_{Error} = SS_{Total} - SS_T - SS_B$$
$$= 683.00 - 369.50 - 273.39 = 40.11$$

Make an ANOVA table (Table 4.15).

**Table 4.15** ANOVA table

| Source | SS value | df | Mean square (MS) | $F_c$ | $F_t$ | Significant/ Not significant |
|---|---|---|---|---|---|---|
| Treatments | 369.50 | 2 | 369.50/2=184.75 | 184.75/6.685=27.64 | 5.14[a] (Table A.4) | Significant |
| Blocks | 273.39 | 3 | 273.39/3=91.13 | 91.13/6.685=13.63 | 4.76[b] (Table A.4) | Significant |
| Error | 40.11 | 6 | 40.11/6=6.685 | | | |
| Total $SS$ | 683.00 | | | | | |

[a] $F_{t(\alpha;\, a,\, (a-1)(b-1))}$
[b] $F_{t(\alpha;\, b,\, (a-1)(b-1))}$

*Step 6.* Make the decision.

Because $F_c = 27.64 > F_t = 5.14$, clearly, the topical antimicrobial preparations are associated with significantly different post-operative infection rates. Also, because $F_c = 13.63 > F_t = 4.76$, the block effect was significant at $\alpha = 0.05$, so the block design was effective.

One can do the same thing via a statistical computer program. Using the MiniTab® computer software's ANOVA function, the data are keyed in as presented in Table 4.16.

**Table 4.16** Data, Example 4.2

| | C1 $x_i$ (infection rate) | C2 Row code = block hospital (1, 2, 3, 4) | C3 Column code = treatment (1, 2, 3) |
|---|---|---|---|
| 1 | 21 | 1 | 1 |
| 2 | 19 | 1 | 2 |
| 3 | 31 | 1 | 3 |
| 4 | 11 | 2 | 1 |
| 5 | 9 | 2 | 2 |
| 6 | 17 | 2 | 3 |
| 7 | 17 | 3 | 1 |
| 8 | 12 | 3 | 2 |
| 9 | 25 | 3 | 3 |
| 10 | 22 | 4 | 1 |
| 11 | 15 | 4 | 2 |
| 12 | 35 | 4 | 3 |

The infection rates are in column 1 (C1), the blocks in column 2 (C2), and the treatments in column 3 (C3). The analysis of variance table created with MiniTab® is

| Source | DF | SS | MS | F | P |
|---|---|---|---|---|---|
| Treatment | 2 | 369.500 | 184.750 | 27.83 | 0.0001 |
| Blocks | 3 | 273.667 | 91.222 | 13.74 | 0.004 |
| Error | 6 | 39.833 | 6.639 | | |
| Total | 11 | 683.00 | | | |

The same results as in a table prepared by hand are achieved by the software program. The $p$-values are those provided using the MiniTab® program. The $p$ value can easily be done by hand, comparing $F_c$ to $F_{tabled}$. One merely finds the first $\alpha$ value for the $F_t$ value, which is greater than $F_c$. For example, if $F_c > F_{t0.05}$, then compare $F_c$ to $F_{t0.01}$. If $F_c > F_{t0.01}$, compare it to $F_{t0.025}$. If $F_c < F_{t0.025}$, then the $p$ value is the last $F_t$ $\alpha$ where $F_c > F_t$. Here, $F_t$ at $\alpha = 0.01$, so the $p$ value is 0.01, or $p \leq 0.01$.

## 4.6 Pair-wise Contrasts

When the treatment effect is significant, differences among treatments exist, but their location is unknown. To determine where the differences exist, a set of contrasts needs to be employed. We will continue to use the Tukey Contrast method. There are $a(a-1)/2 = 3(3-1)/2 = 6/2 = 3$ possible pair-wise contrasts.

$$\bar{x}_1 - \bar{x}_2$$
$$\bar{x}_1 - \bar{x}_3$$
$$\bar{x}_2 - \bar{x}_3$$

where $s_x = \sqrt{MS_E/b} = \sqrt{6.685/4} = 1.2928$ (the number of blocks, "$b$," is used in the denominator).

As with the completely randomized design, if $|\bar{x}_i - \bar{x}_j| > q_{t(\alpha; a, f)} s_x$, where $a$ is the number of treatments, and $f$ is the degrees of freedom of the error term, which is $(a-1)(b-1) = (3-1)(4-1) = 2 \times 3 = 6$. $q_{t(0.05; 3; (a-1)(b-1))} = q_{t(0.05, 3, 6)} = 4.76$ (from Table A.3). $qs_x = 4.76 (1.2928) = 6.15$. If $|\bar{x}_i - \bar{x}_j| > 6.11$, the pairs different from one another at $\alpha = 0.05$.

| | |
|---|---|
| $\|\bar{x}_1 - \bar{x}_2\| = \|17.75 - 13.75\| = 4 < 6.15$ | Not Significant |
| $\|\bar{x}_1 - \bar{x}_3\| = \|17.75 - 27.07\| = 9.25 > 6.15$ | Significant |
| $\|\bar{x}_2 - \bar{x}_3\| = \|13.75 - 27.0\| = 13.25 > 6.15$ | Significant |

$\bar{x}_1$ and $\bar{x}_2$ are not significantly different from each other, but $\bar{x}_1$ and $\bar{x}_3$, as well as $\bar{x}_2$ and $\bar{x}_3$, are at $\alpha = 0.05$. Recall that $\bar{x}_1$ is 4% CHG, $\bar{x}_2$ is alcohol + CHG, and $\bar{x}_3$ is PVP-I. Hence, the 4% CHG and the alcohol + CHG are not significantly different from each other at $\alpha = 0.05$, but the 4% CHG and the alcohol + CHG are both significantly different from the povidone iodine.

## 4.7 100 $(1 - \alpha)$ Confidence Intervals

The $1 - \alpha$ confidence interval for each treatment mean is calculated as:

$$\mu_t = \bar{x}_t \pm t_{(\alpha/2; (a-1)(b-1))} \sqrt{\frac{MS_E}{b}}$$

Let us determine the $1 - \alpha$ confidence intervals for the $\bar{x}_1$, $\bar{x}_2$, and $\bar{x}_3$ that predict $\mu_1$, $\mu_2$, and $\mu_3$, the population parameters. Recall that

$$\bar{x}_1 = 17.75$$
$$\bar{x}_2 = 13.75$$

$$\bar{x}_3 = 27.00$$
$$MS_E = s^2 = 6.685$$
$$a = 3; b = 4$$

Let us use $\alpha = 0.10$, where $\alpha/2 = 0.05$; $t_{(0.05,\ 6)} = 1.943$, from Table A.1, the Student's $t$ table.

$$\mu_1 = \bar{x}_1 \pm t_{(\alpha/2;\ (a-1)(b-1))}\left(\sqrt{\frac{MS_E}{b}}\right)$$

$$\mu_1 = 17.75 \pm 1.943\left(\sqrt{\frac{6.685}{4}}\right)$$

$$17.75 \pm 2.512$$
$$15.24 \le \mu_1 \le 20.26$$

$$\mu_2 = \bar{x}_2 \pm t_{(\alpha/2;\ (a-1)(b-1))}\left(\sqrt{\frac{MS_E}{b}}\right)$$

$$\mu_2 = 13.75 \pm 1.943\left(\sqrt{\frac{6.685}{4}}\right)$$

$$13.75 \pm 2.512$$
$$11.24 \le \mu_2 \le 16.26$$

$$\mu_3 = \bar{x}_3 \pm t_{(\alpha/2;\ (a-1)(b-1))}\left(\sqrt{\frac{MS_E}{b}}\right)$$

$$27 \pm 2.512$$
$$24.49 \le \mu_3 \le 29.51$$

A 90% confidence interval graph can be made of these intervals to determine if the confidence intervals overlap (Fig. 4.4). If they do, they are equivalent; if not, they are different at $\alpha = 0.10$.

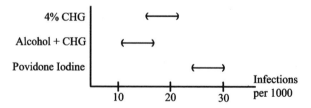

Fig. 4.4 Overlapping confidence intervals, Example 4.2

A statistical software package can also produce the set of $1 - \alpha$ confidence intervals. Figure 4.4 is the output the MiniTab® confidence interval set of these data. The MiniTab® software produces the following Fig. 4.5.

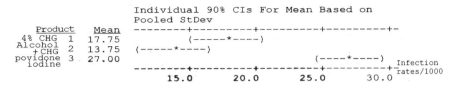

```
                       Individual 90% CIs For Mean Based on
                       Pooled StDev
   Product     Mean    --------+---------+---------+---------+-
4% CHG    1    17.75                 (-----*----)
Alcohol
  + CHG    2    13.75    (-----*----)
povidone  3    27.00                               (----*----)
  iodine                                                       Infection
                       --------+---------+---------+---------+ rates/1000
                          15.0       20.0      25.0      30.0
```

**Fig. 4.5** Individual confidence intervals, Example 4.2

## 4.8 Sample Size Calculation

The sample size calculation that was used in the completely randomized design is also used here. The $s^2 = MS_E$ will generally be smaller than the $MS_E$ in a completely randomized design.

$$n \geq \frac{as^2(z_{\alpha/2} + z_\beta)}{\delta^2}$$

where

$n$ = sample size per treatment group
$a$ = number of treatments to be compared
$z_{\alpha/2}$ = normal tabled value for $\alpha$ (Table A.2)
$z_\beta$ = normal tabled value for $\beta$ (Table A.2)
$\delta^2$ = minimum detectable level required
$s^2 = MS_E$ (from prior experience)

The detection level, $\delta$, can also be determined using the same formula as was used in the completely randomized design.

$$\delta = \sqrt{\frac{as^2(z_{\alpha/2} + z_\beta)}{n}}$$

In conclusion, when performing analysis of variance procedures, use your microbiological reasoning to drive your decision-making process. Also, plan to keep the samples for each group the same.

# Chapter 5

## Regression and Correlation Analysis

An important component in microbiological testing is time, usually presented in terms of exposure time. A good example is the exposure of microorganisms to certain topical antimicrobials over discrete time periods to determine the rate of inactivation, known as the $D$-value, which is the time to reduce the initial microbial population by one $\log_{10}$ – that is, a tenfold reduction.

A regression analysis is appropriate when two variables, $y$ and $x$, are related. $y$ is termed the dependent variable (population of microorganisms), and $x$, the independent variable (an incremental time component). The dependent variable, $y$, is the quantitative measured response when the independent variable, $x$, is set at various levels or points. For example, take the case of a certain microbial population level of *Staphylococcus aureus*, the dependent variable, when exposed to a 10% povidone iodine solution for different time intervals, the independent variable. The independent variable (exposure time) might be set, say, at 10 min, 20 min, and 30 min, and the dependent variable, $y$, is the microbial population counts at the 10-, 20-, and 30-min points of exposure. Regression analysis "fits" a linear function between the two variables, $y$ and $x$. Suppose that when $x = 1$, $y = 2$; when $x = 2$, $y = 4$; when $x = 3$, $y = 6$; when $x = 4$, $y = 8$; and when $x = 5$, $y = 10$.

A function of this relationship is graphed (Fig. 5.1).

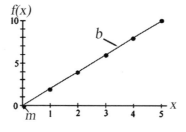

**Fig. 5.1** $f(x)$ = Function of $y$

D.S. Paulson, *Biostatistics and Microbiology*, doi: 10.1007/978-0-387-77282-0_5,
© Springer Science + Business Media, LLC 2008

The dots connected by a line is termed the function of $x$, or $f(x)$. In algebra, the function equation is $f(x) = m + bx$, where $m$ is the $f(x)$ intercept when $x = 0$. In statistics, the equation, $\hat{y} = a + bx$, is used, where $b$ = rise/run = $y_2 - y_1/x_2 - x_1$ = $\Delta y/\Delta x$ = slope of the regression line, $\hat{y}$ = predicted value of $y$, and $a = y$ intercept when $x = 0$.

As in algebra, $b$ has the same basic concept as $m$, the slope of the regression line, except generally, $y$ values corresponding to $x$ values do not result in a perfect linear relationship. Hence, $b$, in statistics, is estimated by the method of least squares. That method determines the fit between $y$ and $x$ that produces the smallest error term. This will be discussed more fully later.

Keeping things simple for the moment, where $\hat{y} = 2x$, the $y$ intercept passes through the origin; or when $y = 0$, $x = 0$. But most times, the $y$ intercept is not 0, but $a$, when $x = 0$. Take, for example, $\hat{y} = a + bx$, in which $a = 3$ and $b = 0.5$. The equation is $\hat{y} = 3 + 0.5x$. When $x = 0$ in this equation, $y = 3$ (Fig. 5.2). When $x = 3$, $\hat{y} = 3 + 0.5(3) = 4.5$, and when $x = 7$, $\hat{y} = 3 + 0.5(7) = 6.5$.

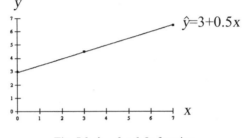

**Fig. 5.2** $\hat{y} = 3 + 0.5x$ function

However, as previously cautioned, only rarely will the $x$ and $y$ values fit to a straight line perfectly, like this example. Instead, both $a$ and $b$ must be estimated by a statistical process. An example of such a case is presented in Fig. 5.3.

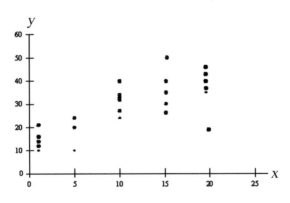

**Fig. 5.3** Collected discrete data points over a time period

In this case, the data points vary at each measured $x$ point.

What is the slope (rise/run) here? When $x = 0$, what does $y$ equal? One could take a straight edge and try to fit a line through the values, but there is a much more accurate way: the method of least squares, which assures that, of all the different data combinations possible, the one calculated produces the smallest overall error term.

## 5.1 Least Squares Equation

The regression procedure begins by calculating the slope, $b$, using the least squares equation.

$$\text{slope} = b = \frac{\sum xy - n\bar{x}\bar{y}}{\sum x^2 - n\bar{x}^2}$$

Next, the $y$ intercept, or $a$, is calculated: $y \text{ intercept} = a = \bar{y} - b\bar{x}$.

Let's use an example to demonstrate the process (Example 5.1). Suppose the following data were collected in a sequential time exposure of $1 \times 10^6$ *Escherichia coli* to povidone iodine for 15 s, 30 s, 1 min, and 5 min, three replicates at each time point ($n = 15$). A requirement of linear regression is that the $y$, $x$ data must have a linear relationship, or produce a straight line. In order to do this, all $y$ values have been transformed to $\log_{10}$ scale, which is customary in microbiology. We also need $xy$ and $x^2$ columns to calculate $b$ (Table 5.1).

**Table 5.1** Example 5.1 Data

| $n$ | $y$ (Colony counts [$\log_{10}$ scale]) | $x$ (Exposure time) | $xy$ | $x^2$ |
|---|---|---|---|---|
| 1 | 6.0 | 0.00 | 0.000 | 0.0000 |
| 2 | 5.8 | 0.00 | 0.000 | 0.0000 |
| 3 | 6.5 | 0.00 | 0.000 | 0.0000 |
| 4 | 5.1 | 0.25 | 1.275 | 0.0625 |
| 5 | 5.2 | 0.25 | 1.300 | 0.0625 |
| 6 | 5.6 | 0.25 | 1.400 | 0.0625 |
| 7 | 4.8 | 0.50 | 2.400 | 0.2500 |
| 8 | 4.7 | 0.50 | 2.350 | 0.2500 |
| 9 | 5.1 | 0.50 | 2.550 | 0.2500 |
| 10 | 4.4 | 1.00 | 4.400 | 1.0000 |
| 11 | 4.6 | 1.00 | 4.600 | 1.0000 |
| 12 | 4.5 | 1.00 | 4.500 | 1.0000 |
| 13 | 2.3 | 5.00 | 11.500 | 25.0000 |
| 14 | 2.1 | 5.00 | 10.500 | 25.0000 |
| 15 | 2.5 | 5.00 | 12.500 | 25.0000 |
| | $\bar{y} = 4.6133$ | $\bar{x} = 1.35$ | $\sum xy = 59.275$ | $\sum x^2 = 78.9375$ |

$$b = \frac{\sum xy - n\bar{x}\bar{y}}{\sum x^2 - n\bar{x}^2} = \frac{59.275 - (15)(1.35)(4.6133)}{78.9375 - 15(1.35)^2} = -0.6618$$

$$a = \bar{y} - b\bar{x} = 4.6133 - (-0.6618)1.35 = 5.5067$$

So, the regression equation is $\hat{y} = 5.5067 - 0.6618x$. When $x = 0$, the $y$ intercept is $a$, or 5.5067 $\log_{10}$. The slope is expressed as a "$b$" change on the $y$ axis for one increment of change on the $x$ axis. That is, for every minute of exposure, $x$, the population, $y$, was reduced by 0.6618 $\log_{10}$. The negative sign on 0.6618 means the values descend over time.

The data evaluated must be linear, forming a straight line, for the equation to be valid in predicting the $y$ values, so it is wise to plot the data points and the estimated regression line. To check for linearity, one can plot the $y$, $x$ data with pencil and paper or use a statistical software package. We will use the MiniTab software to plot the data (Fig. 5.4). Something looks suspicious, because where $x = 0$, $\hat{y}$ is predicted to be 5.5067. Yet, when $x = 0$, the actual data points are 6.0, 5.8, and 6.5, which indicates a significant underestimation of the data by the regression equation.

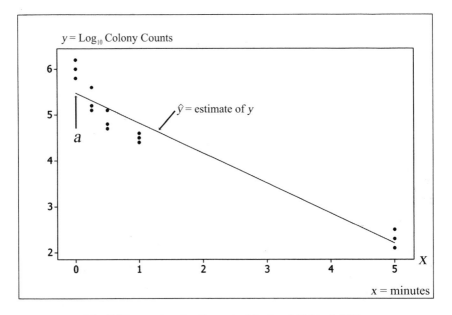

**Fig. 5.4** Regression plot: Example 5.1, $\hat{y} = 5.5067 - 0.6618x$

Perhaps the easiest thing to do here to make the regression line more representative of the data is to drop the $x = 0$ values. Many times in regression, nonlinear patterns are not as obvious as in Fig. 5.4, so a more precise method of detection must be used in conjunction with the $y$: $\hat{y}$ vs. $x$ data plot. That method employs the residual, $e$. The $e$ is merely the difference between the actual $y$ value and the predicted $\hat{y}$

value; that is, $e = y - \hat{y}$. Table 5.2 shows the $y$, $x$, $\hat{y}$, and $e$ data. When plotted against $x$ values, $e$ values should be patternless and centered at 0 (Fig. 5.5).

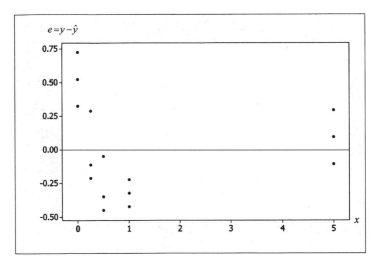

**Fig. 5.5** Scatterplot of $e$ and $x$, Example 5.1

**Table 5.2** Diagnostic regression data values, Example 5.1

| $n$ | $y$ | $x$ | $\hat{y}$ | $e = y - \hat{y}$ |
|---|---|---|---|---|
| 1 | 6.0 | 0.00 | 5.47607 | 0.523934 |
| 2 | 5.8 | 0.00 | 5.47607 | 0.323934 |
| 3 | 6.5 | 0.00 | 5.47607 | 1.023930 |
| 4 | 5.1 | 0.25 | 5.31260 | -0.212597 |
| 5 | 5.2 | 0.25 | 5.31260 | -0.112597 |
| 6 | 5.6 | 0.25 | 5.31260 | 0.287403 |
| 7 | 4.8 | 0.50 | 5.14913 | -0.349128 |
| 8 | 4.7 | 0.50 | 5.14913 | -0.449128 |
| 9 | 5.1 | 0.50 | 5.14913 | -0.049128 |
| 10 | 4.4 | 1.00 | 4.82219 | -0.422190 |
| 11 | 4.6 | 1.00 | 4.82219 | -0.222190 |
| 12 | 4.5 | 1.00 | 4.82219 | -0.322190 |
| 13 | 2.3 | 5.00 | 2.20669 | 0.093314 |
| 14 | 2.1 | 5.00 | 2.20669 | -0.106686 |
| 15 | 2.5 | 5.00 | 2.20669 | 0.293314 |

Clearly, the $e$ data are not patternless (random about the $x$ values), nor are they centered around the $e = 0$.[*]

---

[*]The entire data set of the $e$ values, when summed equals zero, so as a whole, they are centered at 0. What are not centered are the $y_i$ values for specific $x_i$ values.

So, to correct this, the $x = 0$ data and corresponding values of $y$ are removed from the data set. A new regression calculation for $b$ and $a$ is performed, providing $b = -0.59890$, $a = 5.2538$, and a new regression equation, $\hat{y} = 5.2538 - 0.5989x$. The $y$, $\hat{y}$ vs. $x$ data from this second iteration are plotted in Fig. 5.6.

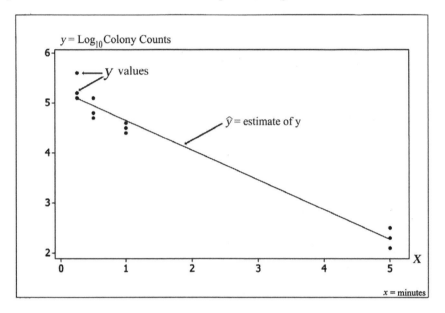

**Fig. 5.6** Regression plot with $x = 0$ values removed: Example 5.1, $\hat{y} = 5.2538 - 0.5989x$

The new regression plot, with $x = 0$ data, and the corresponding $y$ values removed, shows marked improvement in the linearity of the data. Even so, they are certainly not perfect, but rarely will a data set be so.

The $y$, $x$, $\hat{y}_i$, and $e_i$ data are presented in Table 5.3.

**Table 5.3** Diagnostic regression data, Example 5.1, without zero-time values

| $n$ | $y$ | $x$ | $\hat{y}$ | $e = y - \hat{y}$ |
|---|---|---|---|---|
| 1 | 5.1 | 0.25 | 5.09644 | 0.003560 |
| 2 | 5.2 | 0.25 | 5.09644 | 0.103560 |
| 3 | 5.6 | 0.25 | 5.09644 | 0.503560 |
| 4 | 4.8 | 0.50 | 4.94778 | -0.147784 |
| 5 | 4.7 | 0.50 | 4.94778 | -0.247784 |
| 6 | 5.1 | 0.50 | 4.94778 | 0.152216 |
| 7 | 4.4 | 1.00 | 4.65047 | -0.250471 |
| 8 | 4.6 | 1.00 | 4.65047 | -0.050471 |
| 9 | 4.5 | 1.00 | 4.65047 | -0.150471 |
| 10 | 2.3 | 5.00 | 2.27197 | 0.028028 |
| 11 | 2.1 | 5.00 | 2.27197 | -0.171972 |
| 12 | 2.5 | 5.00 | 2.27197 | 0.228028 |

## 5.2 Strategy for Linearizing Data

The situation just encountered is very common in regression. The data simply do not fit a straight line. Fortunately, many data sets can be adjusted by re-expressing the scale. Many times, one does not know the correct mathematical re-expression to use. Four common data patterns are presented in Fig. 5.7 that can be easily re-expressed in a linearized relationship.

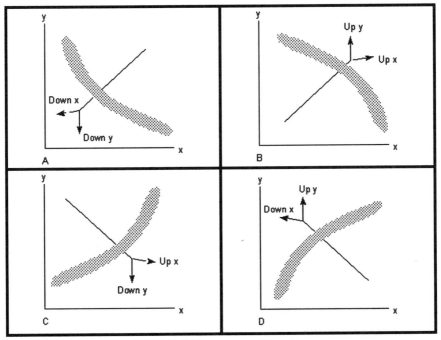

**Fig. 5.7 (a–d)** Four common data patterns

Key Point

Relax; this is not hard.

Down $x$ = reduce power scale of $x$          Up $x$ = increase power scale of $x$
Down $y$ = reduce power scale of $y$          Up $y$ = increase power scale of $y$

If the actual $x$, $y$ data look like Graph A, the data can be linearized by decreasing the power of either the $x$ or $y$ data, even of both. If the data resemble Graph B, the data can be linearized by increasing the power of the $x$ or the $y$ data, even of both. If the data resemble Graph C, the data can be linearized by increasing the power of $x$ or decreasing the power of $y$. If the data resemble Graph D, they can be linearized by increasing the power of $y$ or decreasing the power of $x$.

## 5.3 The Power Scale

| Power | $x$ | $y$ re-expression |
|:---:|:---:|:---:|
| ⋮ | ⋮ | ⋮ |
| 3 | $x^3$ | $y^3$ |
| 2 | $x^2$ | $y^2$ |
| neutral→  1 | $x^1$ | $y^1$ |
| ½ | $\sqrt{x}$ | $\sqrt{y}$ |
| 0 | $\log x$ | $\log y$ |
| −½ | $\dfrac{-1}{\sqrt{x}}$ | $\dfrac{-1}{\sqrt{y}}$ |
| −1 | $\dfrac{-1}{x}$ | $\dfrac{-1}{y}$ |
| −2 | $\dfrac{-1}{x^2}$ | $\dfrac{-1}{y^2}$ |
| −3 | $\dfrac{-1}{x^3}$ | $\dfrac{-1}{y^3}$ |
| ⋮ | ⋮ | ⋮ |

Up in power is >1, and down in power is <1. All one needs to do is iteratively pick a power, transform the data using the power re-expression, and then replot to determine visually if the data appear to be linear.

Generally, it is better to transform the $y$ data, instead of the $x$ values. This is because, when reporting the data, it is often harder for the reader to comprehend the data when the $x$ values are transformed than when the $y$ data are transformed.

## 5.4 Using Regression Analysis

Usually, a microbiologist wants not only to model a data set, but also, to predict $y$ values (response values) based on one or more $x$ values with confidence that the predicted values are close to the measured values. There are two basic approaches for this process: one is to calculate a 100 $(1 - \alpha)$ confidence interval for the *average* ($\bar{\hat{y}}$) of the $y$ values predicted, and the other is a 100 $(1 - \alpha)$ confidence interval for the *specific* $\hat{y}$ value predicted. Although $\bar{\hat{y}}$ and $\hat{y}$ will be the same value, the width of the 100 $(1 - \alpha)$ confidence interval on each will differ. A confidence interval on $\bar{y}$, on the average, is narrower than is a confidence interval on a specific value. Predictions, when using either of these approaches, should avoid extrapolation; that is, predicting $y$ values outside the range of the collected $x$ data points. Why? It is because the data may not be linear beyond the data set collected. Most $\hat{y}$ response predictions will be interpolated values within the range of the $x$ values, and not necessarily an $x$ value used to produce the regression equation (Fig. 5.8).

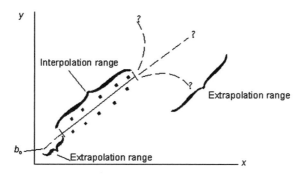

**Fig. 5.8** Extrapolation and interpolation of regression data

## 5.5 Predicting the Average $\hat{y}$ from an $x$ Value

This is the more commonly used of the two prediction processes. First, $\hat{y} = a + bx$ is determined from the data set. Second, the value of an $x$ used to predict the $y$ value is plugged in the regression formula to predict $\hat{y}$. Third, a 100 $(1 - \alpha)$ confidence interval is constructed around the $\hat{y}$. The 100 $(1 - \alpha)$ confidence interval formula is $\hat{y} \pm t_{\alpha/2, n-2} s_y$, where

$$s_y = \sqrt{\frac{\sum(y - \hat{y})^2}{n - 2}\left[\frac{1}{n} + \frac{(x - \bar{x})^2}{\sum(x - \bar{x})^2}\right]}.$$

Let us use the data in Example 5.1 without the 0 time points, as presented in Table 5.3 (repeated here for convenience).

**Table 5.3** Diagnostic regression data, Example 5.1, without zero-time values

| $n$ | $y$ | $x$ | $\hat{y}$ | $e = y - \hat{y}$ |
|---|---|---|---|---|
| 1 | 5.1 | 0.25 | 5.09644 | 0.003560 |
| 2 | 5.2 | 0.25 | 5.09644 | 0.103560 |
| 3 | 5.6 | 0.25 | 5.09644 | 0.503560 |
| 4 | 4.8 | 0.50 | 4.94778 | -0.147784 |
| 5 | 4.7 | 0.50 | 4.94778 | -0.247784 |
| 6 | 5.1 | 0.50 | 4.94778 | 0.152216 |
| 7 | 4.4 | 1.00 | 4.65047 | -0.250471 |
| 8 | 4.6 | 1.00 | 4.65047 | -0.050471 |
| 9 | 4.5 | 1.00 | 4.65047 | -0.150471 |
| 10 | 2.3 | 5.00 | 2.27197 | 0.028028 |
| 11 | 2.1 | 5.00 | 2.27197 | -0.171972 |
| 12 | 2.5 | 5.00 | 2.27197 | 0.228028 |

Recall that $\hat{y} = 5.2538 - 0.5989x$. Let us predict $y$ when $x = 3$ min. $\hat{y} = 5.2538 - 0.5989(3) = 3.4612$.

Next, we compute the $1 - \alpha$ confidence interval on the average $\hat{y}$ when $x = 3$. Let us use $\alpha = 0.05$ for a confidence interval of 95%.

$$\mu = \hat{y} \pm t_{(\alpha/2, n-2)} s_{\bar{y}}$$

$$t_{(0.05/2, 12-2)} = t_{(0.025, 10)} = 2.228 \text{ , from Table A.1 (the Student's } t \text{ table).}$$

$$\frac{\sum(y-\hat{y})^2}{n-2} = \frac{0.5410}{10} = 0.0541$$

$$\sum(x-\bar{x})^2 = \sum(0.25-1.6875)^2 + (0.25-1.6875)^2 + ... + (5.00-1.6875)^2 + (5.00-1.6875)^2$$

$$\sum(x-\bar{x})^2 = 44.7656$$

$$s_{\bar{y}} = \sqrt{\frac{\sum(y_i-\hat{y})^2}{n-2}\left[\frac{1}{n}+\frac{(x-\bar{x})^2}{\sum(x-\bar{x})^2}\right]} = \sqrt{0.0541\left[\frac{1}{12}+\frac{(3.0-1.6875)^2}{44.7656}\right]}$$

$$s_{\bar{y}} = 0.0812$$

$$\mu = \hat{y} \pm t_{(\alpha/2, n-2)} s_{\bar{y}}$$

$$\mu = 3.4612 \pm 2.228(0.0812)$$

$$3.2803 \le \mu \le 3.6422$$

$$3.28 \le \mu \le 3.64$$

So, when predicting the true average, $\mu$, from $x = 3$, 95 out of 100 times, it will be contained within the interval, $3.28 \le \mu \le 3.64$ at $\alpha = 0.05$.

## 5.6 Predicting a Specific $\hat{y}$ Value from an $x$ Value

The prediction of a specific $\hat{y}$ value is similar to the prediction of $\hat{y}_i$, an average value. The first step is to compute the regression equation, $\hat{y} = a + bx$. Next, choose the $x$ from which to predict $\hat{y}$. Let us use $x = 3$.

$$\hat{y} = 5.2538 - 0.5989(3) = 3.4612 .$$

The prediction 100 $(1 - \alpha)$ confidence interval is $Y = \hat{y} \pm t_{(\alpha/2, n-2)} s_y$, where

$$s_y = \sqrt{\frac{\sum(y-\hat{y})^2}{n-2}\left[1+\frac{1}{n}+\frac{(x-\bar{x})^2}{\sum(x-\bar{x})^2}\right]}.$$

The difference between $s_y$ and $s_y$ is that the 1 is added to $(1/n + (x - \bar{x}^2))/\sum(x - \bar{x})^2$ in predicting $s_y$. This is because, in predicting a specific $\hat{y}$ value, the confidence interval will be greater than for predicting an average $\hat{y}$. Let us set $\alpha = 0.05$ and $t_{(0.05/2, 10)} = 2.228$, as before.

$$s_y = \sqrt{0.0541\left[1+\frac{1}{12}+\frac{(3.0-1.6875)^2}{44.7656}\right]} = 0.2464$$

$$Y = \hat{y} \pm t_{(\alpha/2, n-2)} s_y$$

$$Y = 3.4612 \pm 2.228(0.2464)$$

So, 95 out of 100 times, when predicting a $\hat{y}$ value from $x = 3$, the actual $\hat{y}$ value will be contained within the interval, $2.91 \leq Y \leq 4.01$ at $\alpha = 0.05$. This is clearly wider than $\mu$, $3.28 \leq \mu \leq 3.64$, because there is more uncertainty predicting a specific value, $Y$, than predicting the true average value, $\mu$.

## 5.7 Correlation

Directly related to regression analysis is correlation. Correlation measures the degree to which two variables are related or associated. For example, take the equation, $\hat{y} = 1 + 2x$. Table 5.4 presents $x_i$ and $y_i$ values, as well as $\hat{y}_i$ and $e_i$ values. Notice $y = \hat{y}$ in every case, so $e_i = 0$ in every case. There is no variability or error.

**Table 5.4** Correlation of $x_i$ and $y_i$ values

| $x$ | $y$ | $\hat{y}$ | $e = y - \hat{y}$ |
|-----|-----|-----------|-------------------|
| 1 | 3 | 3 | 0 |
| 2 | 5 | 5 | 0 |
| 3 | 7 | 7 | 0 |
| 4 | 9 | 9 | 0 |
| 5 | 11 | 11 | 0 |
| 6 | 13 | 13 | 0 |

Plotting this equation results in perfect correlation (Fig. 5.9).

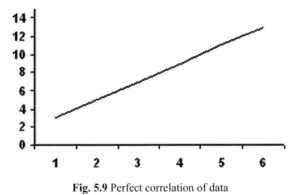

**Fig. 5.9** Perfect correlation of data

Perfect correlation occurs when the predicted $\hat{y}$ s and the actual $y$s are the same value, so $y - \hat{y} = 0$. Perfect correlation is 1. Conversely, when there is no relationship between the predicted $\hat{y}$ and the actual $y$ values, $\hat{y} \neq a + bx$; instead, $\hat{y} = \bar{y}$ (Fig. 5.10).

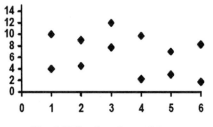

**Fig. 5.10** Random data points

Here, there is only a series of random points, so the correlation is 0. The value, $\bar{y}$, is then used to represent the data, because there is no slope.

Correlation is a very useful tool in regression, telling how well the predicted regression line function, $\hat{y}$, explains the $y_i$ data (Fig. 5.11). Fig. 5.11a shows data that have stronger correlation than those in Fig. 5.11b.

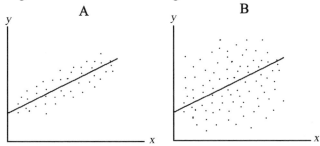

**Fig. 5.11** Correlation in regression

## 5.8 Correlation Coefficient: *r*

The most widely used value in correlation analysis is the correlation coefficient, $r$. The $r$ value ranges between $-1$ and $+1$, negative if the regression slope is negative, and positive if the regression slope is positive (Fig. 5.12).

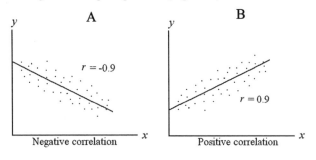

**Fig. 5.12** Negative and positive correlation

This simply means that $r$ is negative if the $y$ values get smaller as the $x$ values get larger (Fig. 5.12a), and $r$ is positive if the $y$ values gets larger as the $x$ values get larger (Fig. 5.12b). The closer the value of $r$ is to $-1$ or 1, the stronger the relationship between the $y$ and $x$ values is. The closer the $r$ value is to 0, the weaker the relationship is. The correlation coefficient is calculated as:

$$r = \frac{\sum xy - \frac{(\sum x)(\sum y)}{n}}{\sqrt{\left[\sum x^2 - \frac{(\sum x)^2}{n}\right]\left[\sum y^2 - \frac{(\sum y)^2}{n}\right]}}.$$

Let us perform the calculation of $r$ using the data in Example 5.1, both with the $n = 15$ values, and then with an $n$ of 12 values, with the $y$ values corresponding to the $x_i = 0$ removed (Table 5.5).

**Table 5.5** Diagnostic regression data, Example 5.1

| $n$ | $y$ | $x$ | $xy$ | $y^2$ | $x^2$ |
|---|---|---|---|---|---|
| 1 | 6.0 | 0.00 | 0.000 | 36.0 | 0.000 |
| 2 | 5.8 | 0.00 | 0.000 | 33.64 | 0.000 |
| 3 | 6.5 | 0.00 | 0.000 | 42.25 | 0.000 |
| 4 | 5.1 | 0.25 | 1.275 | 26.01 | 0.062 |
| 5 | 5.2 | 0.25 | 1.300 | 27.04 | 0.062 |
| 6 | 5.6 | 0.25 | 1.400 | 31.36 | 0.062 |
| 7 | 4.8 | 0.50 | 2.400 | 23.04 | 0.250 |
| 8 | 4.7 | 0.50 | 2.350 | 22.09 | 0.250 |
| 9 | 5.1 | 0.50 | 2.550 | 26.01 | 0.250 |
| 10 | 4.4 | 1.00 | 4.400 | 19.36 | 1.000 |
| 11 | 4.6 | 1.00 | 4.600 | 21.16 | 1.000 |
| 12 | 4.5 | 1.00 | 4.500 | 20.25 | 1.000 |
| 13 | 2.3 | 5.00 | 11.500 | 5.29 | 25.000 |
| 14 | 2.1 | 5.00 | 10.500 | 4.41 | 25.000 |
| 15 | 2.5 | 5.00 | 12.500 | 6.25 | 25.000 |
| | $\sum y = 69.2$ | $\sum x = 20.25$ | $\sum xy = 59.275$ | $\sum y^2 = 344.16$ | $\sum x^2 = 78.94$ |

A negative value is assigned to $r$, referring to the regression slope, $b$, being negative.

$$r = \frac{59.28 - \frac{(20.25)(69.2)}{15}}{\sqrt{\left[78.94 - \frac{(20.25)^2}{15}\right]\left[344.16 - \frac{(69.2)^2}{15}\right]}}$$

$$= \frac{-34.14}{\sqrt{(51.60)(24.92)}} = \frac{-34.14}{35.86} = -0.95$$

A negative sign is placed before 0.95, because $r$ carries the sign of $b$, which is $-0.6618$. This is a relatively high correlation, in that the value, $r$, is close to $-1$.

Now let us see what happens when the $x_i = 0$ values are removed, as well as the corresponding $y_i$ values? We already know they are adding variability to the regression line not being exactly linear (see Fig. 5.4). Table 5.6 presents those summary data.

**Table 5.6** Diagnostic regression data, Example 5.1, without zero-time values

| $n$ | $y$ | $x$ | $xy$ | $y^2$ | $x^2$ |
|-----|-----|-----|------|-------|-------|
| 1 | 5.1 | 0.25 | 1.275 | 26.01 | 0.062 |
| 2 | 5.2 | 0.25 | 1.300 | 27.04 | 0.062 |
| 3 | 5.6 | 0.25 | 1.400 | 31.36 | 0.062 |
| 4 | 4.8 | 0.50 | 2.400 | 23.04 | 0.250 |
| 5 | 4.7 | 0.50 | 2.350 | 22.09 | 0.250 |
| 6 | 5.1 | 0.50 | 2.550 | 26.01 | 0.250 |
| 7 | 4.4 | 1.00 | 4.400 | 19.36 | 1.000 |
| 8 | 4.6 | 1.00 | 4.600 | 21.16 | 1.000 |
| 9 | 4.5 | 1.00 | 4.500 | 20.25 | 1.000 |
| 10 | 2.3 | 5.00 | 11.500 | 5.29 | 25.000 |
| 11 | 2.1 | 5.00 | 10.500 | 4.41 | 25.000 |
| 12 | 2.5 | 5.00 | 12.500 | 6.25 | 25.000 |
| | $\Sigma y = 50.9$ | $\Sigma x = 20.25$ | $\Sigma xy = 59.275$ | $\Sigma y^2 = 232.27$ | $\Sigma x^2 = 78.94$ |

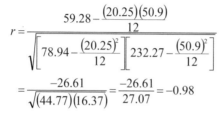

$$r = \frac{59.28 - \dfrac{(20.25)(50.9)}{12}}{\sqrt{\left[78.94 - \dfrac{(20.25)^2}{12}\right]\left[232.27 - \dfrac{(50.9)^2}{12}\right]}}$$

$$= \frac{-26.61}{\sqrt{(44.77)(16.37)}} = \frac{-26.61}{27.07} = -0.98$$

Again, a negative value is assigned to $r$, because $b$ is a negative number.

The value of $r = -0.98$ represents very high negative correlation. That is, the regression slope descends ($y$ data become smaller as the $x$ increases). Yet, how does one actually interpret $r$, other than whether it is closer to 1 or 0? What does an $r$ of 0.88 or 0.92 signify? Nothing – except that 0.92 is larger than 0.88. To make a more specific determination, we need to use the value, $r^2$, known as "the coefficient of determination."

## 5.9 Coefficient of Determination: $r^2$

The coefficient of determination, $r^2$, is an expression more useful than the correlation coefficient, $r$, because it can be interpreted directly. An $r^2 = 0.90$ means that 90% of the data variability can be explained by the regression equation. This is very valuable to know. In the previous example, where $r = -0.96$, $r^2 = 0.92$, meaning only 92% of the variability in the data was explained by the regression equation. When the $x_i = 0$ values are omitted from the equation, $r = -0.98$, and $r^2 = 0.96$. This means that 96% of the variability in the data is explained by the regression equation. The $r^2$ is always a positive value.

## 5.10 Predicting an $x$ Value from a $y$ Value

There are times when a microbiologist will find it advantageous to predict $x$ from a $y$. For example, in time-kill kinetic studies, often one wants to know the time it will take to kill a certain number of bacteria, fungi, or viruses. This process is called the D-value determination. This is just the reverse of what we have done in the previous sections of this chapter, where we estimated $y$ values based on $x$ values. Now we estimate $x$ values from $y$ values.

The computational procedure is straight-forward. First, one determines the regression parameters, $a$ and $b$, as before, via the regression equation $\hat{y} = a + bx$. Once the regression equation has been determined, one rearranges the terms in the equation to predict $\hat{x}$: $\hat{x} = (y - a)/b$. By itself, the $\hat{x}$ value predicted is only partially useful, so construction of a $100(1 - \alpha)$ confidence interval is also recommended. The $100(1 - \alpha)$ confidence interval is of the form, $\hat{x} \pm t_{(\alpha/2, n-2)} s_x$. For estimates of a specific $x$ value from a specific $y$ value, $s_x$ is computed as

$$s_x = \sqrt{\frac{s^2}{b^2}\left[1 + \frac{1}{n} + \frac{(x - \bar{x})^2}{\sum(x - \bar{x})^2}\right]}.$$

For the estimation of the average $\hat{\bar{x}}$ value from a specific $y$ value, $s_{\bar{x}}$ is computed as

$$s_{\bar{x}} = \sqrt{\frac{s^2}{b^2}\left[\frac{1}{n} + \frac{(x - \bar{x})^2}{\sum(x - \bar{x})^2}\right]}.$$

Let us look at an example, Example 5.2. Suppose a microbiologist exposes $1 \times 10^7$ *Bacillus subtilis* spores to a high level disinfectant for set times to determine a lethality curve and, from that, wants to predict exposure times required to kill specific inoculation levels of spores. The collected data are presented in Table 5.7.

**Table 5.7** Collected data, Example 5.2

| $n$ | Exposure time ($x$) | Log$_{10}$ colony counts ($y$) |
|---|---|---|
| 1 | 0 | $1.0 \times 10^7$ |
| 2 | 0 | $1.0 \times 10^7$ |
| 3 | 0 | $1.0 \times 10^7$ |
| 4 | 1 | $5.31 \times 10^6$ |
| 5 | 1 | $6.21 \times 10^6$ |
| 6 | 1 | $9.98 \times 10^6$ |
| 7 | 2 | $6.15 \times 10^5$ |
| 8 | 2 | $5.35 \times 10^5$ |
| 9 | 2 | $3.98 \times 10^5$ |
| 10 | 3 | $7.21 \times 10^4$ |
| 11 | 3 | $1.09 \times 10^5$ |
| 12 | 3 | $5.35 \times 10^4$ |
| 13 | 4 | $6.73 \times 10^3$ |
| 14 | 4 | $5.98 \times 10^3$ |
| 15 | 4 | $4.79 \times 10^3$ |
| 16 | 5 | $6.21 \times 10^2$ |
| 17 | 5 | $5.81 \times 10^2$ |
| 18 | 5 | $7.99 \times 10^2$ |

The $y_i$ values, in scientific notation, must be linearized by $\log_{10}$ transformation (Table 5.8), and $a$ and $b$ values estimated to attain the regression equation, $\hat{y} = a + bx$.

**Table 5.8** $\log_{10}$ transformed data, Example 5.2

| Row | $x$ | $y$ | $\log_{10} y$[a] |
|---|---|---|---|
| 1 | 0 | 10000000 | 7.00000 |
| 2 | 0 | 10000000 | 7.00000 |
| 3 | 0 | 10000000 | 7.00000 |
| 4 | 1 | 5310000 | 6.72509 |
| 5 | 1 | 6210000 | 6.79309 |
| 6 | 1 | 9980000 | 6.99913 |
| 7 | 2 | 615000 | 5.78888 |
| 8 | 2 | 535000 | 5.72835 |
| 9 | 2 | 398000 | 5.59988 |
| 10 | 3 | 72100 | 4.85794 |
| 11 | 3 | 109000 | 5.03743 |
| 12 | 3 | 53500 | 4.72835 |
| 13 | 4 | 6730 | 3.82802 |
| 14 | 4 | 5980 | 3.77670 |
| 15 | 4 | 4790 | 3.68034 |
| 16 | 5 | 621 | 2.79309 |
| 17 | 5 | 581 | 2.76418 |
| 18 | 5 | 799 | 2.90255 |
| | $\sum x = 45$ | | $\sum y = 93.00302$ |

[a]$\log_{10} y$, from now on, will be designated $y$.

$$\bar{y} = \frac{\sum y}{n} = \frac{93.00302}{18} = 5.16683$$

$$\bar{x} = \frac{\sum x}{n} = \frac{45}{18} = 2.5$$

$$b = \frac{\sum xy - n\bar{x}\bar{y}}{\sum x^2 - n\bar{x}^2}$$

$$\sum xy = (0 \cdot 7) + (0 \cdot 7) + ... + (5 \cdot 2.76418) + (5 \cdot 2.90255) = 186.0620$$

$$\sum x^2 = 0^2 + 0^2 + 0^2 + ... + 5^2 + 5^2 + 5^2 = 165.000$$

$$b = \frac{186.062 - 18(2.5)(5.17)}{165 - 18(2.5)^2} = -0.8874$$

$$b = -0.8874$$

$$a = \bar{y} - b\bar{x}$$

$$a = 5.16683 - (-0.8874)(2.5)$$

$$a = 7.38533$$

So,     $\hat{y} = a + bx = 7.39 - 0.887x$

Next, rearrange the regression terms to find $\hat{x}$.

$$\hat{x} = \frac{y - a}{b} = \frac{y - 7.39}{-0.887}$$

Suppose the microbiologist wants to predict the specific time, $\hat{x}$, where $y = 6$, with a 95% confidence interval.

$$\hat{x} = \frac{6 - 7.39}{-0.887} = 1.57 \text{ min exposure}$$

## 5.11 Confidence Interval for a Specific $\hat{x}$

$$\hat{x} \pm t_{(\alpha/2, n-2)} s_x$$

$$s_x = \sqrt{\frac{s^2}{b^2} \left[ 1 + \frac{1}{n} + \frac{(x - \bar{x})^2}{\sum(x - \bar{x})^2} \right]}$$

$$s^2 = \frac{\sum(y - \hat{y})^2}{n - 2}$$

To perform this calculation, we need to know what the $\hat{y}_i$ and $(y - \hat{y})^2$ values are. Table 5.9 provides these values.

**Table 5.9** $x$, $\hat{y}$, $y - \hat{y}$, and $(y - \hat{y})^2$ values, Example 5.2

| $n$ | $x$ | $y$ | $\hat{y}$ | $y - \hat{y}$ | $(y - \hat{y})^2$ |
|---|---|---|---|---|---|
| 1 | 0 | 7.00000 | 7.38533 | −0.38533 | 0.14848 |
| 2 | 0 | 7.00000 | 7.38533 | −0.38533 | 0.14848 |
| 3 | 0 | 7.00000 | 7.38533 | −0.38533 | 0.14848 |
| 4 | 1 | 6.72509 | 6.49793 | 0.22716 | 0.05160 |
| 5 | 1 | 6.79309 | 6.49793 | 0.29516 | 0.08712 |
| 6 | 1 | 6.99913 | 6.49793 | 0.50120 | 0.25120 |
| 7 | 2 | 5.78888 | 5.61053 | 0.17835 | 0.03181 |
| 8 | 2 | 5.72835 | 5.61053 | 0.11782 | 0.01388 |
| 9 | 2 | 5.59988 | 5.61053 | −0.01065 | 0.00011 |
| 10 | 3 | 4.85794 | 4.72313 | 0.13481 | 0.01817 |
| 11 | 3 | 5.03743 | 4.72313 | 0.31430 | 0.09878 |
| 12 | 3 | 4.72835 | 4.72313 | 0.00522 | 0.00003 |
| 13 | 4 | 3.82802 | 3.83573 | −0.00771 | 0.00006 |
| 14 | 4 | 3.77670 | 3.83573 | −0.05903 | 0.00348 |
| 15 | 4 | 3.68034 | 3.83573 | −0.15539 | 0.02415 |
| 16 | 5 | 2.79309 | 2.94833 | −0.15524 | 0.02410 |
| 17 | 5 | 2.76418 | 2.94833 | −0.18415 | 0.03391 |
| 18 | 5 | 2.90255 | 2.94833 | −0.04578 | 0.00210 |

$$\sum(y - \hat{y})^2 = 1.08594$$

$$s^2 = \frac{\sum(y - \hat{y})^2}{n - 2} = \frac{1.08594}{18 - 2} = 0.06787$$

Also, $\sum(x - \bar{x})^2 = (0 - 2.5)^2 + (0 - 2.5)^2 + ... + (5 - 2.5)^2 + (5 - 2.5)^2 = 52.5$

$$s_x = \sqrt{\frac{0.06787}{(-0.887)^2} \left[ 1 + \frac{1}{18} + \frac{(1.57 - 2.5)^2}{52.5} \right]} = 0.304$$

$$X = \hat{x} \pm t_{(\alpha/2,\,n-2)} s_x$$

$$t_{(0.05/2;\,18-2)} = t_{(0.025;\,16)} = 2.12 \ \ (\text{Table A.1})$$

$$\hat{x} \pm 2.12 s_x = 1.57 \pm 2.12(0.304)$$

$$0.93 \leq X \leq 2.21$$

So, the spore population can be taken down 6 $\log_{10}$ with an exposure of 0.93–2.21 min. In practice, the microbiologist would use an exposure time greater than 2.21 min.

## 5.12 Confidence Interval for the Average $\bar{x}$ Value

The only thing that needs to be recalculated is $s_{\bar{x}}$ .

$$s_{\bar{x}} = \sqrt{\frac{s^2}{b^2}\left[\frac{1}{n} + \frac{(x-\bar{x})^2}{\sum(x-\bar{x})^2}\right]} = \sqrt{\frac{0.06787}{(-0.887)^2}\left[\frac{1}{18} + \frac{(1.57-2.5)^2}{52.5}\right]} = 0.0788$$

$$\mu = \hat{x} \pm t_{(\alpha/2,\,n-2)} s_{\bar{x}} = 1.57 \pm 2.12(0.0788)$$

$$1.40 \leq \mu \leq 1.74 \ \ \text{at} \ \alpha = 0.05 \ .$$

On the average, a 6.0 $\log_{10}$ population reduction would be achieved after an exposure between 1.40 and 1.74 min.

## 5.13 *D*-Value Calculation

The $D$ value is defined as the time required to reduce an initial microbial population by 1 $\log_{10}$. That value is simply $|1/b|$. The $b$ value for this example is –0.887.

$$D = \left|\frac{1}{-0.887}\right| = 1.13 \ \text{min}$$

The confidence intervals are exactly as computed for the specific $x$ and average $x$ values.

# Chapter 6

## Qualitative Data Analysis

## 6.1 Binomial Distribution

In many instances, microbiologists collect data that are binary. That is, data can occur in one of two possible outcomes, such as 0/1, +/–, growth/no growth, or pass/fail. The frequencies of each outcome are tabulated, relative to the number of trials, providing a data set ranging from 0 to 1.0. For example, let's take the number of trypticase soy broth tubes that are positive (+) for microbial growth over a 72-h incubation. Suppose ten tubes were used, and eight tubes were positive for growth. Then the proportion of positive growth is the number of positives ÷ total number = 8 ÷ 10 = 0.80. The value, "$p$," is usually designated as the proportion of successes, in this case, positive (+) growth. The proportion of no-growth can also be calculated. The proportion of no-growth is $q = 1 - p$, or $1.0 - 0.80 = 0.20$. Incidentally, what one terms "success" and "failure" is arbitrary.

There are many applications appropriate to using the binomial distribution, such as contingency tables to determine association in two-sample sets, but for microbiologists, probably the most practical application is using binary data to compute proportions (e.g., of successes or failures), which can then be represented in terms of the normal distribution. This makes analysis of binomial data a process to which we can apply methods already discussed.

The mean and standard deviation of a binomial data set are used to approximate the normal distribution. Two versions are used – one for estimating the actual events, and the other for estimating proportions.

### 6.1.1 Version I: Mean, Variance, and Standard Deviation Estimates for Predicting Outcome Events

The true population mean, $\pi$, is estimated by $p$. Where $n$ number of trials, the mean = $p$ = proportion of successes = number of successes = $n$, and vise-versa, number of successes = $np$. In the above example, if $n = 10$ and $p = 0.80$, the average or expected

D.S. Paulson,*Biostatistics and Microbiology*, doi: 10.1007/978-0-387-77282-0_6,  
© Springer Science + Business Media, LLC 2008

number of successes $= np = 10(0.80) = 8$, or 8 tubes. The variance of this data set is estimated as $s^2 = npq$, and the standard deviation as $s = \sqrt{npq}$ .

In our example, where $n = 10$ tubes, $p = 0.80$ successes, and $q = 1 - p = 1 - 0.8 = 0.20$ failures, the variance is $s^2 = 10(0.80)(0.20) = 1.6$, or 1.6 tubes. The standard deviation is $s = \sqrt{1.6} = 1.26$, or 1.26 tubes.

## 6.1.2 Version II: Mean, Variance, and Standard Deviation Estimates for Predicting Proportions or Percentages

The true population mean, $\pi$, is estimated by $p$, or proportion of successes: $\pi = p =$ number of successes $\div$ number of trials $= 8 \div 10 = 0.80$, as in our earlier example. The variance, $s^2$, is estimated as $p(1 - p) = 0.80(0.2) = 0.16$. The standard deviation, $s$, is estimated by $\sqrt{p(1-p)}$, and the standard error of the mean, $s_{\bar{x}}$, by $s_{\bar{x}} = \sqrt{p(1-p)/n}$ .

It is important to bear in mind that binomial, or categorical data are nominal data. That is, these data do not carry numerical values that can be ranked, but only differentiated. They signify the degree of one outcome, termed proportion of successes, out of two possible outcomes, success or failure. For example, an inoculated broth tube may be positive for growth of *Staphylococcus aureus*, and that growth may range from 1 to $10^9$ microorganisms, but all we can do is state "growth" or "no growth." Hence, nominal data are weaker than ordinal data (values that can be ranked) and much weaker than interval-ratio data. This means larger sample sizes are required to have statistical power comparable to those tests employing interval-ratio data. Fortunately, many times, it is easier to collect large samples of binary data than of interval-ratio data.

## 6.2 Confidence Interval Estimation

The $100(1 - \alpha)$ confidence interval for a binomial data set using a normal distribution approximation is:

$$\pi = p \pm z_{\alpha/2} \sqrt{\frac{p(1-p)}{n}}$$

where $\pi$ is the population value of the proportion of successes.

We will use the normal distribution $z$ table, because the Student's $t$ table customarily is not used with binomial data. However, in order to use the $z$ table, there are two restrictions that should be met: $np \geq 5$ and $nq \geq 5$. Let's look at an example.

*Example 6.1.* A microbiologist performed a minimum inhibitory concentration (MIC) test using *Staphylococcus aureus* (MRSa) to challenge a new synthetic $\beta$-lactamase-resistant antibiotic. Fifty (50) tubes were used to measure growth/no growth at each dilution level. At a dilution level of 32:1, there were 15 positive tubes and 35 negative tubes.

$p = 15/50 = 0.30$ = proportion of successes (no growth)

$q = 1 - p = 0.70$ = proportion of failures (growth)

The microbiologist wants to construct the confidence interval for the proportion of successes at $\alpha = 0.05$ for Type I error. From Table A.1, $z_{\alpha/2} = z_{0.05/2} = 1.96$*.

Key Point

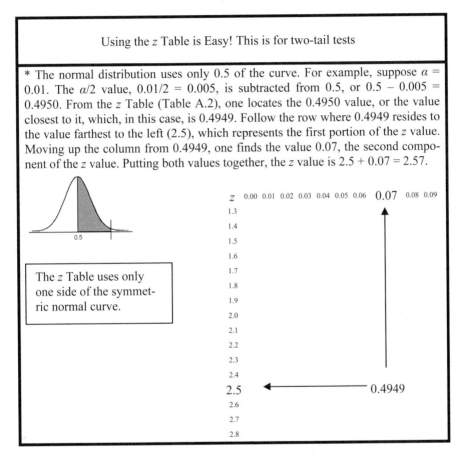

Using the z Table is Easy! This is for two-tail tests

* The normal distribution uses only 0.5 of the curve. For example, suppose $\alpha = 0.01$. The $\alpha/2$ value, $0.01/2 = 0.005$, is subtracted from 0.5, or $0.5 - 0.005 = 0.4950$. From the z Table (Table A.2), one locates the 0.4950 value, or the value closest to it, which, in this case, is 0.4949. Follow the row where 0.4949 resides to the value farthest to the left (2.5), which represents the first portion of the z value. Moving up the column from 0.4949, one finds the value 0.07, the second component of the z value. Putting both values together, the z value is $2.5 + 0.07 = 2.57$.

The z Table uses only one side of the symmetric normal curve.

For $\alpha = 0.05$, $z = 1.96$.

$$\pi \pm z_{\alpha/2}\sqrt{\frac{p(1-p)}{n}} = 0.30 \pm 1.96 \left( \sqrt{\frac{(0.30)(0.70)}{50}} \right)$$

$$= 0.30 \pm 0.127$$

The 95% confidence interval for the population proportion of successes is:

$$0.173 \le \pi \le 0.427$$

Remember to check for the appropriateness of using the z table.

$$np = 50(0.30) = 15 > 5, \text{ and } nq = n(1-p) = 50(0.70) = 35 > 5$$

Both $np$ and $nq$ are greater than 5, so the z table is okay to use.

From the proportions of the high and low ends of the confidence interval, one can easily convert to actual outcome values. For example, if one expects the same proportion to hold for other evaluations, and one cultures 275 tubes, how many will be positive? This can be calculated by using the proportion data from the 95% confidence interval just computed.

The lower 95% confidence interval is 0.173, and the upper boundary is 0.427. The mean value, $pn$, is 0.30 (275) = 82.5. That is, about 83 tubes, on the average, will be negative. The lower boundary is simply the lower proportion, $p_L$, times $n$, or $p_L n$ = 0.173 (275) = 47.58, and the upper boundary is $p_U n$ = 0.427 (275) = 117.42. The 95% confidence interval estimate for the actual occurrences, then, is $47 \leq \pi \leq 118$. Hence, one would expect the true average to be contained in the interval, 47–118, 95% of the time.

### Key Point

#### Note on Confidence Intervals

There are times when one will use the binomial distribution, and $p$ will be very close to 0 or 1; other times, $p$ will be closer to 0.5. Two different approaches to the confidence interval estimation process are necessary for these situations.

In the above example, $p$ = 0.30 is relatively close to 0.5, so many statisticians suggest using a Yates adjustment factor, which is $\pm 1/2n$.

## 6.2.1 Confidence Intervals on Proportions that are not Extreme (Not Close to 0 or 1): The Yates Adjustment

$$\pi = p \pm z_{\alpha/2}\sqrt{\frac{p(1-p)}{n}} \pm \frac{1}{2n}$$

$$p - z_{\alpha/2}\sqrt{\frac{p(1-p)}{n}} - \frac{1}{2n} \leq \pi \leq p + z_{\alpha/2}\sqrt{\frac{p(1-p)}{n}} + \frac{1}{2n}$$

Using data from Example 6.1,

$$p = 0.30, \quad q = 1 - p = 0.70, \quad s_{\bar{x}} = \sqrt{\frac{p(1-p)}{n}} = \sqrt{\frac{0.30(0.70)}{50}} = 0.0648,$$

$z_{\alpha/2} = z_{0.05/2} = 1.96$ (Table A.2), and $\dfrac{1}{2(n)} = \dfrac{1}{2(50)} = 0.01$.

$$p + z_{\alpha/2}\sqrt{\frac{p(1-p)}{n}} + \frac{1}{2n} = 0.30 + 1.96(0.0648) + 0.01 = 0.437$$

$$p - z_{\alpha/2}\sqrt{\frac{p(1-p)}{n}} - \frac{1}{2n} = 0.30 - 1.96(0.0648) - 0.01 = 0.183$$

The $100(1 - \alpha)$, or $100(1 - 0.05)$ = 95% confidence interval is $0.183 \leq \pi \leq 0.437$ at $\alpha = 0.05$.

## 6.2.2 Confidence Intervals on Proportions that are Extreme (Close to 0 or 1)

When $p$ or $q$ values are close to 0 or 1, the $p$ value should be assigned the value nearest to 0, so the Poisson distribution, or the rare event statistic, is used. Note that when $p$ is close to 0, $q$ will be close to 1, and when $p$ is close to 1, $q$ will be close to 0. The $100(1 - \alpha)$ confidence interval is $\pi = p \pm z_{\alpha/2}\sqrt{\dfrac{p}{n}}$ .

*Example 6.2.* A microbiologist working on a Homeland Security project for developing a handheld device for detection of *Bacillus anthracis* spores evaluated the ability of the device to detect the spores in visible powders from five different surfaces types. The overall correct detection level was 997 correct ÷ 1000 trials = 0.997. Hence, $p = 0.997$, and $q = 1 - 0.997 = 0.003$. In this case, $p$ and $q$ are interchanged in order to use the Poisson distribution. Hence, $p = 0.003$. Compute a 99% confidence interval ($\alpha = 0.01$).

$$\pi = p \pm z_{\alpha/2}\sqrt{\frac{p}{n}}$$

$$\sqrt{\frac{p}{n}} = \sqrt{\frac{0.003}{1000}} = 0.00173 \text{, and } z_{0.01/2} = z_{0.005} = 2.57 \text{ (from Table A.2)}.$$

$$\pi = p \pm z_{\alpha/2}\sqrt{\frac{p}{n}} = 0.003 \pm 2.57\sqrt{\frac{0.003}{1000}}$$

This gives $0.003 \pm 0.00445$, or $-0.00145 \leq \pi \leq 0.00745$. The 99% confidence interval is $0 \leq \pi \leq 0.007$.[*]

The interval, $0 \leq \pi \leq 0.007$ is the 99% confidence interval for not detecting *B. anthracis* spores when they are present. This is more commonly known as the false negative rate. To convert back to $p$ = success, simply subtract the lower and upper confidence interval points from 1 and reverse the order. In other words, confidence interval lower becomes confidence interval higher.

$1 - 0 = 1$, and $1 - 0.007 = 0.993$
$0.993 \leq \pi_{\text{success}} \leq 1.0$

The evaluation would most than likely focus on the lower confidence point being 0.993, or 99.3% correct identification with an expected level of 0.997 or 99.7%.

## 6.3 Comparing Two Samples

There are occasions when a microbiologist may want to compare two binomial sample sets. In previous chapters, we examined comparisons of a sample set to a standard value, and of two samples for a difference – a one-tail or two-tail test. Let us now look at the first case, as it applies to binomial data.

### 6.3.1 Proportions: One Sample Compared to a Standard Value

It is straightforward to compare a sample proportion to a standard value. To perform this, we use the form:

---

[*] Because the confidence interval is negative, 0 is used; the low-end value cannot be less than 0.

$$z_c = \frac{p - \acute{s}}{\sqrt{\dfrac{p(1-p)}{n}}}$$

where $p$ = proportion of successes, $x_i/n$ ($x_i$ = successes), $n$ = number of trials, $\acute{s}$ = standard value, and $z_c$ = calculated normal value, which will be compared to a value from the normal $z$ distribution (Table A.2).

*Example 6.3.* In an EPA hard-surface disinfectant test, the penicylinders used for testing must yield from their surfaces at least $1.0 \times 10^5$ colony-forming units (CFUs) per penicylinder. A microbiologist wants to know if a test product can kill all of the challenge microorganisms within 1 h of exposure. In order to do this, only 0.01, or 1% of the penicylinders can test positive for growth following product exposure. The microbiologist will conduct a pilot disinfectant study using 35 inoculated penicylinders to decide if a full study is warranted.

---

### Key Point

Design of this study can be addressed using the six-step procedure.

---

*Step 1.* Formulate the hypothesis.

In this study, the proportion of positives can be no more than 0.01, or 1% (this is noted as $\acute{s}$, for standard proportion). Hence, we need a study that does not exceed 0.01, so it is an upper-tail test.

$H_0$: Proportion of positives $\leq 0.01$
$H_A$: Proportion of positives $> 0.01$

*Step 2.* Select the sample size and $\alpha$ level.

We do not know what the test sample size should be for certain. However, the microbiologist believes the sample size of $n = 35$ is adequate. The microbiologist sets $\alpha = 0.05$.

*Step 3.* Write out the test statistic.

When $np > 5$, the test statistic is:

$$z_c = \frac{p - \acute{s}}{\sqrt{\dfrac{p(1-p)}{n}}}$$

where $p$ = observed proportion of positives, $\acute{s}$ = standard "tolerance" limit of positives, and $n$ = sample size.

If $np \not> 5$, $z_c = \dfrac{p - \acute{s}}{\sqrt{\dfrac{p}{n}}}$.

*Step 4.* State the acceptance/rejection rule.

Because this is an upper-tail test, if $z_c > z_t$, reject $H_0$ at $\alpha$ (Fig. 6.1). Here, $\alpha = 0.05$. Because this is not a two-tail test, we do not divide $\alpha$ by 2; we simply subtract 0.05 from 0.5. The result, 0.45, in Table A.2, is about $z_t = 1.64$. So, we reject $H_0$ if $z_c > 1.64$.

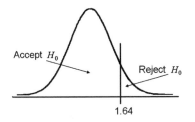

**Fig. 6.1** Upper-tail test, Example 6.3, Step 4

*Step 5.* Conduct the experiment.

The microbiologist gets three positives out of 35.

$p = 3/35 = 0.086$, $1 - p = 0.914$, and $n = 35$.

$$z_c = \frac{0.086 - 0.01}{\sqrt{\dfrac{0.086(0.914)}{35}}} = 1.6037$$

*Step 6.* Make the decision.

Because $z_c = 1.60 < z_t = 1.64$, one cannot reject $H_0$ at $\alpha = 0.05$.

The microbiologist notes that the actual number of positives was proportionally greater than 0.01. Even though there is not enough evidence to reject $H_0$ and claim that the rate of positives does not exceed 0.01, concluding this without caution would be foolish. Furthermore, the microbiologist realizes that $np$ is not greater than 5, so the Poisson distribution should have been used for the calculation. That calculation is

$$z_c = \frac{p - \dot{s}}{\sqrt{\dfrac{p}{n}}} = \frac{0.086 - 0.01}{\sqrt{\dfrac{0.086}{35}}} = 1.53$$

This did not solve the microbiologist's quandary, because using the Poisson calculation makes the statistic more conservative, i.e., it is more difficult to reject $H_0$.

### 6.3.2 Confidence Interval Approach

The test could also be conducted using a 95% confidence interval, as described previously. Here, because $p$ is close to 0, so we will use the Poisson version. The confidence interval approach uses a two-tail analysis: $\pi = p \pm z_{\alpha/2}\sqrt{p/n}$ and $\alpha/2 = 0.05/2 = 0.025$.

Using Table A.2, subtracting $\alpha/2$ from $0.5 = 0.5 - 0.025 = 0.475$. The $z_t$ value in Table A.2 = 1.96.

$$\pi = 0.086 \pm 1.96\sqrt{0.086/35}$$

$$-0.0112 \leq \pi \leq 0.1832$$

Because we cannot have a negative proportion, we use zero.

$0 \leq \pi \leq 0.183$, as diagrammed in Fig. 6.2.

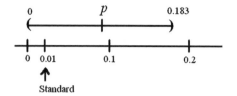

**Fig. 6.2** Confidence interval approach diagram

Because the confidence interval, 0–0.183, includes the standard, 0.01, one cannot reject $H_0$ at $\alpha$. Yet, the better way to use this interval is to see that the upper bound well exceeds $\dot{s} = 0.01$. In fact, the model would call 18% positive not significant. This should concern the microbiologist, in that the test product will not pass the EPA requirement, 0.01 positives, for a hard-surface disinfectant. Hence, the microbiologist would probably want to see an upper limit on the $1 - \alpha$ confidence interval that does not exceed 0.01.

---

### Key Points

#### Process for Comparing One Sample to a Standard

*Step 1.* Write out the hypotheses.

| Lower-Tail | Upper-Tail | Two-Tail |
|---|---|---|
| $H_0: p_c \geq \dot{s}$ | $H_0: p_c \leq \dot{s}$ | $H_0: p_c = \dot{s}$ |
| $H_A: p_c < \dot{s}$ | $H_A: p_c > \dot{s}$ | $H_A: p_c \neq \dot{s}$ |

where $p_c = p_{\text{calculated}}$ and $\dot{s} =$ standard or tolerance proportion

*Step 2.* Set $n$ and $\alpha$.

*Step 3.* Write statistical formula.

$$z_c = \frac{p_c - \dot{s}}{\sqrt{\dfrac{p(1-p)}{n}}}, \text{ where } pn > 5; \text{ if } pn \not> 5, \ z_c = \frac{p_c - \dot{s}}{\sqrt{\dfrac{p}{n}}}$$

*Step 4.* Write the decision rule.

| | | |
|---|---|---|
| $z_t$ at $-\alpha$ | $z_t$ at $\alpha$ | $z_t$ at $-\alpha/2$    $z_t$ at $\alpha/2$ |
| If $-z_c < z_t$, reject $H_0$ at $\alpha$. Using the z table, 0.5 − $\alpha$, then find $z_t$, which will be a negative value. | If $z_c > z_t$, reject $H_0$ at $\alpha$. Using the z table, 0.5 − $\alpha$, then find $z_t$, which will be a positive value. | If $z_c > z_t$ at $\alpha/2$ or $-z_c < -z_t$ at $\alpha/2$, reject $H_0$ at $\alpha$. Using the z table, 0.5 − $\alpha/2$, then find $z_t$, which will take both positive and negative values. |

*Step 5.* Perform the experiment.

*Step 6.* Make the decision.

## 6.4 Comparing Two Sample Proportions

A two-sample proportion test is a very useful statistic when the collected data are nominal scale. The two samples are designated Group 1 and Group 2, and the sample proportions, $p_1$ and $p_2$, are calculated: $p_1$ = (number of successes for group 1) ÷ (number of trials for group 1), and $p_2$ = (number of successes for group 2) ÷ (number of trials for group 2).

Next, the two proportions are combined to give a pooled value, or $p_p$.

$$p_p = \frac{n_1 p_1 + n_2 p_2}{n_1 + n_2} = \frac{(\text{number of successes for group 1}) + (\text{number of successes for group 2})}{(\text{total trials for group 1}) + (\text{total trials for group 2})}$$

Then, the pooled standard deviation is estimated:

$$s_p = \sqrt{p_p(1-p_p)\left(\frac{1}{n_1} + \frac{1}{n_2}\right)}$$

Finally, the test statistic, $z_{\text{calculated}}$, is computed as:

$$z_c = \frac{p_1 - p_2}{\sqrt{p_p(1-p_p)\left(\frac{1}{n_1} + \frac{1}{n_2}\right)}}, \text{ when } np > 5.$$

In this case, both $n_1 p_1$ and $n_2 p_2$ must be greater than 5. If not, use the formula:

$$z_c = \frac{p_1 - p_2}{\sqrt{p_p\left(\frac{1}{n_1} + \frac{1}{n_2}\right)}}$$

*Example 6.4.* In developing a diagnostic test device to detect $>10^2$ CFU *Escherichia coli* (O157/H7) per gram of hamburger meat (40% fat), two sampling card readers have been developed. Card Reader 1 is less expensive to purchase than Card Reader 2, so unless Card Reader 1 is less sensitive than Card Reader 2, it will be selected for use. Two dilutions of the ground meat will be tested, and the dilution that provides proportional values closest to 0.5 will be used for comparative analysis.

*Step 1.* State the hypotheses.

In this evaluation, the microbiologist wants to make sure that Card Reader 1 does not have more error in detecting *E. coli* than does Card Reader 2. Let $p$ = correct detection for O157/H7 *E. coli* and $1 - p$, or $q$, the number of incorrect detections. $p_1$ = the proportion of correct selections for Card Reader 1 and $p_2$ = the proportion of correct selections for Card Reader 2. If $p_1$ cannot be significantly less than $p_2$, then the alternative test is $p_1 < p_2$.

$H_0: p_1 \geq p_2$
$H_A: p_1 < p_2$ (the proportion of correct reads from Card Reader 1 is less than that from Card Reader 2)

*Step 2.* Set $n_1$ and $n_2$, as well as $\alpha$.

Twenty (20) replicates will be used for both $n_1$ and $n_2$, and $\alpha = 0.05$.

*Step 3.* Write out the test statistic to be used.

$$z_c = \frac{p_1 - p_2}{\sqrt{p_p(1-p_p)\left(\frac{1}{n_1} + \frac{1}{n_2}\right)}}$$

*Step 4.* State the decision rule.

If $z_c < z_t = -1.64$,* reject $H_0$. There is significance evidence to claim that Card Reader 1 has more detection error than Card Reader 2 at $\alpha = 0.05$ (Fig. 6.3).

Review

| *Where did the $-1.64$ come from? |
|---|
| Using the normal distribution table (Table A.2), subtract $\alpha$ from 0.5. <br> $\quad 0.5 - 0.05 = 0.450$ <br> Next, from Table A.2, find the value 0.450, and find the left-most value of that row (1.6) and the upper-most value of that column (0.04). Combine them; 1.6 + 0.04 = 1.64. Because this is a lower-tail test, the 1.64 is negative, so $z_t = -1.64$. |

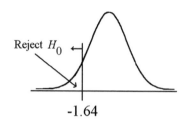

Reject $H_0$

$-1.64$

**Fig. 6.3** Decision rule, Example 6.4, Step 4

*Step 5.* Perform experiment. The associated data (two dilution levels) are presented in Table 6.1.

**Table 6.1** Example 6.4 data

| $n$ | Population $10^3$ Card Reader 1 | Card Reader 2 | Population $10^2$ Card Reader 1 | Card Reader 2 |
|---|---|---|---|---|
| 1 | + | + | 0 | 0 |
| 2 | + | + | 0 | + |
| 3 | + | + | + | 0 |
| 4 | + | + | + | + |
| 5 | 0 | + | + | 0 |
| 6 | + | + | + | + |
| 7 | + | + | 0 | lost |
| 8 | + | + | + | + |
| 9 | + | + | 0 | + |
| 10 | + | + | 0 | + |
| 11 | + | + | + | 0 |
| 12 | + | + | + | 0 |
| 13 | + | + | 0 | + |
| 14 | + | + | 0 | + |
| 15 | + | + | 0 | + |
| 16 | + | + | + | 0 |
| 17 | + | + | + | + |
| 18 | + | + | 0 | + |
| 19 | + | + | + | + |
| 20 | + | + | + | + |

$$p_1 = \frac{19}{20} = 0.95 \qquad p_2 = \frac{20}{20} = 1.00 \qquad p_1 = \frac{11}{20} = 0.55 \qquad p_2 = \frac{13}{19} = 0.684$$

$+$ = correct selection by Card Reader (confirmed positive); $0$ = incorrect negative reading by Card Reader (confirmed positive)

Results from the $10^3$ dilution were not used for the analysis because $p_1$ and $p_2$ were close to 1; the $10^2$ data were used. For Card Reader 2, one sample was lost.

$$p_p = \frac{n_1 p_1 + n_2 p_2}{n_1 + n_2} = \frac{20(0.55) + 19(0.684)}{20 + 19} = \frac{11 + 13}{39} = 0.615$$

$$z_C = \frac{p_1 - p_2}{\sqrt{p_p(1 - p_p)(1/n_1 + 1/n_2)}} = \frac{0.55 - 0.684}{\sqrt{0.615(1 - 0.615)(1/20 + 1/19)}} = -0.8596$$

*Step 6.* Make the decision.

Because $z_c = -0.8596 \not< z_t = -1.64$, one cannot reject $H_0$ at $\alpha = 0.05$. Card Reader 1 does not provide significantly less accurate data. However, using one's microbiological knowledge, one sees that Card Reader 2 was the better detector at both the $10^2$ and $10^3$ levels. This would tip the microbiologist to look into the situation in greater depth. Just because a statistic cannot demonstrate a difference does not mean a difference worth detecting is not present.

Key Point

| Guide for Two-Sample Proportion Test Using the Six-Step Procedure |
| --- |

*Step 1.* State the hypothesis.

| Lower-Tail | Upper-Tail | Two-Tail |
| --- | --- | --- |
| $H_0: p_1 \geq p_2$ | $H_0: p_1 \leq p_2$ | $H_0: p_1 = p_2$ |
| $H_A: p_1 < p_2$ | $H_A: p_1 > p_2$ | $H_A: p_1 \neq p_2$ |

*Step 2.* Set $n$ and $\alpha$.

The sample sizes must satisfy $np > 5$ for both group 1 and group 2 and, if possible, $n_1 = n_2$. If $np \not> 5$, use the Poisson $z_c$ formula.

*Step 3.* Write out the test statistic to use.

If $np > 5$ for both groups, use

$$z_c = \frac{p_1 - p_2}{\sqrt{p_p(1 - p_p)\left(\dfrac{1}{n_1} + \dfrac{1}{n_2}\right)}}$$

If $np \not> 5$ for at least one group, use

$$z_c = \frac{p_1 - p_2}{\sqrt{p_p\left(\dfrac{1}{n_1} + \dfrac{1}{n_2}\right)}}$$

*Step 4.* Make the decision rule.

| Reject $H_0$ if $z_c <$ $z_t$ at $\alpha$. $z_t = 0.5 - \alpha$. Then find $z_t$ in Table A.2, which will be a negative value. | Reject $H_0$, if $z_c > z_t$ at $\alpha$. $z_t = 0.5 - \alpha$. Then find $z_t$ in Table A.2, which will be a positive value. | Reject $H_0$ if $-z_c < -z_t$ or $z_c > z_t$ at $\alpha/2$. $z_t = 0.5 - \alpha/2$. Then find $z_t$, which will take both negative and positive values. |
| --- | --- | --- |

*Step 5.* Perform the experiment.

*Step 6.* Make the decision.

## 6.5 Equivalence Testing: Proportions

Recall from previous chapters that, in equivalence testing, the test hypothesis is equivalence, not difference. When microbiologists use a conventional statistical hypothesis test, as we have been doing in this chapter up until now, and conclude that two samples are not significantly different at a specified $\alpha$ level, the $H_0$ hypothesis is not rejected, but this does not mean the samples are equivalent. One can only state a significant difference between the samples was not detected. The sample data could be very different, but the sample size may have been too small and/or the variability too great to detect a significant difference. This is not the case in equivalence testing. If the test is underpowered, the statistic reverts to stating there is not enough information to conclude equivalence.

### Key Point

> To conclude that two samples are not different using a standard test hypothesis is not the same as saying they are equivalent. To do that, one must use an equivalence test.

### 6.5.1 Equivalence Testing: One Proportion Sample Compared to a Standard

Technically, a two-tail test is used for equivalence, and a single-tail test for either a nonsuperiority or a noninferiority test, depending upon the direction. Also, remember that, in standard statistical hypothesis testing, testing is for a difference. It is the alternative, or $H_A$ hypothesis that is the focus of the test, and the null hypothesis is the hypothesis of no difference. Conversely, in equivalence testing, it is the null, or $H_0$ hypothesis that postulates a difference. If the test value is not large enough, the $H_A$ hypothesis of equivalence is rejected. What constitutes equivalence is a set or specified difference, $\Delta$. Often, the $\Delta$ level is set at 20% of the standard, as in bioequivalence studies for drug products. That is, if sample data are not more or less than 20% of the standard value (for the equivalence test), the sample is considered equivalent to the standard; the $H_A$ hypothesis of equivalence is accepted. If it is a directional test, and the sample is not less than 20% of a standard, it would be considered noninferior, and if not more than 20% of a standard, nonsuperior.

### Key Point

> Fortunately, the six-step procedure can be used to evaluate for nonsuperiority, noninferiority, or equivalence.

*Step 1.* State the hypothesis.

| Noninferiority | Nonsuperiority | Equivalence |
|---|---|---|
| $H_0$: $d \leq \Delta$ (inferior) | $H_0$: $d \geq \Delta$ (superior) | $H_0$: $-d \leq -\Delta$ and $d \geq \Delta$ |
| $H_A$: $d > \Delta$ (noninferior) | $H_A$: $d < \Delta$ (nonsuperior) | $H_A$: $-\Delta < d < \Delta$ (equivalence) |

NOTE: $\pi$ is a theoretical or established proportion, $\Delta$ is the set limit value, and $d = p - \pi$, where $p =$ proportion of successes.

*Step 2.* Select the sample size, and state $\alpha$.
*Step 3.* Write out the test statistic to use.

| Sample not inferior to standard value | Sample not superior to standard value | Samples equivalent to range values $\Delta_L$ and $\Delta_U$ |
|---|---|---|
| $z_c = \dfrac{d - (-\Delta)}{s_d}$ | $z_c = \dfrac{\Delta - d}{s_d}$ | $z_c = \dfrac{d - (-\Delta)}{s_d}$ and $z_c = \dfrac{\Delta - d}{s_d}$ |

For all conditions, $s_d = \sqrt{\dfrac{\pi(1 - \pi)}{n}}$.

*Step 4.* Write out the acceptance/rejection criteria.

| Reject $H_0$, that sample proportion is less than the standard, if $z_c > z_t$. | Reject $H_0$, that sample proportion is greater than the standard, if $z_c > z_t$ | Reject $H_0$, if both $z_c$ calculations are greater than $z_t$. This is a one-tail test performed two times. |
|---|---|---|

*Step 5.* Perform the experiment.
*Step 6.* Make the decision.

*Example 6.5.* A microbiologist wants to claim that sutures E-407-1 are noninferior to (not worse than) standard sutures, 757-A1, in tensile strength. To meet the set strength of the standard, each test suture must remain unbroken after 45 kg of continuous weight for 24 h. The test sutures must have a success rate that is within 20% of the standard success rate, $\pi$, of 95%.
*Step 1.* State the hypothesis and set the $\Delta$ level.
   $\Delta = 20\%(\pi)$
Noninferior Test
   $H_0: d \leq \Delta$. The test sutures are worse than the standard.
   $H_A: d > \Delta$. The test sutures are not worse than (inferior to) the standard.
*Step 2.* Select the sample size, and set $\alpha$.
   $n = 100$, and $\alpha = 0.05$, values set by the microbiologist
*Step 3.* Write out the test statistic.
   $z_c = \dfrac{d - (-\Delta)}{s_d}$, where $d = p - \pi$, $\Delta = (20\%)(\pi)$, and $s_d = \sqrt{\dfrac{\pi(1 - \pi)}{n}}$
*Step 4.* State the decision rule.
   If $z_c > z_t$, reject $H_0$; the new sutures are not inferior to the standard at $\alpha$. Using Table A.2, $z_{0.05} = 0.5 - 0.05 = 0.450$, which corresponds to 1.645. If $z_c > 1.645$, reject $H_0$.
*Step 5.* Perform the experiment.
   Of the 100 sutures tested, $p = 97$ successes.
   $p = 97/100 = 0.97$, so $d = 0.97 - 0.95 = 0.02$

$$s_d = \sqrt{\frac{0.95(1-0.95)}{100}} = 0.0218 \text{ , and } \Delta = (0.2)(0.95) = 0.19$$

$$z_c = \frac{0.02 - (-0.19)}{0.0218} = 9.633$$

*Step 6.* Make the decision.

Because $z_c = 9.633 > z_t = 1.645$, there is strong evidence to claim noninferiority of the sample sutures to the standard at $\alpha = 0.05$. The test sutures are not worse than the standard ones at $\alpha = 0.05$.

## 6.5.2 Confidence Interval Approach

The same six-step procedure can be done with the confidence interval. However, Steps 3–6 differ.

*Step 3.* The test statistic is:

$$\Delta \geq d + z_t s_d$$

*Step 4.* State the decision rule.

If $\Delta \geq d + z_t s_d$, the test sutures are not inferior to the standard.

*Step 5.* Perform the experiment.

$$0.19 \geq 0.02 + 1.645 \ (0.0218)$$
$$0.19 \geq 0.056$$

So, $H_0$ is rejected at $\alpha$; there is significant evidence to conclude noninferiority.

## 6.5.3 Nonsuperiority

*Example 6.6.* Suppose a microbiologist wants to assure that a substitute topical antimicrobial used to treat peripherally-inserted central catheters is not less effective than a standard and does not result in increased infection rates. This can be viewed as a nonsuperiority test in that superiority, here, means increased infection rates. If the $H_0$ hypothesis is not rejected, it means that the infection rate is greater than with the standard. The standard is associated with an infection rate of $\pi = 0.23$.

*Step 1.* State the hypothesis and set the $\Delta$ level.

$H_0$: $d \geq \Delta$. The test product is associated with higher rates of infection than is the standard.

$H_A$: $d < \Delta$. The test product provides no worse infection prevention than does the standard preparation.

*Step 2.* Select the sample size, and set $\alpha$.

$n = 32$, and $\alpha = 0.10$

*Step 3.* Write out the test statistic.

$$z_c = \frac{\Delta - d}{s_d} \text{, where } \Delta = 20\% \text{ of } \pi, \ d = p - \pi$$

$$d = 0.156 - 0.230 = -0.074$$

$$s_d = \sqrt{\frac{\pi(1-\pi)}{n}}$$

$\Delta = 0.2(0.23) = 0.046$, and $s_d - 0.0744$

$$z_c = \frac{\Delta - d}{s_d} = \frac{0.046 - (-0.074)}{0.0744} = 1.63$$

*Step 4.* State the decision rule.

Reject $H_0$ if $z_c > z_t$. $z_t = z_{0.10} = 0.5 - 0.10 = 0.40 = 1.285$ (Table A.2).

*Step 5.* Perform the experiment.

Out of 32 test cases, five were positive, so $p = 5/32 = 0.156$.

*Step 6.* Make the decision.

Because $z_c = 1.63 > z_t = 1.285$, there is significant evidence to reject $H_0$. The substitute antimicrobial is no worse in infection prevention than is the standard at $\alpha = 0.10$. The confidence interval approach is:

Reject $H_0$ if $\Delta \geq d + z_t s_d$.

Where $\Delta = 0.046 \geq -0.074 + 1.285 (0.0744)$, $0.046 > 0.022$.

So, one can reject $H_0$; the test product produces infection prevention not worse than that of the standard.

## 6.5.4 Two-Tail Test: Equivalence

This process will apply only if there is a standard range of $\Delta$; that is, $\pm\Delta$. For example, the standard may be $\pm x\%$ of a set value. Suppose the microbiologist in Example 6.5 wants to demonstrate equivalence of the new suture to the standard suture in terms of tensile strength. Let us say that the success rate of the new sutures must be $\pm 20\%$ of that of the standard success rate of 95%. Hence, $\Delta = 0.2\pi = 0.2 (0.95) = 0.19$.

*Step 1.* State the hypotheses.

$H_0$: $-d \leq -\Delta$, and $d \geq \Delta$

$H_A$: $-\Delta < d < \Delta$

*Step 2.* Select the sample size, and set $\alpha$.

$n = 100$, and $\alpha = 0.05$

*Step 3.* State the test formulas. There are two single-sided (one-tail) tests.

For the lower-side, $z_{c_{(Lower)}} = z_{c_L} = \dfrac{d - (-\Delta)}{s_d}$

For the upper-side, $z_{c_{(Upper)}} = z_{c_U} = \dfrac{\Delta - d}{s_d}$

*Step 4.* State the decision rule.

If $z_{c_L} > z_{tabled}$ and $z_{c_U} > z_{tabled}$, reject $H_0$; the substitute and the standard are equivalent at $\alpha = 0.10$.

$z_t = 1.645$ for $\alpha = 0.05$ (Table A.2)

*Step 5.* Perform the experiment.

$$p = \frac{97}{100} = 0.97 \,, d = 0.97 - 0.95 = 0.02, \Delta = 0.19, \text{ and } s_d = 0.0218$$

$$z_{c_L} = \frac{\Delta - (-d)}{s_d} = \frac{0.19 + 0.02}{0.0218} = 9.633$$

$$z_{c_U} = \frac{\Delta - d}{s_d} = \frac{0.19 - 0.02}{0.0218} = 7.798$$

*Step 6.* Make the decision.

Both the upper and lower values of $z_c$ must be greater than $z_t = 1.645$ in order to reject $H_0$. In this example, they are. Hence, the tensile strength of the new sutures is equivalent to that of the standard sutures.

### 6.5.5 Confidence Interval

If $\pm\Delta$ falls outside the confidence interval, $d - z_t s_d < \Delta < d + z_t s_d$, the $H_0$ hypothesis is rejected.

$\alpha = 0.05$, and $\Delta = 0.19$.

$d \pm z_t s_d = 0.02 \pm 0.036$, and $-0.016 < 0.19 \not< 0.056$.

Hence, $H_0$ is again rejected.

## 6.6 Two-Sample Equivalence: Proportions

Equivalence, nonsuperiority, and noninferiority tests can be applied to comparing two sample sets to each other. Many times, the two samples compared will be a test and a reference or control group, although this certainly does not have to be the case.

As noted earlier, equivalence is accepted, in many cases, if the test group does not differ from the control group by more than 20% (i.e., ±20%), in terms of average values. For example, if a reference product fill is between 100 mL and 120 mL, and if the test product average is within that range, equivalence in fill-volume is concluded between the test and control products.

Usually, a more formal statistical procedure is applied. In this case, the difference between the test and the control, say $p_1$ and $p_2$, respectively, is compared to the specified tolerance limits. The basic formula is $d = p_1 - p_2$, and the test statistics are dependent on equivalence, nonsuperiority, and noninferiority.

$$z_c = \frac{\Delta - d}{s_d} \text{ or } \frac{d - (-\Delta)}{s_d}$$

where $\Delta$ is the specified difference allowable (±20%, in this case) between $p_1$ and $p_2$

$$p_1 = \frac{\text{number of successes}}{\text{number of total trials of group 1}} = \frac{x_1}{n_1}$$

$$p_2 = \frac{\text{number of successes}}{\text{number of total trials of group 2}} = \frac{x_2}{n_2}$$

$$s_d = \sqrt{\left(\frac{1}{n} + \frac{1}{n}\right)\left(\frac{n_1 p_1 + n_2 p_2}{n_1 + n_2}\right)\left(1 - \frac{n_1 p_1 + n_2 p_2}{n_1 + n_2}\right)}$$

All three tests – equivalence, noninferiority, and nonsuperiority – can be performed.

Let us write out the form of the statistics using the six-step procedure.

*Step 1.* State the specified allowable difference, $\Delta$, that the test $(p_1)$ can be from the reference $(p_2)$ and still be equivalent. The test hypotheses are:

| Noninferior | Nonsuperior | Equivalent |
|---|---|---|
| $H_0: d \leq \Delta$ | $H_0: d \geq \Delta$ | $H_0: -d \leq -\Delta$, and $d \geq \Delta$ |
| $H_A: d > \Delta$ | $H_A: d < \Delta$ | $H_A: -\Delta < d < \Delta$ |

*Step 2.* Select sample sizes for both $n_1$ and $n_2$, which ideally will be the same. Set $\alpha$.

*Step 3.* Write out the test statistic.

| Noninferior | Nonsuperior | Equivalent |
|---|---|---|
| $z_c = \dfrac{d-(-\Delta)}{s_d}$ | $z_c = \dfrac{\Delta - d}{s_d}$ | $z_c = \dfrac{d-(-\Delta)}{s_d}$ and $z_c = \dfrac{\Delta - d}{s_d}$ |

and for all tests, $s_d = \sqrt{\left(\dfrac{1}{n_1}+\dfrac{1}{n_2}\right)\left(\dfrac{n_1 p_1 + n_2 p_2}{n_1 + n_2}\right)\left(1-\dfrac{n_1 p_1 + n_2 p_2}{n_1 + n_2}\right)}$

*Step 4.* Write out the acceptance/rejection criteria.

| Reject $H_0$ if | Reject $H_0$ if | Reject $H_0$ if $z_{c_L} > z_t$ and |
|---|---|---|
| $z_c > z_{t(\alpha; n_1 + n_2 - 2)}$ | $z_c > z_{t(\alpha; n_1 + n_2 - 2)}$ | $z_{c_U} > z_t$ |

*Step 5.* Conduct the experiment.
*Step 6.* Make the decision.

*Example 6.7.* A microbiologist uses *Clostridium difficile* spores from an outside laboratory source to evaluate the effectiveness of hospital room disinfectants. Lately, the standard spores have been harder to purchase from the source. Because the hospital quality control historically has been based on the resistance of the standard spores, before switching to a new laboratory source, the microbiologist wants to assure equivalence in spore resistance to a 0.05% peracetic acid solution. This will be measured in terms of $D$-values, or the time that 0.05% peracetic acid reduces the initial spore load by 1.0 $\log_{10}$. Each initial spore population level must be at least 5.0 $\log_{10}$.

*Step 1.* State the hypotheses. This is clearly a bioequivalence study, a two-tail test.
   $H_0$: $-d \leq -\Delta$ and $d \geq \Delta$
   $H_A$: $-\Delta < d < \Delta$
$\Delta$ is $\pm 0.50$ $\log_{10}$. The unknown population difference is estimated by $d$. There can be no more difference between $p_1$ and $p_2$ than $\pm 0.50$ $\log_{10}$.

*Step 2.* State $n_1$, $n_2$, and $\alpha$.
   $n_1 = n_2 = 15$, and $\alpha = 0.05$

*Step 3.* Write out the test statistic (there are two one-sided tests).
$$z_c = \frac{d-(-\Delta)}{s_d} \text{ and } z_c = \frac{\Delta - d}{s_d}$$

where $s_d = \sqrt{\left(\dfrac{1}{n_1}+\dfrac{1}{n_2}\right)\left(\dfrac{n_1 p_1 + n_2 p_2}{n_1 + n_2}\right)\left(1-\dfrac{n_1 p_1 + n_2 p_2}{n_1 + n_2}\right)}$.

*Step 4.* State the decision rule.
   If $z_{c_L}$ and $z_{c_U}$ are larger than $z_t$, reject $H_0$ at $\alpha$.
   $z_\alpha = z_{0.05} = 1.645$ (Table A.2)

*Step 5.* Perform the experiment. Fifteen sub-samples of each *Clostridium difficile* spore sample were tested. All samples were greater than $1 \times 10^5$ spores per mL, based on dilution of a stock spore suspension ($1 \times 10^9$ spores per mL) and verified by baseline colony counts. A "+" indicates that a value was within 0.50 $\log_{10}$ ($\pm 0.25$) of the $D$-value of a 1.0 $\log_{10}$ reduction. A "−" designates that it was not (Table 6.2).

**Table 6.2** Example 6.7 data

| $n$ | Group 1 | Group 2 (standard) |
|---|---|---|
| 1 | + | − |
| 2 | + | + |
| 3 | − | + |
| 4 | + | − |
| 5 | + | − |
| 6 | + | + |
| 7 | − | + |
| 8 | − | + |
| 9 | + | + |
| 10 | + | − |
| 11 | + | − |
| 12 | − | + |
| 13 | + | − |
| 14 | − | − |
| 15 | − | − |

$$p_1 = \frac{9}{15} = 0.60 \text{, and } p_2 = \frac{7}{15} = 0.4667$$

$$s_d = \sqrt{\left(\frac{1}{15}+\frac{1}{15}\right)\left(\frac{15(0.60)+15(0.4667)}{15+15}\right)\left(1-\frac{15(0.60)+15(0.4667)}{15+15}\right)}$$

$$s_d = 0.1822$$

$$d = p_1 - p_2 = 0.60 - 0.4667 = 0.1333$$

$$z_{c_L} = \frac{d-(-\Delta)}{s_d} = \frac{0.1333+0.5}{0.1822} = 3.4759$$

$$z_{c_U} = \frac{\Delta-d}{s_d} = \frac{0.5-0.1333}{0.1822} = 2.0126$$

*Step 6.* Make the decision.

Because $z_{c_L} = 3.4759 > z_t = 1.645$ and $z_{c_U} = 2.0126 > z_t = 1.645$, one rejects $H_0$ at $\alpha = 0.05$. There is significant evidence that the two groups are equivalent.

The confidence interval approach is as follows:

$$d \pm z_\alpha s_d = 0.1333 \pm 1.645\,(0.1822)$$
$$= 0.1333 \pm 0.2997$$
$$-0.5 < -0.1664 < d < 0.4330 < 0.5$$

Because both the lower and upper values of the $(1 - 2\alpha) = 1 - 2(0.05) = 90\%$ of the confidence intervals are contained within the upper and lower bounds of $\Delta$, Group 1 and Group 2 are equivalent at $2\alpha = 0.10$.

Suppose the microbiologist wanted to ensure that Group 1 was not inferior to Group 2, the standard. This means the unknown population proportion for the test group (Group 1), in effect, is larger than the unknown population proportion for the test group standard (Group 2) by more than the lower set limit, $-\Delta$. If this is so, then the alternative hypothesis, $H_A$, of noninferiority is concluded.

*Step 1.* State the hypotheses.
$H_0: d \leq \Delta$
$H_A: d > \Delta$
where $\Delta = 0.05$, and $d = p_1 - p_2$

*Step 2.* State $n_1$, $n_2$, and $\alpha$.
$n_1 = n_2 = 15$, and $\alpha = 0.05$.

*Step 3.* Write out the test statistic.
$$z_c = \frac{d - (-\Delta)}{s_d}$$

*Step 4.* State the decision rule.
$z_t = z_{0.05} = 1.645$ (Table A.2)
If $z_c > z_t = 1.645$, reject $H_0$ at $\alpha = 0.05$.

*Step 5.* Perform the experiment.
$$z_c = \frac{d - (-\Delta)}{s_d}$$
$$d = p_1 - p_2 = 0.60 - 0.4667 = 0.1333$$
$$z_c = \frac{d - (-\Delta)}{s_d} = \frac{0.1333 - (-0.5)}{0.1822} = \frac{0.1333 + 0.5}{0.1822} = 3.4759$$

*Step 6.* Make the decision.
Because $z_c = 3.4759 > z_t = 1.645$, one must reject $H_0$ at $\alpha = 0.05$. The new group (Group 1) is not inferior to Group 2 at $\alpha = 0.05$.
The confidence interval approach is as follows:
$$d - z_a s_d = 0.1333 - 1.645(0.1822)$$
$$= 0.1333 - 0.2997$$
$$= -0.1664$$
If $-\Delta < d - z_t(s_d)$, reject $H_0$ at $\alpha$. Because $-0.5 < -0.1644$, the $H_0$ hypothesis is rejected at $\alpha = 0.05$.
Suppose the microbiologist wants to assure that Group 1 $(p_1)$ is not superior to Group 2 $(p_2)$ by $\Delta$.

*Step 1.* State the hypotheses.
$H_0: d \geq \Delta$
$H_A: d < \Delta$

*Step 2.* State $n_1$, $n_2$, and $\alpha$.
$n_1 = n_2 = 15$, and $\alpha = 0.05$.

*Step 3.* Write out the test statistic.
$$z_c = \frac{\Delta - d}{s_d}$$

*Step 4.* State the decision rule.
$z_t = z_a = 1.645$ (Table A.2)
If $z_c > z_t = 1.645$, reject $H_0$ at $\alpha = 0.05$.

*Step 5.* Perform the experiment.
$$d = p_1 - p_2 = 0.60 - 0.4333 = 0.1333$$
$$z_c = \frac{\Delta - d}{s_d} = \frac{0.5 - 0.1333}{0.1822} = 2.0126$$

*Step 6.* Make the decision.

Because $z_c = 2.0126 > z_t = 1.645$, one must reject $H_0$ at $\alpha = 0.05$. $p_1$ is nonsuperior to $p_2$.

The confidence interval approach is as follows:

If $\Delta > d + z_\alpha s_d$, reject $H_0$

$0.1333 + 1.645\ (0.1822) = 0.4330$

$\Delta = 0.5 > 0.4330$, so $p_1$ is not superior to $p_2$.

# Chapter 7

## Nonparametric Statistical Methods

We have discussed many statistical test methods to this point. Much of our attention has been focused on tests for analyzing interval-ratio data, which are quantitative in nature. These include *t*-tests, Analysis of Variance (ANOVA), and regression analysis. We have also discussed statistical methods using nominal data, sometimes called binary data, because they can have only two outcomes: success/failure, positive/negative, or growth/no growth. In this last chapter, we will again deal with interval-ratio data and nominal data and will also deal with data that fall between these two extremes of the spectrum – data that can ranked. These data are termed ordinal data and, like nominal data, are qualitative in nature. Table 7.1 summarizes these distinctions.

**Table 7.1** Three data scale types

| Nominal data | Ordinal data | Continuous data |
|---|---|---|
| Binary values (1,0) representing growth/no growth, success/failure, positive/negative; these can be distinguished, but not ranked | Distinguished by multiple attributes and can be ranked in order (viz., better or worse) | Interval-ratio data; these data represent the highest level of differentiation of data and are continuous quantitative measurements, without gaps. |

Data in all three of these scales of measurement can be analyzed using nonparametric statistical methods. Nonparametric statistical methods are favored by many in research, for they do not require the often rigorous assumptions that parametric methods require. Parametric statistics are designed to estimate parameters such as the population mean and variance. In order to be useful measurements, both the mean and variance must be from data that are normally distributed. That is, the data are

D.S. Paulson, *Biostatistics and Microbiology*, doi: 10.1007/978-0-387-77282-0_7, 121
© Springer Science + Business Media, LLC 2008

symmetrical around the mean (Fig. 7.1a). This symmetry gives the data the bell-shaped curve illustrated throughout this book. Parametric statistics also require that the data be unimodal, or have only one peak, instead of two or more (Fig. 7.1b). Parametric statistical methods are prone to bias when the data are not symmetrical about the mean (Fig. 7.1c, d), as well as when extreme values are present in the data set (Fig. 7.1e). These problems are especially prevalent in studies having relatively small sample sizes usually necessary in real-world practice, instead of the larger ones used in statistical theory.

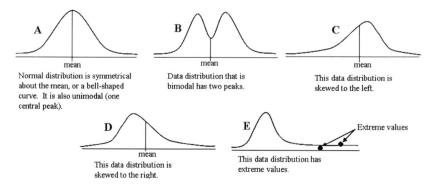

**Fig. 7.1a–e** Data distribution curves

Many times, a microbiologist must be satisfied with a sample size of 10 or even fewer. When the data distribution requirements just discussed cannot be assured, nonparametric statistical methods are useful. However, not just anything goes – all nonparametric methods, like the parametric ones, require that the data have been collected at random.

By convention, two types of statistical methods are included under the heading of nonparametric: 1) truly nonparametric procedures and 2) distribution-free statistical methods. As previously stated, nonparametric statistics, strictly speaking, are not concerned with estimating population parameters, $\mu$ and $\sigma$. These include $\chi^2$ goodness-of-fit tests and tests for randomness of a data set. Distribution-free tests do not depend on the underlying data distribution – normal, skewed, uniform, or bimodal.

Yet, there is a price – sometimes heavy – to pay. Nonparametric statistical methods generally lack the power of parametric statistics, so they require greater true differences to exist between tested groups in order to detect a significant difference than do their counterpart parametric statistics. That is, they err on the Type II side, concluding no significant difference exists when one really does. This is a problem when performing small studies to ascertain proof of concept or to find a product group that may show promise but needs further development. One solution is to set the alpha error level higher; i.e., instead of using $\alpha = 0.05$, use $\alpha = 0.10$ or $0.15$.

We will discuss each nonparametric test method, first using the nominal data scale; next, ordinal data will be covered, and then interval data.

## 7.1 Comparing Two Independent Samples: Nominal Scale Data

### 7.1.1 Comparing Two Independent Samples: 2 × 2 Chi Square Test

For comparing two independent samples, a 2 × 2 Chi Square ($\chi^2$) contingency table is very commonly applied. One could also use a two-sample proportion $t$-test, covered in Chapter 6, given there are enough collected data points; $n > 30$. For work with the 2 × 2 contingency table, we do not convert count data to proportion data. Basically, the 2 × 2 design uses two treatment conditions performed on two sample groups[*] (Table 7.2).

**Table 7.2** 2×2 Contingency table

|  |  | Treatment condition | | |
|---|---|---|---|---|
|  |  | A | B |  |
| Sample group | 1 | $a$ | $b$ | $a + b$ |
|  | 2 | $c$ | $d$ | $c + d$ |
|  |  | $a + c$ | $b + d$ | $N$ |

The statistical assumptions of this test are easy to make. The selection of test units assigned to Groups A and B must be random, and each treatment group independent in relation to the other. We will use the six-step procedure in the following example.

*Step 1.* Formulate the test hypotheses. Three test conditions can be evaluated.

| Lower-Tail Test | Upper-Tail Test | Two-Tail Test |
|---|---|---|
| $H_0: A \geq B$ | $H_0: A \leq B$ | $H_0: A = B$ |
| $H_A: A < B$ | $H_A: A > B$ | $H_A: A \neq B$ |

*Step 2.* Select the sample sizes and $\alpha$ level.

*Step 3.* Write out the test statistic. Both the lower-tail and upper-tail tests use the normal distribution approximation:

$$z_c = \frac{\sqrt{N}(ad - bc)^2}{\sqrt{(a+b)(c+d)(a+c)(b+d)}} \quad (z \text{ Distribution [Table A.2]}).$$

The two-tail test uses the $\chi^2$ (Chi Square) distribution:

$$\chi_c^2 = \frac{N(ad - bc)^2}{(a+b)(c+d)(a+c)(b+d)} \quad (\text{Chi Square [Table A.5]}).$$

$$\chi_t^2 = \chi_{(\alpha,1)}^2 \text{ with 1 degree of freedom.}$$

*Step 4.* State the decision rule.

---

[*] The treatment condition is usually A = growth, B = no growth, or A = +, B = –, etc.

| Lower-Tail Test | Upper-Tail Test | Two-Tail Test |
|---|---|---|
| Reject $H_0$ if $z_c < -z_{t(\alpha)}$ | Reject $H_0$ if $z_c > z_{t(\alpha)}$ | Reject $H_0$ if $\chi_c^2 > \chi_{t(\alpha,1)}^2$ |
| | | (Table A.5) |

*Step 5*. Perform experiment.
*Step 6*. Make decision.

### 7.1.1.1 Two-Tail Test

*Example 7.1*. A microbiologist wants to know if bacteriological media from two different geographic locations that use their local waters to mix ingredients will provide different growth data. A number of sample runs are performed over the course of 3 months, collecting the number of positive and negative growth outcomes for the same microorganism, *Clostridium difficile*.

*Step 1*. State the hypothesis.

To determine if the media differ on the basis of geographical locations, the microbiologist will use a two-tail test: $A$ = geographical region A; $B$ = geographical region B.

$H_0$: $A = B$
$H_A$: $A \neq B$

*Step 2*. Set the sample sizes, $n$, and the $\alpha$ level.

The microbiologist would like to set the sample sizes from both labs the same, but must rely on data provided by each laboratory manager. However, $\alpha$ will be set at 0.10.

*Step 3*. Write out the test statistic.

This is a $\chi^2$ (Chi Square) test for difference (two-tail).

$$\chi_c^2 = \frac{N(ad-bc)^2}{(a+b)(c+d)(a+c)(b+d)}$$

*Step 4*. State decision rule.

If $\chi_c^2 > \chi_t^2$, reject $H_0$. The growth results from the two labs are significantly different at $\alpha = 0.10$.

$\chi_t^2 = \chi_{(0.10,1)}^2 = 2.706$ (From Table A.5)

If $\chi_c^2 > 2.706$, reject $H_0$.

*Step 5*. Perform the experiment (Table 7.3).

**Table 7.3** 2×2 Contingency table

| | Location A | | | Location B | |
|---|---|---|---|---|---|
| Sample | $n$ | Failures | Sample | $n$ | Failures |
| 1 | 20 | 2 | 1 | 17 | 3 |
| 2 | 30 | 1 | 2 | 31 | 7 |
| 3 | 41 | 3 | 3 | 25 | 1 |
| 4 | 17 | 5 | 4 | 45 | 1 |
| 5 | 32 | 2 | 5 | 27 | 6 |
| | 140 | 13 | | 145 | 18 |

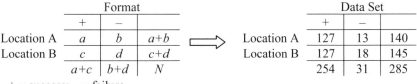

+ = success; – = failure

$$\chi_c^2 = \frac{N(ad-bc)^2}{(a+b)(c+d)(a+c)(b+d)} = \frac{285[(127\cdot18)-(13\cdot127)]^2}{140\cdot145\cdot254\cdot31} = 0.719$$

*Step 6.* Make decision.

Because $\chi_c^2 = 0.719 \not> \chi_t^2 = 2.706$, one cannot reject $H_0$ at $\alpha = 0.10$. There is no detectable difference in the growth data from the two different locations when $\alpha = 0.10$.

### 7.1.1.2 Lower-Tail Test

Suppose the number of negatives at Location A was 13, and at Location B, it was 51.

|  | Location A | Location B |
|---|---|---|
|  | $n = 140$ | $n = 145$ |
|  | Negatives = 13 | Negatives = 51 |

*Step 1.* Determine the hypotheses.
   $H_0$: $A \geq B$
   $H_A$: $A < B$ (Fewer negatives at Location A than at Location B; the key is what is expected as "fewer" must go in A of the contingency table.)
*Step 2.* Set the alpha level; $\alpha = 0.10$.
*Step 3.* Write out the test statistic.

$$z_c = \frac{\sqrt{N}(ad-bc)^2}{\sqrt{(a+b)(c+d)(a+c)(b+d)}}$$

*Step 4.* Write out the acceptance rule.
   If $z_c < z_t$, reject $H_0$ at $\alpha$.
   $0.5 - 0.10 = 0.40 \approx -1.282 = -z_t$ (Table A.2).
   Remember that, because this is a lower-tail test, $z$ is negative. If $z_c < -1.282$, reject $H_0$ at $\alpha = 0.10$.
*Step 5.* Perform the experiment (Table 7.4).

**Table 7.4** 2×2 Contingency table

|  | Location | | |
|---|---|---|---|
|  | A | B |  |
| Failures (–) | 13 (*a*) | 51 (*b*) | 64 (*a+b*) |
| Successes (+) | 127 (*c*) | 94 (*d*) | 221 (*c+d*) |
|  | 140 (*a+c*) | 145 (*a+b*) | 285 N |

$$t_c = \frac{\sqrt{N}(ad-bc)^2}{\sqrt{(a+b)(c+d)(a+c)(b+d)}} = \frac{\sqrt{285}[(13 \cdot 94)-(51 \cdot 127)]}{\sqrt{64 \cdot 221 \cdot 140 \cdot 145}} = -5.236$$

*Step 6.* Make the decision.

Because $t_c = -5.236 < -1.282$, reject $H_0$ at $\alpha = 0.10$. There is strong evidence that media Location A has significantly fewer failures than does Location B.

### 7.1.1.3 Upper-Tail Test

Suppose the microbiologist wants to evaluate media successes instead of failures.

$H_A: A > B$ will be the alternate hypothesis.

*Step 1.* State the hypotheses.

$H_0: A \leq B$

$H_A: A > B$ (Location A media has more successes than does that of Location B.)

*Step 2.* Let $\alpha = 0.10$.

*Step 3.* Write out the test statistic.

$$t_c = \frac{\sqrt{N}(ad-bc)^2}{(a+b)(c+d)(a+c)(b+d)}$$

*Step 4.* State the decision rule.

If $t_c > t_t = 1.282$ (Table A.2), reject $H_0$ at $\alpha = 0.10$.

*Step 5.* Perform the experiment.

Key Point

For the positives, $ad - bc > 0$, for an upper-tail test to occur, this requires $a$ to be the successes.

|  | Location | | |
|---|---|---|---|
|  | A | B |  |
| Successes | 127 (*a*) | 94 (*b*) | 221 (*a+b*) |
| Failures | 13 (*c*) | 51 (*d*) | 64 (*c+d*) |
|  | 140 (*a+c*) | 145 (*b+d*) | 285 (*N*) |

$$t_c = \frac{\sqrt{285}(127 \times 51)-(94 \times 13)}{\sqrt{221 \times 64 \times 140 \times 145}} = 5.236$$

*Step 6.* Make the decision.

Because $t_c = 5.236 > 1.282$, reject $H_0$ at $\alpha = 0.10$.

### 7.1.2 Comparing Two Related Samples: Nominal Scale Data

Paired samples or observations are samples that are blocked according to similarities. This strategy, when applied correctly, makes the statistic more powerful. In a specific example, say for the rate of lung cancer among smokers and nonsmokers in a very specific population, 45- to 60-year-old African males living in Louisiana, the pairing would be the specific population, age group, and race. Each smoker would be paired with a like nonsmoker, and lung cancer rates would be compared over time. Table 7.5 shows the schema. The trick is that each group (smokers/nonsmokers) is paired.

**Table 7.5** Paired data

|  |  | Smokers | | |
|---|---|---|---|---|
|  |  | Cancer | No Cancer |  |
| Nonsmokers | Cancer | $a$ | $b$ | $a+b$ |
|  | No Cancer | $c$ | $d$ | $c+d$ |
|  |  | $a+c$ | $b+d$ | $N$ |

where

$a$ = Pairs of smokers and nonsmokers with cancer[*]
$b$ = Pairs of nonsmokers with cancer and smokers without cancer
$c$ = Pairs of nonsmokers without cancer and smokers with cancer
$d$ = Pairs of smokers and nonsmokers without cancer

Suppose ten pairs of subjects are used, smokers and nonsmokers, and followed over time – that is, a follow-up, or prospective study. Additionally, a retrospective study could be conducted (though weaker) by pairing after the fact smokers and nonsmokers who have similarities that constitute the pairing rationale, and who have or have not developed cancer. The key to this design is in the "pairing" strategy. Let + represent subjects with cancer, and – represent those with no cancer (Table 7.6).

**Table 7.6** Smoker/nonsmoker pairing schema

| Pair | Smoker | Nonsmoker |
|---|---|---|
| 1 | + | + |
| 2 | + | + |
| 3 | + | – |
| 4 | + | – |
| 5 | – | + |
| 6 | – | + |
| 7 | – | + |
| 8 | – | – |
| 9 | – | – |
| 10 | – | – |

A paired design also is applicable when comparing status of a variable over time, such as before treatment vs. after treatment. Here, each person is his/her own control – that is, each subject is paired with him/herself. For example, in a skin irritation evaluation, subjects enter the study with irritated or nonirritated hands, are treated, and then evaluated again (Fig. 7.2).

---

[*] Each pair is one sample.

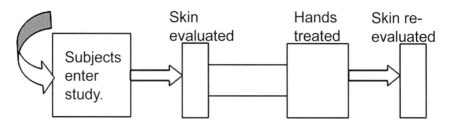

**Fig. 7.2** Time sequence schema

The paired design is:

|  |  | Posttreatment | |
|---|---|---|---|
|  |  | Irritated | Nonirritated |
| Pretreatment | Irritated | $a$ | $b$ |
|  | Nonirritated | $c$ | $d$ |

The paired $2 \times 2$ contingency table test to use is McNemar's Statistic, which is simply $z_c = b - c / \sqrt{b + c}$ , and easily can be completed using the six-step procedure. The analysis is performed only as a two-tail test to detect a difference.

*Example 7.2.* Suppose a microbiologist was assigned to evaluate the nosocomial incidence of *Staphylococcus aureus* colonization in the nasal cavities of patients staying at least 48 h in a large teaching hospital. $P_A$ = patients entering the hospital; $P_B$ = the same patients leaving after 48 h. Each patient is evaluated only against him/herself, so this is a pre-post design, wherein each person is his or her own control.

*Step 1.* State the hypotheses.

$H_0$: $P_A = P_B$; colonization of nasal cavities by *Staphylococcus aureus* is the same for patients entering and leaving the hospital. In other words, they do not acquire the staph colonization in the hospital.

$H_A$: $P_A \neq P_B$; above statement is not true. Patients entering and leaving the hospital have different *S. aureus* colonization rates.

*Step 2.* Set sample size, $n$, at 100, and $\alpha = 0.05$.

*Step 3.* Write out test statistic.

$$z_c = \frac{b - c}{\sqrt{b + c}}$$

*Step 4.* State decision rule.

If $|z_c| > |z_t|$, reject $H_0$ at $\alpha = 0.05$.

$z_t$ is from Table A.2 (z Distribution table)

To find $z_t$, $z_t = 0.5 - (\alpha/2) = 0.5 - (0.05/2) = 0.5 - 0.025 = 0.4750$. The z values corresponding to $0.4750 = 1.96$ or $-1.96$ and 1.96 for a two-tail test. So, if $|z_c| > |1.96|$, reject $H_0$ at $\alpha = 0.05$.

*Step 5.* Perform the experiment.

$N = 100$ = total number of patients evaluated.

$a = 7$ = the number of patients entering and leaving the hospital without *S. aureus* colonization.

$b = 27$ = the number of patients entering the hospital without S. *aureus* but leaving with S. *aureus* colonization.

$c = 3$ = the number of patients entering the hospital with *S. aureus* but leaving without *S. aureus* colonization.

$d = 63$ = the number of patients both entering and leaving the hospital with *Staphylococcus aureus* colonization.

|  | Posthospitalization | |
|---|---|---|
|  | Without *S. aureus* | With *S. aureus* |
| Prehospitalization Without *S. aureus* | 7 (*a*) | 27 (*b*) |
| With *S. aureus* | 3 (*c*) | 63 (*d*) |

$$z_c = \frac{b-c}{\sqrt{b+c}} = \frac{27-3}{\sqrt{27+3}} = 4.38$$

*Step 6.* Make the decision.

Because $z_c = 4.38 > 1.96$, reject $H_0$ at $\alpha = 0.05$. There is clearly a difference in *Staphylococcus aureus* colonization of the nasal cavities among those entering and leaving the hospital. Because this was a positive test ($b > c$), there is evidence that patients acquire *Staphylococcus aureus* colonization of their nares at a significant rate during their hospitalization.

## 7.1.3 Comparing More than Two Independent Samples: Nominal Scale Data

The Chi Square ($\chi^2$) test for association can be used for comparing more than two samples groups. It is done by comparing the "expected" and "observed" frequencies. The collected data are arranged in rows and columns in the form of an ($r \times c$) contingency table (Table 7.7).

**Table 7.7** Row × column contingency table

|  | Column (categories) | | | | Total |
|---|---|---|---|---|---|
|  | 1 | 2 | . . . | $c$ |  |
| 1 | $n_{11}$ | $n_{12}$ | . . . | $n_{1c}$ | $n_{1.}$ |
| Row 2 | $n_{21}$ | $n_{22}$ | . . . | $n_{2c}$ | $n_{2.}$ |
| (levels) . | . | . | | . | . |
| . | . | . | . . . | . | . |
| . | . | . | | . | . |
| $r$ | $n_{r1}$ | $n_{r2}$ | . . . | $n_{rc}$ | $n_{r.}$ |
|  | $n_{.1}$ | $n_{.2}$ | . . . | $n_{.c}$ | $n_{..} = N$ |

A contingency table is constructed from the observed data, and expected frequency data are also computed. The observed frequencies, "$0_{ij}$," are what actually occurred; the expected frequencies, "$E_{ij}$," are what is expected if the data are completely random events.

The $\chi^2$ test statistic calculated is:

$$\chi_c^2 = \sum_{i=1}^{r}\sum_{j=1}^{c}\frac{(0_{ij} - E_{ij})^2}{E_{ij}},$$

where $E_{ij} = n_i.(n_{.j})/n..$

The tabled $x^2$ value is

$$\chi_t^2 = \chi_{(\alpha;[(r-1)(c-1)])}^2$$

where $r$ = number of rows, and $c$ = number of columns.

If $\chi_c^2 > \chi_t^2$, reject $H_0$ at $\alpha = 0.05$.

The entire process can be completed easily using the six-step procedure. Only a two-tail test can be determined. Let us work an example.

*Example 7.3.* A microbiologist wants to evaluate test kits made by four different manufacturers for reliability in detecting target organisms at three microbial inoculum levels, low, medium, and high. Low levels are defined as inoculum levels less than or equal to $1 \times 10^4$ colony-forming units (CFUs)/mL, medium levels are greater than $1 \times 10^4$, but less than or equal to $1 \times 10^9$, and high levels are greater than $1 \times 10^9$ CFUs/mL. The observed reliability number is $p/n$, or the number of correct determinations over the total number of trials at that inoculum level.

*Step 1.* State the test hypotheses.

$H_0$: No difference exists between the test kits' reliabilities in determining inoculum levels.

$H_A$: At least one test kit differs from the others in reliability.

*Step 2.* Specify $\alpha$ and $n_i$ and $n_j$.

Ten tests ($n_{ij} = 10$) will be completed using each of three inoculum levels. There are four test kit types to be evaluated, so the total sample size is $10 \times 4 \times 3 = 120$. Let $\alpha = 0.05$.

*Step 3.* Write out the test statistic.

$$\chi_c^2 = \sum_{i=1}^{r}\sum_{j=1}^{c}\frac{(0_{ij} - E_{ij})^2}{E_{ij}}$$

where

$$0_{ij} = \frac{\text{number of successes}}{\text{number of trials}}$$

*Step 4.* Make the decision rule.

If $\chi_c^2 > \chi_t^2$, reject $H_0$ at $\alpha = 0.05$.

$$\chi_t^2 = \chi_{0.05;[(r-1)(c-1)]}^2 .$$

There are four products, or four columns, and three levels, or three rows. So the $(r-1)(c-1)$ degrees of freedom value is $r - 1 = 3 - 1 = 2$, $c - 1 = 4 - 1 = 3$, and $(r-1)(c-1) = 2 \times 3 = 6$.

$$\chi_{t(0.05;\,6)}^2 = 12.592 \quad \text{(Chi Square Table, Table A.5).}$$

If $\chi_c^2 > 12.592$, reject $H_0$ at $\alpha = 0.05$.

*Step 5.* Perform the experiment.

A contingency table was constructed as $P_i$ = successes ÷ total number of trials at that level.

| | | Test Kit | | | |
|---|---|---|---|---|---|
| | | 1 | 2 | 3 | 4 |
| | Low $(\leq 10^4)$ | $\frac{2}{10} = 0.2$ | $\frac{4}{10} = 0.4$ | $\frac{3}{10} = 0.3$ | $\frac{5}{10} = 0.5$ |
| Inoculum Level | Medium $(>10^4 \leq 10^9)$ | $\frac{7}{10} = 0.7$ | $\frac{6}{10} = 0.6$ | $\frac{7}{10} = 0.7$ | $\frac{7}{10} = 0.7$ |
| | High $(>10^9)$ | $\frac{10}{10} = 1.0$ | $\frac{9}{10} = 0.9$ | $\frac{9}{10} = 0.9$ | $\frac{10}{10} = 1.0$ |

Sometimes, for computation purposes and avoiding too much "round-off" error, it is good to use whole numbers. Hence, each value of $P_i$ is multiplied by 10.

| | | Test Kit Observed Values | | | | Row Totals |
|---|---|---|---|---|---|---|
| | | 1 | 2 | 3 | 4 | $n_i.$ |
| | $\leq 10^4$ | 2 | 4 | 3 | 5 | 14 |
| Inoculum Level | $>10^4 \leq 10^9$ | 7 | 6 | 7 | 7 | 27 |
| | $>10^9$ | 10 | 9 | 9 | 10 | 38 |
| Column Totals | $n._j$ | 19 | 19 | 19 | 22 | 79 |

Next, the expected contingency table is made. $E_{ij} = \dfrac{n_i.(n._j)}{n..}$

| | Test Kit Expected Values | | | |
|---|---|---|---|---|
| | 1 | 2 | 3 | 4 |
| $\leq 10^4$ | 3.37* | 3.37 | 3.37 | 3.90 |
| $>10^4 \leq 10^9$ | 6.49 | 6.49 | 6.49 | 7.52 |
| $>10^9$ | 9.14 | 9.14 | 9.14 | 10.58 |

$$E_{11} * \frac{19 \cdot 14}{79} = 3.37$$

Using the data from the tables of observed and expected values, $\chi_c^2$ can be determined.

$$\Sigma\Sigma = \frac{(0_{ij} - E_{ij})^2}{E_{ij}} = \frac{(2-3.37)^2}{3.37} + \frac{(4-3.37)^2}{3.37} + \frac{(3-3.37)^2}{3.37} + \frac{(5-3.90)^2}{3.90}$$

$$+ \frac{(7-6.49)^2}{6.49} + \frac{(6-6.49)^2}{6.49} + \frac{(7-6.49)^2}{6.49} + \frac{(7-7.52)^2}{7.52}$$

$$+ \frac{(10-9.14)^2}{9.14} + \frac{(9-9.14)^2}{9.14} + \frac{(9-9.14)^2}{9.14} + \frac{(10-10.58)^2}{10.58} = 1.30$$

$$\chi_c^2 = 1.30$$

*Step 6.* Make the decision.

Because $\chi_c^2 = 1.30 \neq 12.592 = \chi_t^2$, one cannot reject $H_0$ at $\alpha = 0.05$. The manufacturers are not significantly different from one another.

## 7.1.4 Comparing More than Two Related Samples: Nominal Scale Data

In situations in which sample groups are related – blocked design – the Cochran Test for related observations is very useful. For example, skin irritation studies, such as the repeat insult patch test (RIPT), use individual subjects each with multiple products applied. The blocked or related factor in this case is each subject. In another example, when bioreactors are used to grow bacteria in their natural biofilm state, the same reactor and microorganism are used, but with different attachment surfaces. If one compares several surfaces (more than two) for bacterial attachment equality, the Cochran Test will be appropriate.

The Cochran's related sample test, also known as Cochran's Q Test, requires that all $c$ treatments are applied to each of the $k$ blocks (e.g., all products are applied to each subject block). However, the samples are randomized within each block. This design is analogous to the randomized block ANOVA design in parametric statistics. The Cochran's Q test statistic is

$$Q = \frac{c(c-1)\sum_{j=1}^{c} c_j^2 - (c-1)N^2}{cN - \sum_{i=1}^{r} r_i^2},$$

which is a $\chi^2$ test with $c - 1$ degrees of freedom, where $c$ = number of treatments.

The data are arranged in a contingency table (Table 7.8).

**Table 7.8** Contingency table, Cochran's Q Test

| Block | | Treatment 1 | 2 | 3 | ... | $c$ | Row totals |
|---|---|---|---|---|---|---|---|
| | 1 | $n_{11}$ | $n_{12}$ | $n_{13}$ | ... | $n_{1c}$ | $r_1$ |
| | 2 | $n_{21}$ | $n_{22}$ | $n_{23}$ | ... | $n_{2c}$ | |
| | 3 | $n_{31}$ | $n_{32}$ | $n_{33}$ | ... | $n_{3c}$ | |
| | . | . | . | . | | . | . |
| | . | . | . | . | ... | . | . |
| | . | . | . | . | | . | . |
| | $r$ | $n_{r1}$ | $n_{r2}$ | $n_{r3}$ | ... | $n_{rc}$ | $r_n$ |
| Column totals | | $c_1$ | $c_2$ | $c_3$ | ... | $c_c$ | $N$ |

*Example 7.4.* A microbiologist wants to know if five different healthcare personnel handwash products differ in causation of skin irritation. Ten individuals have five Finn chambers applied to their backs, each containing an aliquot of a specific handwash product. Following 24-h exposures, the chambers are removed and the skin graded: 0 = no irritation, and 1 = irritation. The data analysis can be approached using the six-step procedure.

*Step 1.* State the hypotheses.

$H_0$: Each of the five products is equivalent in causation of skin irritation.

$H_A$: The above is not true.

*Step 2.* Set $n$ and $\alpha$. $n = 10$ for each block to receive all 5 treatments; $\alpha = 0.05$.

*Step 3.* Write out the test statistic. $Q = \dfrac{c(c-1)\sum_{j=1}^{c} c_j^2 - (c-1)N^2}{cN - \sum_{i=1}^{r} r_i^2}$

*Step 4.* State the decision rule. $\chi_t^2 = \chi_{(\alpha;\,c-1)}^2 = \chi_{(0.05;\,4)}^2 = 9.488$ (Table A.5).

If $Q > 9.488$, reject $H_0$ at $\alpha = 0.05$.

*Step 5.* Run the experiment.

Product/Treatment

| Block/Subject | 1 | 2 | 3 | 4 | 5 | $r_i$. |
|---|---|---|---|---|---|---|
| 1 | 1 | 1 | 1 | 0 | 1 | 4 |
| 2 | 1 | 0 | 0 | 1 | 0 | 2 |
| 3 | 0 | 0 | 1 | 0 | 0 | 1 |
| 4 | 0 | 0 | 0 | 0 | 0 | 0 |
| 5 | 0 | 1 | 1 | 1 | 1 | 4 |
| 6 | 0 | 1 | 0 | 1 | 1 | 3 |
| 7 | 0 | 1 | 1 | 1 | 1 | 4 |
| 8 | 0 | 1 | 1 | 1 | 1 | 4 |
| 9 | 1 | 0 | 1 | 1 | 1 | 4 |
| 10 | 0 | 1 | 0 | 1 | 1 | 3 |
| Column totals | 3 | 6 | 6 | 7 | 7 | 29 |

$c_j$

$$Q = \frac{c(c-1)\sum_{29}^{c} c_j^2 - (c-1)N^2}{cN - \sum r_i^2} = \frac{5(5-1)(3^2 + 6^2 + 6^2 + 7^2 + 7^2) - (5-1)29^2}{5(29) - (4^2 + 2^2 + 1^2 + 0^2 + 4^2 + 3^2 + 4^2 + 4^2 + 4^2 + 3^2)}$$

$$Q = 5.14$$

*Step 6.* Make decision.

Because $Q = 5.14 \not> 9.488$, one cannot reject $H_0$ at $\alpha = 0.05$. The products produce equivalent irritation.

## 7.2 Ordinal Scale Data: Rankable

To use ordinal statistical methods, the data must be rankable; for example, 1 = best, 2 = satisfactory, and 3 = poor.

### 7.2.1 Comparing Two Independent Sample Sets: Ordinal Data

The Mann-Whitney U Test is a very useful nonparametric statistic analogous to the parametric two-sample independent $t$-test. Two sample groups are compared, Group A and Group B. In using the Mann-Whitney U Statistic, upper-, lower- and two-tail tests can be performed. The Mann-Whitney U test statistic is:

$$M_c = \sum_{i=1}^{n} R_i \frac{n_A(n_A+1)}{2}$$

where $M_c$ = test statistic calculated, $\sum_{i=1}^{n} R_i$ = the sum of the ranked numeric values for Group A (only Group A values are used, but values for both Group A and Group B are ranked), and $n_A$ = sample size of Group A.

The six-step procedure is as follows.

*Step 1.* State the hypotheses.

| Lower-Tail Test | Upper-Tail Test | Two-Tail Test |
|---|---|---|
| $H_0: A \geq B$ | $H_0: A \leq B$ | $H_0: A = B$ |
| $H_A: A < B$ | $H_A: A > B$ | $H_A: A \neq B$ |

*Step 2.* Specify $\alpha$ and $n$. The sample sizes for the two groups do not have to be the same.

*Step 3.* Write out the test statistic.

$$M_c = \sum_{i=1}^{n} R_i - \frac{n_A(n_A + 1)}{2}$$

*Step 4.* Present the decision rule.

| Lower-Tail Test | Upper-Tail Test | Two-Tail Test |
|---|---|---|
| Reject $H_0$ if $M_c < M_t$, where the tabled value of $M_t$ (Table A.6), in this case, is $M_\alpha$; $M_{t(\alpha; n_A, n_B)}$ | Reject $H_0$ if $M_c > M_t$, where the Mann-Whitney tabled value is $M_{t(1-\alpha)}$; $n_A n_B - M_{(\alpha; n_A, n_B)}$ | Reject $H_0$ if $M_c < M_t = M_{(\alpha/2)}$ or $M_c > M_t = M_{(1-\alpha/2)}$. $M_{(\alpha/2; n_A, n_B)}$ = lower tail, and $M_{(1-\alpha/2)} = n_A n_B - M_{(\alpha/2; n_A, n_B)}$ = upper value |

*Step 5.* Perform the experiment.

*Step 6.* Make the decision.

Let us use the six-step procedure to perform the example, Example 7.5, that follows. All three tests will be performed.

*Step 1.* Formulate the hypothesis.

| Lower tail | Upper tail | Two-tail |
|---|---|---|
| $H_0: x_A \geq x_B$ | $H_0: x_A \leq x_B$ | $H_0: x_A = x_B$ |
| $H_A: x_A < x_B$ | $H_A: x_A > x_B$ | $H_A: x_A \neq x_B$ |

*Step 2.* Select $\alpha$ level and the sample size to be used.

*Step 3.* Write out the test statistic.

$$M_c = \sum_{i=1}^{n} R_i - \frac{n_A(n_A+1)}{2}$$

where $\sum_{i=1}^{n} R_i$ = sum of the ranks of Sample Group A, $n_A$ = sample size of Sample Group A, and $M_c$ = test statistic calculated.

*Step 4.* Make the decision rule.

| Lower tail | Upper tail | Two-tail |
|---|---|---|
| Reject $H_0$ if $M_c < M_{(\alpha)}$ where $M_{(\alpha)} = M_{(\alpha: n_A, n_B)}$ | Reject $H_0$ if $M_c > M_{(1-\alpha)}$ where $M_{(1-\alpha)} = n_A n_B - M_{(\alpha/2: n_A, n_B)}$ | Reject $H_0$, if $M_c < M_{(\alpha/2)}$ or $M_c > M_{(1-\alpha/2)}$ where $M_{(\alpha/2)}$ is the tabled Mann Whitney value at $M_{(\alpha/2: n_A, n_B)}$ and $M_{(1-\alpha/2)} = n_A n_B - M_{(\alpha/2: n_A, n_B)}$ |

Note: Table A.6 (Mann-Whitney Table) is used for all determinations of $M_t$.
*Step 5.* Compute statistic.
*Step 6.* Decision.

*Example 7.5.* Suppose a microbiologist wants to "compare" the nosocomial infection rates from hospitals in two separate geographic regions, A and B, based on data from five hospitals in each region. We will first calculate a two-tail test to determine if the rates differ. Following this, the upper- and lower-tail tests will be demonstrated. Table 7.9 presents the data received over a specified 1-month period, as to the average infection rate in the five hospitals in each region.

**Table 7.9** Data for Example 7.5

| Region A ($n_A$) | Region B ($n_B$) |
|---|---|
| 11.3 | 12.5 |
| 15.2 | 10.6 |
| 19.0 | 10.3 |
| 8.2 | 11.0 |
| 6.8 | 17.0 |
| 11.3 | 18.1 |
| 16.0 | 13.6 |
| 23.0 | 19.7 |
| 19.1 | |
| 10.6 | |

First, it is a good idea to plot these data to make sure there are no trends – increases or decreases in infection rates over time, indicating a time element that must be considered. The time element might include variables of further interest, such as policy changes (e.g., disinfectant usages, etc.), a seasonal effect, an epidemic occurrence (such as the flu), or some other important influence. In this particular example, it is a preliminary analysis during an exploratory stage.

*Step 1.* Formulate the hypothesis. This is a two-tail test.

$$H_0: x_A = x_B$$
$$H_A: x_A \neq x_B$$

where $x_A$ = average Region A nosocomial infection rate in five hospitals, and $x_B$ = average Region B nosocomial infection rate in five hospitals.

*Step 2.* Select $\alpha$ level.

$$\alpha = 0.05, \text{ so } \alpha/2 = 0.025$$

*Step 3.* Write out the test statistic ($t_c$).

$$M_c = \sum_{i=1}^{n} R_i - \frac{n_A(n_A + 1)}{2}$$

Both Regions A and B will be ranked, but only Region A data is summed. In cases of ties, the sequential values equaling the number of ties will be summed and divided by the number of ties. For example, suppose values 7, 8, and 9 were all tied. The same value would be used in slots 7, 8, and 9, which is $(7 + 8 + 9)/3 = 8$.

*Step 4.* State the decision rule. Reject $H_0$ if:

$$M_c < M_{(\alpha/2)} \text{ or } M_c > M_{(1 - \alpha/2)}$$

where ($n_A = 10$) and ($n_B = 8$).

$$M_{(\alpha/2; \, n_A, n_B)} = M_{(0.05/2, \, 10, \, 8)} = 18 \text{ from Table A.6 } (n_1 = n_A \text{ and } n_2 = n_B).$$

$$M_{1 - \alpha/2} = n_A n_B - M_{(\alpha/2; \, n_A, n_B)} = 10 \cdot 8 - 18 = 62.$$

So, if $M_c$ is less than 18 or greater than 62, we reject $H_0$ at $\alpha = 0.05$ (Fig. 7.3).

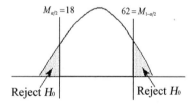

$M_{\alpha/2} = 18$          $62 = M_{1-\alpha/2}$

Reject $H_0$          Reject $H_0$

**Fig. 7.3** Example 7.5, Step 4

*Step 5.* Compute the test statistic, $M_c$.

First find $\sum_{i=1}^{n} R_A$, ranking both A and B (Table 7.10).

**Table 7.10** Rank values of A and B from Table 7.8

| $x_A$ Values | $x_A$ Ranking | | $x_B$ Values | $x_B$ Ranking | |
|---|---|---|---|---|---|
| 6.8 | 1 | | | | |
| 8.2 | 2 | | | | |
| | | | 10.3 | 3 | |
| 10.6 | 4.5 | | 10.6 | 4.5$^a$→ | (4+5)÷2=4.5 |
| | | | 11.0 | 6 | |
| 11.3 | 7.5→ | (7+8)÷2=7.5 | | | |
| 11.3 | 7.5 | | | | |
| | | | 12.5 | 9 | |
| | | | 13.6 | 10 | |
| 15.2 | 11 | | | | |
| 16.0 | 12 | | | | |
| | | | 17.0 | 13 | |
| | | | 18.1 | 14 | |
| 19.0 | 15 | | | | |
| 19.1 | 16 | | | | |
| | | | 19.7 | 17 | |
| 23.0 | 18 | | | | |

$$\sum_{i=1}^{10} R_A = 94.5$$

$^a$ Because there are two values of 10.6, which compete for ranks 4 and 5, both 10.6 values are given the mean value.

$$M_c = \sum_{i=1}^{n} R_A - \frac{n_A(n_A+1)}{2} = 94.5 - \frac{10(10+1)}{2} = 39.5$$

*Step 6.* Because 39.5 is contained in the interval 18 to 62, one cannot reject $H_0$ at $\alpha = 0.05$ (Fig. 7.4).

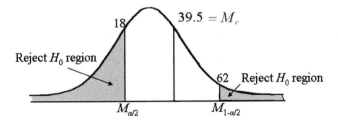

**Fig. 7.4** Example 7.5, Step 6

Let us now compute $M_{1-\alpha}$ and $M_\alpha$ to demonstrate both an upper- and a lower-tail test. Set $\alpha = 0.05$.

### 7.2.1.1 Upper-Tail Test

$H_0: x_A \leq x_B$
$H_A: x_A > x_B$

$$M_{1-\alpha} = n_A \, n_B - M_{(\alpha:\, n_A, n_B)}$$
$$= (10 \times 8) - 21$$
$$= 80 - 21 = 59$$

So, if $M_c > 59$, reject $H_0$ at $\alpha = 0.05$ (Fig. 7.5).

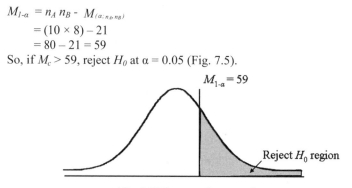

**Fig. 7.5** Diagram of upper-tail test

## 7.2.1.2 Lower-Tail Test

$H_0: x_A \geq x_B$
$H_A: x_A < x_B$
$M_\alpha = M_{(\alpha:\, n_A, n_B)} = 21$

So, if $M_c < 21$, reject $H_0$ at $\alpha = 0.05$ (Fig. 7.6).

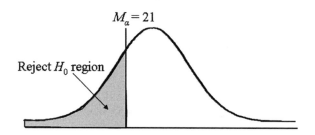

**Fig. 7.6** Diagram of lower-tail test

## 7.2.2 Comparing Two Related Sample Sets: Ordinal Data

For two related data sets, the Sign Test statistic is appropriate and can be applied for lower-, upper-, and two-tail tests. The test compares two groups, A and B, arbitrarily labeled as "+" and "–," on the basis of the sign of the "better" group in relation to the binomial table (Table A.7, Binomial Probability Distribution). "Better" may be less irritating, faster reaction, more potential, etc. One can use the same schema focusing on the worse of the two, also. In skin irritation studies, for example, two products are compared, A and B. Let A = + and B = –; each subject receives both treatments and are visually evaluated in terms of redness (Table 7.11). The treatment that produces the most skin redness is selected, an example of the worst-case situation being the one of interest.

**Table 7.11** Treatment evaluation results

| Subject | Treatment A (+) | Treatment B (−) | Sign of greatest irritation group |
|---|---|---|---|
| 1 | Red | Slightly Red | + |
| 2 | Not Red | Slightly Red | − |
| 3 | Red | Slightly Red | + |
| 4 | Red | Slightly Red | + |
| 5 | Red | Slightly Red | + |
| 6 | Red | Slightly Red | + |
| 7 | Not Red | Red | − |

If the treatments are equivalent, the number of "+" equals the number of "−." That is, $p(+) = p(-)$, where $p$ = proportion = 0.5. Using the six-step procedure to demonstrate,

*Step 1.* State the hypotheses.

| Lower-Tail Test | Upper-Tail Test | Two-Tail Test |
|---|---|---|
| $H_0: p(+) \geq p(-)$ | $H_0: p(+) \leq p(-)$ | $H_0: p(+) = p(-)$ |
| $H_A: p(+) < p(-)$ | $H_A: p(+) > p(-)$ | $H_A: p(+) \neq p(-)$ |

Again, we use $p = 0.5$ here because we are using the binomial table as an indicator of $p(+) = p(-)$, or a completely randomized case where $p(+) = p(-) = p(0.5)$. If the $p = 0.5$ tabled value is exceeded by the actual occurrence value of the + value, then the test is significant for an upper-, lower-, or two-tail test at $\alpha$ in that the probability is not $p = 0.5$.

*Step 2.* State $\alpha$ and $n$.
*Step 3.* Write out the test statistic, which in this case, is simply the use of the binomial table.

| Lower-Tail Test | Upper-Tail Test | Two-Tail Test |
|---|---|---|
| $P[(T_c \leq t \mid n, p)]$ | $P[(T_c \geq n - t \mid n, p)]$ | $P[(T_c \leq t \mid n, p)]$ or $P[(T_c \geq n - t \mid n, p)]$ |

*Step 4.* Present the decision rule.

| Lower-Tail Test | Upper-Tail Test | Two-Tail Test |
|---|---|---|
| Reject $H_0$ if $P[(T_c \leq t \mid n, 0.5)] \leq \alpha$ | Reject $H_0$ if $P[(T_c \geq n - t \mid n, 0.5)] \leq \alpha$ | Reject $H_0$ if $P[(T_c \leq t \mid n, 0.5)] \leq \alpha/2$ or $P[(T_c \geq n - t \mid n, 0.5)] \leq \alpha/2$ |

*Step 5.* Perform the experiment.
*Step 6.* Make the decision.

*Example 7.6.* A research and development microbiologist wants to determine if there is a difference between two test kits in the rapid detection of *Streptococcus pneumoniae*. Test Kit Card A will be designated "+," and Test Kit Card B will be designated "–." The quicker response of the two, in terms of minutes, will be termed the better response and, therefore, the sign representing that card will be used. However, only the "+" values will be enumerated for use in the analysis.

*Step 1.* Formulate the hypothesis.

$H_0$: $p(+) = p(-)$, or Test Kit Card A = Test Kit Card B
$H_A$: $p(+) \neq p(-)$ or Test Kit Card A ≠ Test Kit Card B

*Step 2.* Select $\alpha$ and $n$.

We will use $\alpha = 0.10$ and $n = 20$.

*Step 3.* Specify the test statistic.

Because this is a two-tail test, we use two cut-off levels of significance.
$P[(T_c \leq t|n, 0.5)]$ and $P[(T_c \geq n - t|n, 0.5)]$.

*Step 4.* Make the decision rule.

If $P[(T_c \leq t|n, 0.5)] \leq 0.10/2 = 0.05$, or $P[(T_c \geq n - t|n, 0.5)] \leq 0.10/2 = 0.05$, reject $H_0$ at $\alpha = 0.10$.

$T_c$ = the number of + values

*Step 5.* Perform the experiment.

The collected data are displayed in Table 7.12.

**Table 7.12** Data, Example 7.6

| $n$ | Card A(+) | Card B(–) | Faster |
|---|---|---|---|
| 1 | 3.7 | 3.8 | + |
| 2 | 4.2 | 3.7 | – |
| 3 | 5.1 | 2.7 | – |
| 4 | 3.2 | 8.9 | + |
| 5 | 5.6 | 4.3 | – |
| 6 | 2.7 | 3.2 | + |
| 7 | 6.9 | 2.7 | – |
| 8 | 3.2 | 4.1 | + |
| 9 | 2.1 | 5.8 | + |
| 10 | 5.1 | 2.1 | – |
| 11 | 2.7 | 5.1 | + |
| 12 | 3.7 | 7.2 | + |
| 13 | 5.4 | 3.7 | – |
| 14 | 6.3 | 5.2 | – |
| 15 | 6.1 | 6.9 | + |
| 16 | 3.2 | 4.1 | + |
| 17 | 5.1 | 5.1 | Tie |
| 18 | 5.4 | 4.7 | – |
| 19 | 8.1 | 6.8 | – |
| 20 | 6.9 | 7.2 | + |
| | | | $T_c = 10$ |

Total number of + values = 10.
Because there is a tie, $n$ is reduced by one to $n = 19$.
*Step 6.* Make the decision.

### Key Point

---

$p$ = proportion of +/– values for equality, $p = 0.5$
$n$ = number of pairs evaluated
$T_c$ = number of + values
$t$ = tabled value for lower-tail test
$n - t$ = tabled value for upper-tail test

---

First, find the critical value in the binomial table (Table A.7) for $p = 0.5$ and $n = 19$ that is closest to and less than $\alpha = 0.10/2 = 0.5$. Note that the binomial table comprises discrete values of $t$ ($= r$ in the table), so the levels of $\alpha$ will not be exact, but rather, the nearest value less than the selected $\alpha$. Hence, $P[(T_c \leq 5|19, 0.5)] \leq 0.05$ becomes $P[(T_c \leq 5|19, 0.5)] \leq 0.0222$, and $P[(T_c \geq 19 - 5|19, 0.5)] = P[(T_c > 14|19, 0.5)] \leq 0.0222$.
The $H_0$ hypothesis is rejected for $T_c \leq 5$ or $T_c \geq 14$. Because $T_c = 10$, which is between 5 and 14, one cannot reject $H_0$ at $\alpha = 0.10$.

## 7.2.2.1 Lower-Tail Test

To test the lower-tail:
*Step 1.* State the hypothesis.
$H_0: p(+) \geq p(-)$
$H_A: p(+) < p(-)$
*Step 2.* Select $\alpha$ and $n$.
We will use $\alpha = 0.10$ and $n = 19$, as before.
*Step 3.* State the test statistic.
$P[(T_c > t|n, 0.5)]$
*Step 4.* State the decision rule.
$P[(T_c \leq t|19, 0.5)] \leq 0.10$, or as tabled (Table A.7),
$P[(T_c \leq 7|19, 0.5)] \leq 0.0961$. If $t \leq 7$, reject $H_0$ at $\alpha = 0.10$.
*Step 5.* Perform the experiment.
$T_c = 10$
*Step 6.* Make the decision.
Because $T_c = 10$, which is not less than or equal to 7, one cannot reject $H_0$ at $\alpha = 0.0961$.

## 7.2.2.2 Upper-Tail Test

To test the upper-tail:
*Step 1.* State the hypothesis.
$H_0: p(+) \leq p(-)$
$H_A: p(+) > p(-)$

*Step 2.* Select $\alpha$ and $n$.
We will use $\alpha = 0.10$ and $n = 19$.
*Step 3.* Write out the test statistic.
$P[(T_c \geq n - t|19, 0.5)] \leq 0.10$
*Step 4.* State the decision rule.
$P[(T_c \geq 19 - 7|19, 0.5)] \leq 0.0961$; if $t \geq 12$, reject $H_0$ at $\alpha = 0.10$.
*Step 5.* Perform the experiment.
$T_c = 10$
*Step 6.* Make the decision.
Because $T_c = 10 \not> 12$, one cannot reject $H_0$ at $\alpha = 0.074$.

## 7.2.3 Comparing More than Two Independent Samples: Ordinal or Interval Data

When more than two independent sample sets are compared, the Kruskal-Wallis statistic is perhaps the most widely applied. The Kruskal-Wallis Test is the nonparametric version of the one-factor Analysis of Variance. It is used for completely randomized designs that produce ordinal- or interval-level data, meaning that the test run order between sampled groups is completely random. For example, if there are to be four replicates of three different treatments, the fundamental design is presented in Table 7.13.

**Table 7.13** Treatment design, Kruskal-Wallis Test

|            | 1 | 2 | 3 |
|------------|---|---|---|
|            | 1 | 1 | 1 |
| Replicates | 2 | 2 | 2 |
|            | 3 | 3 | 3 |
|            | 4 | 4 | 4 |

The 12 actual test runs (4 replicates, 3 treatments) are selected at random for evaluation. For example, the complete random sampling schema may be as shown in Table 7.14.

**Table 7.14** Sampling schema, Kruskal-Wallis Test

|            | Treatment #1 Run order | | Treatment #2 Run order | | Treatment #3 Run order | |
|------------|---|----|---|---|---|----|
|            | 1 | 2  | 1 | 3 | 1 | 6  |
| Replicates | 2 | 8  | 2 | 7 | 2 | 10 |
|            | 3 | 12 | 3 | 1 | 3 | 4  |
|            | 4 | 5  | 4 | 9 | 4 | 11 |

Note that replicate 3, treatment 2 is run first (Run Order = 1), replicate 1, treatment 1 is run second, and replicate 3, treatment 1 is the final, or twelfth run.

The Kruskal-Wallis test statistic is:

$$t_c = \frac{12}{N(N+1)}\sum_{i=1}^{k}\frac{R_i^2}{n_i} - 3(N+1)$$

where $R_i^2$ = sum of ranks squared for the $i^{th}$ treatment, $n_i$ = number of treatment replicates, and $N$ = total observations.

Ties can be a problem with the Kruskal-Wallis Test, so if one-quarter or more of the values are tied, a correction factor should be applied to $t_c$, as follows.

$$t_{c(corrected)} = \frac{t_c}{\left(1-\sum_{i=1}^{k}T\right)/(N^3 - N)}$$

where

$T = t^3 - t$ for each treatment sample data set of ties
$t$ = the number of tied observations in a group
$N$ = total number of observations

The six-step procedure is used in this evaluation.

*Step 1.* State the hypothesis; it will always be a two-tail test.

$H_0$: The $k$ populations are equivalent.

$H_A$: At least one of the $k$ populations differs from the others.

*Step 2.* Select $\alpha$ and $n$.

*Step 3.* Write out the test statistic.

$$t_c = \frac{12}{N(N+1)}\sum_{i=1}^{k}\frac{R_i^2}{n_i} - 3(N+1)$$

*Step 4.* State the decision rule.

If $k = 3$ and the sample size of each $k$ group is not more than 5, use Table A.8 (Critical Values of the Kruskal-Wallis Test). If $k > 3$ and/or $n > 5$, use the $\chi^2$ table (Table A.5), with $k - 1$ degrees of freedom. If $t_c >$ the critical tabled value, reject $H_0$ at $\alpha$.

*Step 5.* Perform the experiment.

*Step 6.* Make the decision, based on Step 4.

*Example 7.7.* Three different antimicrobial wound dressings were selected for use in a simulated wound care model. The model substrate was a fresh, degermed pig skin, incubated at 35–37°C, with 3 cm incisions made and inoculated with $10^8$ CFU of *Staphylococcus epidermidis* bacteria in a mix of bovine blood serum. The three wound dressings – A = 25% Chlorhexidine Gluconate, B = Silver Halide, and C = Zinc Oxide – were applied, and one inoculated wound was left untreated to provide a baseline value. Each pig skin was placed in a sealed container with distilled water to provide moisture and incubated for 24 h. Sampling was performed using the cup scrub procedure. The $\log_{10}$ reductions recorded are presented in Table 7.15.

**Table 7.15** $Log_{10}$ reduction data and ranks, Example 7.7*

| $A = 1$ | $R_1$ | $B = 2$ | $R_2$ | $C = 3$ | $R_3$ |
|---|---|---|---|---|---|
| 3.10 | 6.5 | 5.13 | 12.0 | 2.73 | 1.5 |
| 5.70 | 13.0 | 4.57 | 9.0 | 3.51 | 8.0 |
| 4.91 | 10.0 | 3.01 | 4.5 | 3.01 | 4.5 |
| 3.10 | 6.5 | 2.98 | 3.0 | 2.73 | 1.5 |
| 5.01 | 11.0 | | | | |
| | $\sum R_1 = 47$ | | $\sum R_2 = 28.5$ | | $\sum R_3 = 15.5$ |
| | $n_1 = 5$ | | $n_2 = 4$ | | $n_3 = 4$ |

$A$, $B$, $C$ = wound dressing types; $R_i$ = rank of values in each wound dressing group, lowest to highest, across all groups.

Key Point

* All rank data are used and ranked from smallest to largest values. In case of ties, each tied value is assigned an equal value. For example, if values 15, 16, and 17 are tied at 7.9, the ranks for 15, 16, and 17 would be $(15 + 16 + 17)/3 = 16$.

The investigator wants to know if there was a significant difference between the antimicrobial wound dressing in terms of bacterial $log_{10}$ reductions they produce.

Step 1. Formulate hypothesis.

$H_0$: Group A = Group B = Group C in $log_{10}$ reductions

$H_A$: At least one group differs from the other two

Step 2. Set $\alpha$.

The investigator will use $\alpha = 0.10$.

Step 3. Write the test statistic to be used.

$$t_c = \frac{12}{N(N+1)} \sum_{i=1}^{3} \frac{R_i^2}{n_i} - 3(N+1)$$

Step 4. Decision rule.

If $t_c$ > critical value, reject $H_0$. From the Kruskal-Wallis Table (Table A.8), for $n_1 = 5$, $n_2 = n_3 = 4$, and $\alpha = 0.10$, the critical value is 4.6187.

Key Point

Find $n_1$, $n_2$, and $n_3$, as well as the critical value corresponding to the $\alpha$ level. The critical selected value is $\leq \alpha$.

So if $t_c$ > 4.6187, reject $H_0$ at $\alpha = 0.10$. Because nearly one-half of the $log_{10}$ data are ties, if $t_c$ is not significant, we will apply the adjustment formula to see if the correction makes a difference.

$$t_{c(adjusted)} = \frac{t_c}{1 - \sum T(N^3 - N)}$$

*Step 5.* Compute statistic.

$$t_c = \frac{12}{N(N+1)}\sum_{i=1}^{3}\frac{R_i^2}{n_i} - 3(N+1) = \frac{12}{13(14)}\left[\frac{47^2}{5} + \frac{28.5^2}{4} + \frac{15.5^2}{4}\right] - 3(14) = 4.4786 \not> 4.6187$$

Because $t_c$ is not significant, we will use the correction factor adjustment for ties,

$$t_{c(adjusted)} = \frac{t_c}{(1 - \sum t)/(N^3 - N)} =$$

where

$T = t^3 - t$ for each group of sample data
$t$ = number of ties
$T_1 = 2^3 - 2 = 6$; two ties within Group 1
$T_2 = 0$; no ties within Group 2
$T_3 = 2^3 - 2 = 6$; two ties within Group 3

$$t_{c(adjusted)} = \frac{4.4786}{1 - [(6+0+6)/(13^3 - 13)]} = 4.5033$$

*Step 6.* Conclusion.

Because $t_{c(adjusted)} = 4.5033 \not> 4.6187$, one cannot reject $H_0$ at $\alpha = 0.10$. But this is so close that for $\alpha = 0.15$, the $H_0$ hypothesis would be rejected. It is likely, with a larger sample size, a difference would be detected.

Key Point

If the $\chi^2$ Table (Table A.5) had been used, the critical value at $\alpha = 0.10$ would be $\chi^2_{(0.10, 3-1)} = \chi^2_{(0.10, 2)} = 4.605$.

The Kruskal-Wallis statistic can be particularly valuable in analyzing subjective rankings of scores (ordinal data). For example, in skin irritation studies, scores observed may range between 0 and 5, 0 being no irritation and 5 being severe. Suppose three products are evaluated, and the following irritation scores result (Table 7.16).

**Table 7.16** $Log_{10}$ reduction data and ranks, Example 7.7

| Product 1 | Product 2 | Product 3 |
|-----------|-----------|-----------|
| 0 | 1 | 4 |
| 0 | 1 | 4 |
| 1 | 1 | 3 |
| 2 | 3 | 2 |
| 1 | 2 | 0 |
| 1 | 2 | 1 |
| 1 | 0 | 5 |
| 3 | 1 | |
| | 1 | |

This evaluation would be performed exactly like that in Example 7.7, with the tie adjustment formula applied.

## 7.2.4 Multiple Contrasts

In nonparametric statistics, as in parametric statistics, if a difference is detected when evaluating more than two sample groups, one cannot know where it lies – that is, between what groups of samples – without performing a multiple contrast procedure. The Kruskal-Wallis multiple comparison, like the parametric ANOVA contrast, provides a $1 - \alpha$ confidence level for the family of contrasts performed, on a whole.

The procedure is straight-forward. First, one computes the sum-of-rank means, $\overline{R_j}$. Next, a value of $\alpha$ is selected that is generally larger than the customary $\alpha = 0.05$ (e.g., 0.15, 0.20, or even 0.25, depending on the number of groups, $k$). The larger the $k$ value, the more difficult it is to detect differences. From a practical standpoint, try to limit $k$ to 3, or at most, 4. The next step is to find, on the normal distribution table (Table A.2), the value of $z$ that has $\alpha/k(k - 1)$ area to its right and compare the possible $n(n + 1)/2$, $|\overline{R_i} - \overline{R_j}|$ pairs to the test inequalities.

Key Point

| Contrast Test Formula Value |
|---|
| Use $z_{\alpha/k(k-1)}\sqrt{\dfrac{N(N+1)}{12}\left(\dfrac{1}{n_i}+\dfrac{1}{n_j}\right)}$ for sample sizes unequal between groups |
| and $z_{\alpha/k(k-1)}\sqrt{\dfrac{K(N+1)}{6}}$ for sample sizes equal between groups |

The entire contrast design is

$$\text{If } |\overline{R_i} - \overline{R_j}| > z_{\alpha/k(k-1)}\sqrt{\frac{N(N+1)}{12}\left(\frac{1}{n_i}+\frac{1}{n_j}\right)} \text{ reject } H_0.$$

A significant difference exists between $|\overline{R_i} - \overline{R_j}|$ at $\alpha$.

*Example 7.7* (cont.). Let us use Example 7.7 data, even though no significant difference between the groups was detected. This is only a demonstration of the computation of the multiple comparison. However, this time, let us set $\alpha = 0.15$.

$$\overline{R_1} = 47.0/5 = 9.40$$
$$\overline{R_2} = 28.5/4 = 7.13$$
$$\overline{R_3} = 15.5/4 = 3.88$$

$n(n - 1)/2 = 3$ contrasts possible, which are $(\overline{R_1} \ \overline{R_2})$, $(\overline{R_1} \ \overline{R_3})$, and $(\overline{R_2} \ \overline{R_3})$.

### 7.2.4.1 First Contrast

$$|\overline{R}_1 - \overline{R}_2| = |9.40 - 7.13| \le z_{\frac{0.15}{3(2)}} = z_{0.025} \sqrt{\frac{13(14)}{12}\left(\frac{1}{5} + \frac{1}{4}\right)} = z_{0.025}(2.6125)$$

Note: The 0.025 $z$ value is found in Table A.2, where $0.5 - 0.025 \approx 0.4750$. The value, 0.4750, is roughly equivalent to a tabled value of $1.96 = z$. Hence, $2.27 \le 1.96$ (2.6125), or $2.27 \le 5.1205$. Therefore, no difference exists at $\alpha = 0.15$ between chlorhexidine gluconate and silver halide in their antimicrobial effects.

### 7.2.4.2 Second Contrast

$$|\overline{R}_1 - \overline{R}_3| = |9.40 - 3.88| = 5.52$$

Because $5.52 \le 5.1205$ (the same value as the First Contrast), there is no difference between data from Groups 1 and 3 (chlorhexidine gluconate and zinc oxide) at $\alpha = 0.15$.

### 7.2.4.3 Third Contrast

$$|\overline{R}_2 - \overline{R}_3| = |7.13 - 3.88| = 3.25$$

$$(1.96)\sqrt{\frac{13(14)}{12}\left(\frac{1}{4} + \frac{1}{4}\right)} = (1.96)2.7538 = 5.3974$$

$$3.25 \not> 5.3974$$

There is no detectable difference between Groups 2 and 3 (silver halide and zinc oxide) at $\alpha = 0.15$ in their antimicrobial effects.

## 7.2.5 Comparing More than Two Related Samples: Ordinal Data

Analogous to the ANOVA complete block design, in which the replicate samples are related blocks, is the Friedman Test. Here, randomization is not across all observations, but within blocks of observations. If three treatments are evaluated with four replicates, the design looks like the completely randomized design, presented again here as Table 7.17.

**Table 7.17** Completely randomized design (randomization over all $3 \times 4 = 12$ values)

|  | Treatments | | |
|---|---|---|---|
|  | 1 | 2 | 3 |
| Replicates | 1 | 1 | 1 |
|  | 2 | 2 | 2 |
|  | 3 | 3 | 3 |
|  | 4 | 4 | 4 |

However, it is not entirely the same. The replication now occurs within each block, not on the entire experiment (Table 7.18).

**Table 7.18** Randomization in blocks (randomization within each of the four blocks)

|  |  | Treatments | | |
|---|---|---|---|---|
|  |  | 1 | 2 | 3 |
|  | 1 | 1 | 1 | 1 |
| Blocks/replicates | 2 | 2 | 2 | 2 |
|  | 3 | 3 | 3 | 3 |
|  | 4 | 4 | 4 | 4 |

The randomization order may look like Table 7.19.

**Table 7.19** Randomization of the complete block

|  |  | Treatments | | |  |
|---|---|---|---|---|---|
|  |  | 1 | 2 | 3 |  |
|  | 1 | 1 | 1 | 1 | Run Order Block 1 |
| Blocks | 2 | 2 | 2 | 2 | Run Order Block 2 |
|  | 3 | 3 | 3 | 3 | Run Order Block 3 |
|  | 4 | 4 | 4 | 4 | Run Order Block 4 |

Let's look at the first replicate block.

|  | Treatment | | |
|---|---|---|---|
| Block 1 | 1 | 2 | 3 |

The randomization occurs between treatments 1, 2, and 3 within Block Replicate 1. Replicates 2 through 4 follow the same pattern.

The critical point is that the treatments in block 1 are related or more homogenous to one another than to those in block 2. The variation noise between blocks is accounted for by the block effect, not the treatment effect, so the design is more powerful than the Kruskal-Wallis Test, if true blocking occurs. The blocks may be a specific lot, a specific culture, a specific microbiologist, or a specific test subject, for example.

The test statistic is $\chi_c^2 = \dfrac{12}{\ell k(\ell+1)} \sum_j^\ell C_j^2 - 3k(\ell+1)$.

where $\ell$ = number of treatments, $k$ = number of blocks, and $C_j$ = sum of the $jth$ treatment. If there are many ties (one-quarter or more of the values), a modified version of the test statistic should be used.

$$\chi_{(mod)}^2 = \dfrac{\chi_c^2}{1 - \sum_{i=1}^{k} T_i \big/ \left( \ell k(\ell^2 - 1) \right)}$$

where

$T_i = \sum t_i^2 - \sum t_i$ , and $t_i$ is the number of tied observations in the $ith$ block

The six-step procedure can be easily employed.

*Step 1*. State hypotheses.

The Friedman statistic is always two-tail.

$H_0$: The test groups are equivalent.

$H_A$: The above statement is not true.

*Step 2*. Set $\alpha$ and the sample size for each treatment block.

*Step 3*. Write the test statistic to be used.

$$\chi_c^2 = \frac{12}{\ell k(\ell+1)} \sum_{j=1}^{\ell} C_j^2 - 3k(\ell+1)$$

*Step 4*. State the decision rule.

If $\ell$ and $k$ are small ($\ell$ = 3 to 4) and $k$ = 2 – 9, then exact tables can be used (Table A.9, Freidman Exact Table). If not, the $\chi^2$ Table with $\ell$ – 1 degrees of freedom is used (Table A.5).

*Step 5*. Perform the experiment.

*Step 6*. Make decision rule.

*Example 7.8.* A microbiologist wants to compare skin irritation scores from ten test panelists who have swatches of three test product formulations containing different emollient levels attached to the skin of their backs for 96 h to simulate exposure during long-term veinous catheterization. A standard skin score regimen is to be used.

    0 = no irritation

    1 = mild irritation (redness, swelling, and/or chaffing)

    2 = moderate irritation

    3 = frank irritation

    4 = extreme irritation

*Step 1*. State hypotheses. The Friedman Test is always two-tail.

$H_0$: Test formulation 1 = test formulation 2 = test formulation 3

$H_A$: At least one formulation differs from the others.

*Step 2*. Set $\alpha$ = 0.05, and samples (blocks) = 3.

*Step 3*. Write the test statistic. $\chi_c^2 = \frac{12}{\ell k(\ell-1)} \sum_{j=1}^{\ell} C_j^2 - 3k(\ell+1)$.

*Step 4*. State the decision rule. If $\chi_c^2 \geq \chi_t^2$ , reject $H_0$ at $\alpha$. In this evaluation, $\ell$ = number of treatments = 3 and $k$ = 10. In order to use the exact Friedman tables, the number of replicates, $k$, can be no more than 9. Hence, we will use the $\chi^2$ Table, where $\chi_t^2 = \chi_{t(\alpha;\,\ell-1)}^2 = \chi_{t(0.05;\,3-1)}^2 = \chi_{t(0.05;\,2)}^2 = 5.991$ (Table A.5). So, if $\chi_c^2 \geq \chi_t^2 = 5.991$, reject $H_0$ at $\alpha$ = 0.05.

*Step 5*. Perform the experiment. The scores, as follow, resulted from evaluations of the skin of each subject (block) for each of the three treatments.

|  | Treatment | | |
|---|---|---|---|
|  | 1 | 2 | 3 |
| 1 | 2 | 2 | 4 |
| 2 | 1 | 3 | 3 |
| 3 | 0 | 2 | 2 |
| 4 | 2 | 3 | 3 |
| 5 | 3 | 4 | 3 |
| 6 | 2 | 4 | 3 |
| 7 | 1 | 3 | 4 |
| 8 | 3 | 5 | 3 |
| 9 | 2 | 4 | 4 |
| 10 | 0 | 2 | 3 |

(Subject/Block labels rows 1–10)

First, rank the scores in terms of each block. That is, block 1 [2, 2, 4] is now [1.5, 1.5, 3]. Table 7.20 provides these rankings, having accounted for ties.

**Table 7.20** Scores ranked by block, Example 7.8

|  | Treatment | | |
|---|---|---|---|
|  | 1 | 2 | 3 |
| 1 | 1.5 | 1.5 | 3 |
| 2 | 1 | 2.5 | 2.5 |
| 3 | 1 | 2.5 | 2.5 |
| 4 | 1 | 2.5 | 2.5 |
| 5 | 2.5 | 3 | 2.5 |
| 6 | 1 | 3 | 2 |
| 7 | 1 | 2 | 3 |
| 8 | 2.5 | 3 | 2.5 |
| 9 | 1 | 2.5 | 2.5 |
| 10 | 1 | 2 | 3 |
| Total | $C_1 = 13.5$ | $C_2 = 24.5$ | $C_3 = 26.0$ |

(Subject/block labels rows 1–10)

$$\chi_c^2 = \frac{12}{\ell k(\ell+1)}\sum_{j=1}^{\ell}C_j^2 - 3k(\ell+1)$$

$$\chi_c^2 = \frac{12}{3\cdot10\,(3+1)}\left[13.5^2 + 24.5^2 + 26^2\right] - 3\,(10)(3+1) = 25.85$$

Because there are so many ties, the tie correction formula should be used.

$$\chi_{(mod)}^2 = \frac{\chi_c^2}{1-\sum_{i=1}^{k}T_i/\ell k(\ell^2-1)} = \frac{25.85}{1-\left[\dfrac{14}{3\,(10)(9-1)}\right]} = 27.45$$

where $T_i = (2^2 - 2) + (2^2 - 2) + (2^2 - 2) + (2^2 - 2) + (2^2 - 2) + (2^2 - 2) + (2^2 - 2) = 14$.

*Step 6.* Make decision rule.

Because $\chi_{c(mod)}^2 = 27.45 \geq \chi_t^2 = 5.991$, reject $H_0$ at $\alpha = 0.05$. The three product formulations are significantly different in degree of irritating properties at $\alpha = 0.05$.

Key Point

Using Exact Tables

Suppose $k = 9$ and $\ell = 3$. Then, the Exact Friedman ANOVA Table (Table A.9) could be used. Here, $\alpha = 0.05$ is not presented, but $\alpha = 0.057$ is, which corresponds to 6. That is, $\chi_t^2 = 6.00$ at $\alpha = 0.057$. Because $\chi_{c(\text{mod})}^2 > 6.00$, reject $H_0$ at $\alpha = 0.057$.

When $H_0$ is rejected, the microbiologist generally will want to determine where the differences lie. A series of contrasts can be made to determine the difference in the normal $z$ distribution. The contrast decision formula is:

$$\text{If } |C_i - C_j| \geq z \sqrt{\frac{\ell k(\ell+1)}{6}} \text{ , where } z \text{ corresponds to } \alpha/\ell(\ell-1).$$

Letting $\alpha = 0.05$, divide $\alpha$ by the number of contrasts, $\ell(\ell-1)$, where $\ell =$ number of treatment groups. Hence, $3(3-1)=6$, and $0.05/6 = 0.0083$.

$0.5 - 0.0083 = 0.4917$ corresponds to $z = 2.39$ (Table A.2).

So, $|C_1 - C_2| = |13.5 - 24.5| = 11 > 2.39\sqrt{(3(10)(2))/6} = 7.5578$.

Results from test groups 1 and 2 differ.

$C_1 \neq C_2$ at $\alpha = 0.05$

$|C_1 - C_3| = |13.5 - 26.0| = 12.5 > 7.5578$

Results from test groups 1 and 3 differ.

$C_1 \neq C_3$ at $\alpha = 0.05$

$|C_2 - C_3| = |24.5 - 26.0| = 1.5 \not> 7.5578$

Results from test groups 2 and 3 do not differ.

$C_2 = C_3$ at $\alpha = 0.05$.

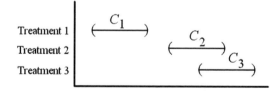

**Fig. 7.7** Treatment comparisons, Example 7.8

$C_1$ is less irritating than both $C_2$ and $C_3$, but $C_2$ and $C_3$ are equivalent (Fig. 7.7).

## 7.3 Interval-Ratio Scale Data

When one is not sure that interval-ratio data meet the requirements for normal distribution (e.g., sample size is small), nonparametric statistical methods offer a valuable alternative option.

### 7.3.1 Comparing Two Independent Samples: Interval-Ratio Data

The Mann-Whitney U test, as explained for application to ordinal data (Section 7.2.1), is a powerful, nonparametric approach to analyzing interval/ratio data. Upper-, lower-, and two-tail tests can be performed. Let us, once again, set up the analysis using the six-step procedure.

*Step 1.* Formulate the hypothesis.

| Two-tail | Upper-tail | Lower-tail |
|---|---|---|
| $H_0: A = B$ | $H_0: A \leq B$ | $H_0: A \geq B$ |
| $H_A: A \neq B$ | $H_A: A > B$ | $H_A: A < B$ |

*Step 2.* Select the $\alpha$ level.
*Step 3.* Write out the test statistic.

$$M_c = R_A - \frac{n_A(n_A + 1)}{2}$$

where $n_A$ = number of replicates in group A and $R_A$ = sum of the ranks in group A.
*Step 4.* Make the decision rule.

| Lower-tail | Upper-tail | Two-tail* |
|---|---|---|
| Reject $H_0$, if $M_c < M_{(\alpha)}$ | Reject $H_0$, if $M_c > M_{(1-\alpha)} = n_A n_B - M_\alpha$ | Reject $H_0$, when $M_c < M_{(\alpha/2)}$ or $M_c > M_{(1-\alpha/2)} = n_A n_B - M_{(\alpha/2)}$ |

* Use the tabled Mann-Whitney value for $M$ where $n_A$ and $n_B$ are the sample sizes of Groups A and B.

See the Mann-Whitney U Test two-sample comparison for ordinal data for examples of upper- and lower-tail Mann-Whitney U tests.
*Step 5.* Compute statistic.
*Step 6.* Conclusion.

*Example 7.9.* In a pharmaceutical manufacturing plant, a quality control scientist measured the dissolution rates of 50 mg tablets produced by two different table presses. She wanted to know if they differed significantly in dissolution rates when dissolved in a pH 5.2 solution held at 37°C in a controlled-temperature water bath.

Table 7.21 presents the dissolution rates in minutes of tablets randomly-sampled from each tablet press and the relative rankings of the rates.

**Table 7.21** Dissolution rates in minutes and rankings, Example 7.9

| $n$ | Tablet press A | Rank A | Tablet press B | Rank B |
|---|---|---|---|---|
| 1 | 2.6 | 3 | 3.2 | 14 |
| 2 | 2.9 | 8.5 | 3.2 | 14 |
| 3 | 3.1 | 11.5 | 3.0 | 10 |
| 4 | 2.7 | 5.5 | 3.1 | 11.5 |
| 5 | 2.7 | 5.5 | 2.9 | 8.5 |
| 6 | 3.2 | 14 | 3.4 | 18.5 |
| 7 | 3.3 | 16.5 | 2.7 | 5.5 |
| 8 | 2.7 | 5.5 | 3.7 | 20 |
| 9 | 2.5 | 1.5 | | $R_B = 102$ |
| 10 | 3.3 | 16.5 | | |
| 11 | 3.4 | 18.5 | | |
| 12 | 2.5 | 1.5 | | |
| | | $R_A = 108$ | | |

*Step 1.* Formulate the hypothesis. The researcher will use a two-tail test to detect a difference.

$H_0$: $A = B$; the dissolution rates of the tablets from Tablet Presses A and B are the same.

$H_A$: $A \neq B$; there is a significant difference in dissolution rates of tablets from the two presses.

*Step 2.* The researcher will use $\alpha = 0.05$, or $\alpha/2 = 0.025$.

*Step 3.* The test statistic to use is:

$$M_c = R_A - \frac{n_A(n_A - 1)}{2}$$

*Step 4.* In this experimental design, it is better to keep $n_A$ and $n_B$ equal in size, but this often is not possible, as is the case here.

$M_{\alpha/2} = 0.05/2 = 0.025$; using Table A.6, $M_{\alpha/2} = 23$ at $n_A = 12$ and $n_B = 8$

$M_{(1-\alpha)} = n_A n_B - M_{(\alpha/2)} = 12 \times 8 - 23 = 73$

$n_A = 12$

$n_B = 8$

$M_{0.025(20, 8)} = 42$ (Table A.6)

If $M_c$ is $< 23$ or $> 73$, the researcher will reject $H_0$ at $\alpha = 0.05$ (Fig. 7.8).

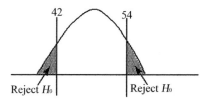

**Fig. 7.8** Example 7.9, Step 4

*Step 5*. Perform analysis.

The ranks for both groups have already been computed in Table 7.15.

$$M_c = R_A - \frac{n_A(n_A + 1)}{2}$$

$$M_c = 108 - \frac{12(12 + 1)}{2} = 30$$

*Step 6*. Because $M_c = 30$, which is not less than 23, we cannot reject $H_0$ at $\alpha = 0.05$. There is no significant difference in the variability of the dissolution rates of the 50 mg tables from presses A and B.

Note: When either sample size is greater than 20, or both are normal, Table A.2 should be used for determination of significance.

$$z_{calculated} = \frac{t_c - \left( \frac{n_A n_B}{2} \right)}{\sqrt{\frac{n_A n_B (n_A + n_B + 1)}{12}}}$$

## 7.3.2 Comparing Two Related or Paired Samples: Interval-Ratio Data

The Wilcoxon matched-pair signed-ranks statistic is a very useful and popular nonparametric statistic for evaluating two related or paired samples when the collected data are interval scale. As with the two-sample matched-pair $t$ test of parametric statistics, the Wilcoxon Test converts the two samples, $x_A$ and $x_B$, to one sample, that we label "$D$" for difference.

$$D_i = x_{A_i} - x_{B_i} \text{ for } i = 1, 2, \ldots n$$

The test is then performed on these sample value differences. The data consist of $n$ values of $D_i$, where $D_i = x_{A_i} - x_{B_i}$. Each pair of measurements, $x_{A_i}$ and $x_{B_i}$, are 1) taken on the same subject or group, etc. (e.g., before/after, pre/post) or 2) taken on subjects or groups, etc. that have been paired meaningfully with respect to important, but non-measured variables (e.g., sex, weight, organ function, etc.). The sampling of each pair, $x_{A_i}$ and $x_{B_i}$, is random.

The Wilcoxon Matched Pair Test can be used for both two-tail and one-tail tests, and the test procedure is straight-forward.

First, the signed value difference, $D_i$, of each $x_A$ and $x_B$ data set is obtained.

$$D_i = x_{A_i} - x_{B_i}$$

Second, the absolute values, $|D_i|$, are ranked from smallest to largest. Finally, the original signs from $D_i = x_{A_i} - x_{B_i}$ are reinstated, and the positive ranks ($R+$) and the negative ranks (R–) are summed. Depending upon the direction, upper or lower tail, one merely sums the $R+$ or R– values to derive a $W_c$ value and compares that value to the Wilcoxon Table value, $W_t$, (d in Table A.10), for a specific significance level at $\alpha$. The Wilcoxon Table has one- and two-tail options and requires the $n$ value (number of $D_i$ values) and the d value (corresponding to the desired $\alpha$ value), which is compared to the $W_c$ value (the sum of the $R+$ or R– values). The tabled values can be used for a number of $D_i$ values up to $n = 25$. For a $n$ larger than 25, a correction factor can be used that allows use of the normal distribution, values of $z$.

The correction factor is:

$$z_c = \frac{W_c - \frac{[n(n+1)]}{4}}{\sqrt{\frac{n(n+1)(2n+1)}{24}}}$$

where $n$ = number of $D_i$ values and $W_c$ = the sum of ranks of the negative or positive $R$ values, depending on direction of the test.

### 7.3.2.1 Ties

There are two types of ties. The first occurs when specific $x_A$, $x_B$ values to be paired are equal so the $D_i$ is 0. All pairs of $x_{A_i} = x_{B_i}$ are dropped from the analysis, and $n$ is reduced 1 for each pair dropped. The other case is where two or more $D_i$s are equal. These ties receive the average rank value. This researcher has not found the tie correction formulas to be of much practical value.

Using the six-step procedure:

*Step 1*. Specify hypothesis.

| Two-tail | Upper tail | Lower tail |
|---|---|---|
| $H_0: x_A = x_B$ | $H_0: x_A \leq x_B$ | $H_0: x_A \geq x_B$ |
| $H_A: x_A \neq x_B$ | $H_A: x_A > x_B$ | $H_A: x_A < x_B$ |

*Step 2*. Specify the $\alpha$ level.

Use $\alpha/2$ for two-tail tests for a specific level of $\alpha$, as always, and $\alpha$ for single tail tests.

*Step 3*. Write out the test statistic to use.

For small sample sizes ($n \leq 25$, which is the number of $D_i$ values), the Wilcoxon Table (Table A.10) can be used directly, requiring only $n$ and the sum of the ranks of $R+$ or $R-$ values, depending on the direction of the test. For larger samples, the correction factor must be used. That is:

$$z_c = \frac{W_c - \frac{[n(n+1)]}{4}}{\sqrt{\frac{n(n+1)(2n+1)}{24}}}$$

*Step 4*. Specify the decision rule.

| Two-Tail | Upper Tail | Lower Tail |
|---|---|---|
| The test is dependent on the sum of the ranks ($W_c$), $R+$ and $R-$, using whichever is the smaller. If $W_C$ is equal to or less than the $d$ tabled value at $\alpha/2$, reject $H_0$ at $\alpha$ | The test uses the sum of the ranks of the negative values, $R- (W_c)$. If $W_c$ is equal to or smaller than the $d$ tabled value at $\alpha$, reject $H_0$. | The test uses the sum of the ranks ($W_c$) of the positive values, $R+$. If $W_c$ is equal to or smaller than the $d$ tabled value at $\alpha$, reject $H_0$. |

*Step 5.* Perform the calculation.
*Step 6.* Formulate the conclusion.

*Example 7.10.* A researcher in a skincare laboratory wants to evaluate a dermatological lotion's ability to reduce atopic dermatitis on the skin of the hands associated with standard work-related tasks of healthcare workers (e.g., skin exposure to hot/cold, to skin cleansers, to wearing surgical/examination gloves over periods of time, and to repeated washing). Prior to the use of the test lotion, the researcher used a Visioscan device to measure skin scaliness on the hands of ten randomly selected healthcare workers. This represented the baseline measurement ($x_A$). After a 24-h period of treatment during which the test dermatological lotion was used three times, the skin on the hands of the subjects was again measured ($x_B$) for degree of skin scaliness. This was an actual in-use study, so healthcare workers went about their usual activities in a normal manner.

On the basis of the before-and-after Visioscan data processed for a Wilcoxon analysis (Table 7.22), the researcher determined if the treatment was effective in reducing skin scaliness at an $\alpha = 0.05$ level of confidence.

**Table 7.22** Visioscan data for degree of skin scaliness, Example 7.10

| Subject | 1 | 2 | 3 | 4 | 5 | 6 | 7 | 8 | 9 | 10 |
|---|---|---|---|---|---|---|---|---|---|---|
| Baseline (pre-treatment $= x_A$) | 54 | 57 | 85 | 81 | 69 | 72 | 83 | 58 | 75 | 87 |
| 24 h. post-treatment $= x_B$ | 41 | 53 | 63 | 81 | 73 | 69 | 75 | 54 | 69 | 70 |
| Difference ($D_i$) | 13 | 4 | 22 | 0 | −4 | 3 | 8 | 4 | 6 | 17 |
| Rank ($D_i$) | 7 | 3 | 9 | − | 3 | 1 | 6 | 3 | 5 | 8 |
| Signed Rank ($R$) | +7 | +3 | +9 | − | −3 | +1 | +6 | +3 | +5 | +8 |

The six-step procedure for this evaluation is:
*Step 1.* Formulate hypothesis.
The researcher wants to determine if the lotion treatment reduced skin scaliness. For that to occur, the $x_B$ values must be significantly lower than the $x_A$ values. Hence, this is an upper tail test ($H_A: x_A > x_B$).

$H_0: x_A \leq x_B$
$H_A: x_A > x_B$     The treatment reduces the scaliness of the skin.
*Step 2.* Specify $\alpha$. Let us set $\alpha$ at 0.05.
Because this is an upper tail test, we sum the ranks of the negative values ($R-$ values) to calculate $W_c$, which is expected to be small, if $H_0$ is to be rejected.
*Step 3.* Decision rule.
The tabled value of $W_t$ is found from Table A.10 (Wilcoxon Table) for $n = 9$ (1 pair value was lost due to a tie) and a one-tail $\alpha = 0.05$. For this table, like the Sign Test, the $\alpha$ value ($\alpha''$ for two tail tests, and $\alpha'$ for one tail tests) is not precisely 0.001,

0.01, 0.025, 0.05, or 0.10, so the researcher uses the tabled $\alpha$ value closest to the specified $\alpha$ value. In this case, with $n = 9$ and $\alpha' = 0.05$, the tabled value is $d = W_t = 9$ at $\alpha = 0.049$. Hence, we reject $H_0$ if $W_c$, the sum of the ranked negative values $(R-)$, is less than or equal to 9 at $\alpha = 0.049$.

*Step 4.* Choose the test statistic.

Because this is an upper tail test, we sum the negative rank values $(R-)$ and compare the result to $W_t = 9$. If the $R-$ value $= W_c \leq 9$, reject $H_0$ at $\alpha = 0.049$.

*Step 5.* Perform analysis.

The sum of the negative rank values $(R-)$ is $3 = W_c$.

*Step 6.* Conclusion.

Because the sum of the negative rank values, $W_c = 3$, is less than the tabled value of 9, we can reject at $\alpha = 0.05$ the $H_0$ hypothesis that the treatment does not significantly reduce skin scaliness.

### 7.3.2.2 Comments

Suppose this had been a two-tail test at $\alpha = 0.05$:

$H_0$: $x_A = x_B$
$H_A$: $x_A \neq x_B$

The sum of negative or positive $R$ values, whichever $R_i$ is smaller, must be less than or equal to 7 at $\alpha = 0.055$ (Table A.10). The sum of $R- = 3$, so we reject $H_0$ at $\alpha = 0.055$, with a $p$ value less than 0.039.

Suppose the test was a lower tail test.

$H_0$: $x_A \geq x_B$
$H_A$: $x_A < x_B$

In this case, we use the sum of the positive ranks. If the sum of the positive ranks is $\leq 9$ with an $n$ of 9, we reject $H_0$ at $\alpha = 0.049$.

$W_c - R+ = 7 + 3 + 9 + 1 + 6 + 3 + 5 + 8 = 42.$
$W_c = 42 > W_t = 9$, so we cannot reject $H_0$ at $\alpha = 0.049$.

### 7.3.3 Independent Samples, $n > 2$: Interval-Ratio Data

Use the Kruskal-Wallis Test, as previously discussed in the ordinal data section.

### 7.3.4 Related Samples, $n > 2$: Interval-Ratio Data

The nonparametric analog to the randomized complete block parametric statistic is the Quade Test, when using interval scale data. Although not as well known as the Kruskal-Wallis or Friedman Tests, the Quade Test often can be an extremely useful tool for the applied statistical researcher.

As in the last two tests discussed, the data display for the Quade Test (interval data) is:

|  | | Treatments | | | |
|--|--|------------|--|--|--|
|  |  | 1 | 2 | ... | $\ell$ |
|  | 1 | $x_{11}$ | $x_{12}$ | ... | $x_{11}$ |
|  | 2 | $x_{21}$ | $x_{22}$ | ... | $x_{21}$ |
| Blocks | . | . | . | . | . |
|  | . | . | . | . | . |
|  | . | . | . | . | . |
|  | $k$ | $x_{k1}$ | $x_{k2}$ | | $x_{k\ell}$ |

### 7.3.4.1 Assumptions

1. The data consist of $k$ mutually-independent blocks of number of treatments, $\ell$.
2. The data are blocked in meaningful ways, such as age, sex, weight, height, same subject, etc.
3. The data within each block can be ranked – data are at least ordinal. However, this microbiologist prefers to use this test with interval data and suggests that, if the data are ordinal, the Friedman test be used.
4. The sample range within each block can be determined. (There is a smallest and largest number in each block; the $x$ values are not all equal.)
5. The blocks, themselves, must be rankable by range.
6. No significant interaction occurs between blocks and treatments.
7. The data are continuous.
    The test hypothesis is a two-tail test.
        $H_0$: The test populations are equivalent.
        $H_A$: They differ in at least one.
Note: Ties do not adversely affect this test, so a correction factor is not necessary.
    The $F$ Distribution Table is used in this test. The $F_T$ value is $F_{\alpha[\ell-1;\,(\ell-1)(k-1)]}$. That is, the numerator degrees of freedom are $(\ell-1)$ and the denominator degrees of freedom are $(\ell-1)(k-1)$.

### 7.3.4.2 Procedure

*Step 1.* Let $R(x_{ij})$ be the rank from 1 to $\ell$ of each block, $i$. For example, in block (row) 1, the individual $\ell$ treatments are ranked. A rank of 1 is provided to the smallest and a rank of $\ell$ to the largest. So step one is to rank all the observations, block-by-block, throughout the $k$ blocks. In case of ties, the average rank is used, as before.

*Step 2.* Using the original $x_{ij}$ values – not the ranks – determine the range of each block. The range in block $i$ = MAX $(x_{ij})$ – MIN $(x_{ij})$. There will be $k$ sample ranges, one for each block.

*Step 3.* Once the ranges are determined, rank the block ranges, assigning 1 to the smallest up to $k$ for the largest. If ties occur, use the average rank. Let $R_1, R_2, \ldots R_k$ be the ranks assigned to the 1, 2, . . .$k$ blocks.

*Step 4.* Each block rank, $R_i$, is then multiplied by the difference between the rank within block $i$, $[R(x_{ij})]$, and the average rank within the blocks, $(\ell+1)/2$, to get the value for $S_{ij}$, which represents the relative size of each observation within the block, adjusted to portray the relative significance of the block in which it appears.

$$S_{ij} \text{ value} = R_i \left[ R x_{ij} - \frac{\ell + 1}{2} \right].$$

Each treatment group sum is denoted by $S'_{ij} = \sum S_{ij}$.

The test statistic is similar to ANOVA.

$$SS_{TOTAL} = \sum_{i=1}^{k} \sum_{j=1}^{\ell} S_{ij}^2$$

If there are no ties in $SS_{TOTAL}$, the equation is:

$$SS_{TOTAL} = \frac{k(k+1)(2k+1)\ell(\ell+1)(\ell-1)}{72}$$

The treatment sum of squares is:

$$SS_{TREATMENT} = \frac{1}{k} \sum_{j=1}^{\ell} S_{ij}^2$$

The test statistic is:

$$F_c = \frac{(k-1)\ SS_{TREATMENT}}{SS_{TOTAL} - SS_{TREATMENT}}$$

Note: If $SS_{TOTAL} = SS_{TREATMENT}$ ($SS_{TOTAL} - SS_{TREATMENT} = 0$), use the "0" as if it was in the critical region and calculate the critical level as $\alpha^* = (1/\ell!)^{k-1}$, where $\ell!$ is $\ell$ factorial or $\ell \times (\ell-1) \times (\ell-2) \ldots (\ell-1+1)$ (e.g., $\ell = 5$; $\ell! = 5 \times 4 \times 3 \times 2 \times 1 = 120$). The decision rule is, using the $F$ Table (Table A.4):

If $F_c > F_t = F_{\alpha[\ell-1;\ (\ell-1)(k-1)]}$, reject $H_0$ at $\alpha$.

Again, the six-step procedure is easily adapted to this statistical procedure.

*Step 1.* Formulate hypothesis, which will always be for a two-tail test.

$H_0$: The groups are equal.

$H_A$: The groups are not equal.

*Step 2.* Choose $\alpha$.

*Step 3.* Write out the test statistic.

$$F_c = \frac{(k-1)SS_{TREATMENT}}{SS_{TOTAL} - SS_{TREATMENT}}$$

*Step 4.* Decision rule.

If $F_c > F_t$, reject $H_0$ at $\alpha$.

*Step 5.* Perform statistic.

*Step 6.* Conclusion.

*Example 7.11.* A microbiologist working with *Pseudomonas aeuroginosa* wants to test the resistance to biofilm formation of several antimicrobial compounds applied to the surface of veinous/arterial catheters. Three different sample configurations of catheter material were introduced in five bioreactors, each of which were considered a block for the analysis. After a 72-h growth period, in a continuous flow nutrient system, the catheter materials were removed, and the microorganism/biofilm levels were enumerated in terms of $\log_{10}$ colony-forming units. The researcher wanted to know if there was a significant difference in microbial adhesion among the products.

| | | Test Catheters | | |
|---|---|---|---|---|
| | | 1 | 2 | 3 |
| Reactors (Blocks) | 1 | 5.03 | 3.57 | 4.90 |
| | 2 | 3.25 | 2.17 | 3.10 |
| | 3 | 7.56 | 5.16 | 6.12 |
| | 4 | 4.92 | 3.12 | 4.92 |
| | 5 | 6.53 | 4.23 | 5.99 |

The collected data were tabulated, as above. Because the study was a small pilot study with few replicate blocks, and the blocked data varied so much, a nonparametric model was selected for the analysis.

*Step 1.* Formulate hypothesis.

$H_0$: Catheter material 1 = 2 = 3 in microbial adherence

$H_A$: At least one catheter material is different

*Step 2.* Specify $\alpha$.

Because this was a small study, $\alpha$ was selected at 0.10.

*Step 3.* Write out the test statistic.

The test statistic to be used is:

$$F_c = \frac{(k-1)\,SS_{TREATMENT}}{SS_{TOTAL} - SS_{TREATMENT}}$$

*Step 4.* Present decision rule.

If $F_c > F_t$, reject $H_0$ at $\alpha = 0.10$. $F_t = F_{\alpha[\ell-1;\,(\ell-1)(k-1)]}$, where numerator degrees of freedom = $\ell - 1 = 3 - 1 = 2$, denominator degrees of freedom = $(\ell - 1)(k - 1) = (3 - 1)(5 - 1) = 8$, and $\alpha = 0.10$. $F_{t[0.10(2,\,8)]} = 3.11$. Therefore, if $F_c > F_t = 3.11$, reject $H_0$ at $\alpha = 0.10$.

*Step 5.* Perform computation; the collected results, ranked within blocks, are presented below.

| | | Test Catheter | | |
|---|---|---|---|---|
| | | 1 | 2 | 3 |
| | 1 | 5.03 (3) | 3.57 (1) | 4.90 (2) |
| Block | 2 | 3.25 (3) | 2.17 (1) | 3.10 (2) |
| (reactor) | 3 | 7.56 (3) | 5.16 (1) | 6.12 (2) |
| | 4 | 4.92 (2.5) | 3.12 (1) | 4.92 (2.5) |
| | 5 | 6.53 (3) | 4.23 (1) | 5.99 (2) |

*Step 6.* Conclusion.

*Step 1.* First rank the within block values in 1, 2, 3, .... order. Above, the ranks are in parentheses to the right of each test value.

*Step 2.* Next, determine range of actual values in each block, which is the high value minus the low value, and record results in a work table (Table 7.17).

Block 1 = 5.03 – 3.57 = 1.46          4 = 4.92 – 3.12 = 1.80

2 = 3.25 – 2.17 = 1.08          5 = 6.53 – 4.23 = 2.30

3 = 7.56 – 5.16 = 2.40

*Step 3.* Next, rank the blocks, and enter the ranks into the work table (Table 7.23).

**Table 7.23** Work table, Example 7.11

| Block | (Step 2) Sample block range | (Step 3) Block rank ($R_i$) | (Step 4) Catheter | | |
|---|---|---|---|---|---|
| | | | 1 | 2 | 3 |
| 1 | 1.46 | 2 | 2 | −2 | 0 |
| 2 | 1.08 | 1 | 1 | −1 | 0 |
| 3 | 2.40 | 5 | 5 | −5 | 0 |
| 4 | 1.80 | 3 | 1.5 | −3 | 1.5 |
| 5 | 2.30 | 4 | 4 | −4 | 0 |
| | | | $S'_{.1} = 13.5$ | $S'_{.2} = -15.0$ | $S'_{.3} = 1.5$ |

*Step 4.* Determine $S_{ij} = R_i \left[ R(x_{ij}) - \dfrac{\ell+1}{2} \right]$ *for each* $x_{ij}$, and enter those values into the work table (Table 7.23).

$$S_{11} = 2(3 - \tfrac{3+1}{2}) = 2$$
$$S_{12} = 2(1 - 2) = -2$$
$$S_{13} = 2(2 - 2) = 0$$
$$S_{21} = 1(3 - 2) = 1$$
$$S_{22} = 1(1 - 2) = -1$$
$$S_{23} = 1(2 - 2) = 0$$
$$S_{31} = 5(3 - 2) = 5$$
$$S_{32} = 5(1 - 2) = -5$$
$$S_{33} = 5(2 - 2) = 0$$
$$S_{41} = 3(2.5 - 2) = 1.5$$
$$S_{42} = 3(1 - 2) = -3$$
$$S_{43} = 3(2.5 - 2) = 1.5$$
$$S_{51} = 4(3 - 2) = 4$$
$$S_{52} = 4(1 - 2) = -4$$
$$S_{53} = 4(2 - 2) = 0$$

*Step 5.* Determine $SS_{TOTAL}$

$$SS_{TOTAL} = \sum_{i=1}^{k} \sum_{j=1}^{\ell} S_{ij}^2$$
$$= 2^2 + (-2)^2 + 0^2 + 1^2 + (-1)^2 + 0^2$$
$$+ 5^2 + (-5)^2 + 0^2 + (-1.5)^2 + (-3)^2 + 1.5^2$$
$$+ 4^2 + (-4)^2 + 0^2$$

$SS_{TOTAL} = 105.50$

$SS_{TREATMENT} = \dfrac{1}{k}\sum_{j=1}^{\ell} S_{ij}^2 = \dfrac{1}{5}\left[13.5^2 + (-15)^2 + 1.5^2\right] = 81.90$

Next, compute $F_c$.

$F_c = \dfrac{(k-1)\ SS_{TREATMENT}}{SS_{TOTAL} - SS_{TREATMENT}} = \dfrac{(5-1)\ 81.90}{105.50 - 81.90} = 13.88$

*Step 6.* Conclusion.

Because $F_c = 13.88 > F_t = 3.11$, the $H_0$ hypothesis is rejected at $\alpha = 0.10$. At least one catheter is different from the others.

### 7.3.4.3 Multiple Contrasts

As before, multiple contrasts are conducted only when $H_0$ is rejected. The computation formula is for all possible $\ell(\ell - 1)/2$ contrast combinations. If $|S_i - S_j|$ is greater than $t_{\alpha/2}\sqrt{\dfrac{2k\ (SS_{TOTAL} - SS_{TREATMENT})}{(\ell - 1)(k - 1)}}$, conclude the difference is significant at $\alpha$.

Let us contrast the catheter products.

$\sqrt{\dfrac{2k\ (SS_{TOTAL} - SS_{TREATMENT})}{(\ell - 1)(k - 1)}} = \sqrt{\dfrac{2(5)\ (105.50 - 81.90)}{2(4)}} = 5.4314$

$t_{0.10/2\ with\ (\ell-1)(k-1)\ df} = t_{(0.05,8)} = 1.86$

So, $t_{\alpha/2}\sqrt{\dfrac{2k\ (SS_{TOTAL} - SS_{TREATMENT})}{(\ell - 1)(k - 1)}} = 1.86(5.4314) = 10.10$

The catheter product contrasts are:

| | | |
|---|---|---|
| 1 vs 2 = $\|13.5 - (-15)\| = 28.5 > 10.10$ | | Significant |
| 1 vs 3 = $\|13.5 - 1.5\| = 12 > 10.10$ | | Significant |
| 2 vs 3 = $\|-15 - 1.5\| = 16.5 > 10.10$ | | Significant |

Each of the catheter products is significantly different from the others, at $\alpha = 0.10$.

# Appendix

**Tables of Mathematical Values**

**Table A.1** Student's $t$ table (percentage points of the $t$ distribution)

| $v$ \ $\alpha$ | .40 | .25 | .10 | .05 | .025 | .01 | .005 | .0025 | .001 | .0005 |
|---|---|---|---|---|---|---|---|---|---|---|
| 1 | .325 | 1.000 | 3.078 | 6.314 | 12.706 | 31.821 | 63.657 | 127.32 | 318.31 | 636.62 |
| 2 | .289 | .816 | 1.886 | 2.920 | 4.303 | 6.965 | 9.925 | 14.089 | 23.326 | 31.598 |
| 3 | .277 | .765 | 1.638 | 2.353 | 3.182 | 4.541 | 5.841 | 7.453 | 10.213 | 12.924 |
| 4 | .271 | .741 | 1.533 | 2.132 | 2.776 | 3.747 | 4.604 | 5.598 | 7.173 | 8.610 |
| 5 | .267 | .727 | 1.476 | 2.015 | 2.571 | 3.365 | 4.032 | 4.773 | 5.893 | 6.869 |
| 6 | .265 | .727 | 1.440 | 1.943 | 2.447 | 3.143 | 3.707 | 4.317 | 5.208 | 5.959 |
| 7 | .263 | .711 | 1.415 | 1.895 | 2.365 | 2.998 | 3.499 | 4.019 | 4.785 | 5.408 |
| 8 | .262 | .706 | 1.397 | 1.860 | 2.306 | 2.896 | 3.355 | 3.833 | 4.501 | 5.041 |
| 9 | .261 | .703 | 1.383 | 1.833 | 2.262 | 2.821 | 3.250 | 3.690 | 4.297 | 4.781 |
| 10 | .260 | .700 | 1.372 | 1.812 | 2.228 | 2.764 | 3.169 | 3.581 | 4.144 | 4.587 |
| 11 | .260 | .697 | 1.363 | 1.796 | 2.201 | 2.718 | 3.106 | 3.497 | 4.025 | 4.437 |
| 12 | .259 | .695 | 1.356 | 1.782 | 2.179 | 2.681 | 3.055 | 3.428 | 3.930 | 4.318 |
| 13 | .259 | .694 | 1.350 | 1.771 | 2.160 | 2.650 | 3.012 | 3.372 | 3.852 | 4.221 |
| 14 | .258 | .692 | 1.345 | 1.761 | 2.145 | 2.624 | 2.977 | 3.326 | 3.787 | 4.140 |
| 15 | .258 | .691 | 1.341 | 1.753 | 2.131 | 2.602 | 2.947 | 3.286 | 3.733 | 4.073 |
| 16 | .258 | .690 | 1.337 | 1.746 | 2.120 | 2.583 | 2.921 | 3.252 | 3.686 | 4.015 |
| 17 | .257 | .689 | 1.333 | 1.740 | 2.110 | 2.567 | 2.898 | 3.222 | 3.646 | 3.965 |
| 18 | .257 | .688 | 1.330 | 1.734 | 2.101 | 2.552 | 2.878 | 3.197 | 3.610 | 3.922 |
| 19 | .257 | .688 | 1.328 | 1.729 | 2.093 | 2.539 | 2.861 | 3.174 | 3.579 | 3.883 |
| 20 | .257 | .687 | 1.325 | 1.725 | 2.086 | 2.528 | 2.845 | 3.153 | 3.552 | 3.850 |
| 21 | .257 | .686 | 1.323 | 1.721 | 2.080 | 2.518 | 2.831 | 3.135 | 3.527 | 3.819 |
| 22 | .256 | .686 | 1.321 | 1.717 | 2.074 | 2.508 | 2.819 | 3.119 | 3.505 | 3.792 |
| 23 | .256 | .685 | 1.319 | 1.714 | 2.069 | 2.500 | 2.807 | 3.104 | 3.485 | 3.767 |
| 24 | .256 | .685 | 1.318 | 1.711 | 2.064 | 2.492 | 2.797 | 3.091 | 3.467 | 3.745 |
| 25 | .256 | .684 | 1.316 | 1.708 | 2.060 | 2.485 | 2.787 | 3.078 | 3.450 | 3.725 |
| 26 | .256 | .684 | 1.315 | 1.706 | 2.056 | 2.479 | 2.779 | 3.067 | 3.435 | 3.707 |
| 27 | .256 | .684 | 1.314 | 1.703 | 2.052 | 2.473 | 2.771 | 3.057 | 3.421 | 3.690 |
| 28 | .256 | .683 | 1.313 | 1.701 | 2.048 | 2.467 | 2.763 | 3.047 | 3.408 | 3.674 |
| 29 | .256 | .683 | 1.311 | 1.699 | 2.045 | 2.462 | 2.756 | 3.038 | 3.396 | 3.659 |
| 30 | .256 | .683 | 1.310 | 1.697 | 2.042 | 2.457 | 2.750 | 3.030 | 3.385 | 3.646 |
| 40 | .255 | .681 | 1.303 | 1.684 | 2.021 | 2.423 | 2.704 | 2.971 | 3.307 | 3.551 |
| 60 | .254 | .679 | 1.296 | 1.671 | 2.000 | 2.390 | 2.660 | 2.915 | 3.232 | 3.460 |
| 120 | .254 | .677 | 1.289 | 1.658 | 1.980 | 2.358 | 2.617 | 2.860 | 3.160 | 3.373 |
| $\infty$ | .253 | .674 | 1.282 | 1.645 | 1.960 | 2.326 | 2.576 | 2.807 | 3.090 | 3.291 |

$v$ = degrees of freedom

**Table A.2** z-table (normal curve areas [entries in the body of the table give the area under the standard normal curve from 0 to z])

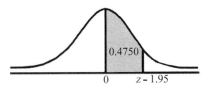

| z | .00 | .01 | .02 | .03 | .04 | .05 | .06 | .07 | .08 | .09 |
|---|------|------|------|------|------|------|------|------|------|------|
| **0.0** | .0000 | .0040 | .0080 | .0120 | .0160 | .0199 | .0239 | .0279 | .0319 | .0359 |
| **0.1** | .0398 | .0438 | .0478 | .0517 | .0557 | .0596 | .0636 | .0675 | .0714 | .0753 |
| **0.2** | .0793 | .0832 | .0871 | .0910 | .0948 | .0987 | .1026 | .1064 | .1103 | .1141 |
| **0.3** | .1179 | .1217 | .1255 | .1293 | .1331 | .1368 | .1406 | .1443 | .1480 | .1517 |
| **0.4** | .1554 | .1591 | .1628 | .1664 | .1700 | .1736 | .1772 | .1808 | .1844 | .1879 |
| **0.5** | .1915 | .1950 | .1985 | .2019 | .2054 | .2088 | .2123 | .2157 | .2190 | .2224 |
| **0.6** | .2257 | .2291 | .2324 | .2357 | .2389 | .2422 | .2454 | .2486 | .2517 | .2549 |
| **0.7** | .2580 | .2611 | .2642 | .2673 | .2704 | .2734 | .2764 | .2794 | .2823 | .2852 |
| **0.8** | .2881 | .2910 | .2939 | .2967 | .2995 | .3023 | .3051 | .3078 | .3106 | .3133 |
| **0.9** | .3159 | .3186 | .3212 | .3238 | .3264 | .3289 | .3315 | .3340 | .3365 | .3389 |
| **1.0** | .3413 | .3438 | .3461 | .3485 | .3508 | .3531 | .3554 | .3577 | .3599 | .3621 |
| **1.1** | .3643 | .3665 | .3686 | .3708 | .3729 | .3749 | .3770 | .3790 | .3810 | .3830 |
| **1.2** | .3849 | .3869 | .3888 | .3907 | .3925 | .3944 | .3962 | .3980 | .3997 | .4015 |
| **1.3** | .4032 | .4049 | .4066 | .4082 | .4099 | .4115 | .4131 | .4147 | .4162 | .4177 |
| **1.4** | .4192 | .4207 | .4222 | .4236 | .4251 | .4265 | .4279 | .4292 | .4306 | .4319 |
| **1.5** | .4332 | .4345 | .4357 | .4370 | .4382 | .4394 | .4406 | .4418 | .4429 | .4441 |
| **1.6** | .4452 | .4463 | .4474 | .4484 | .4495 | .4505 | .4515 | .4525 | .4535 | .4545 |
| **1.7** | .4554 | .4564 | .4573 | .4582 | .4591 | .4599 | .4608 | .4616 | .4625 | .4633 |
| **1.8** | .4641 | .4649 | .4656 | .4664 | .4671 | .4678 | .4686 | .4693 | .4699 | .4706 |
| **1.9** | .4713 | .4719 | .4726 | .4732 | .4738 | .4744 | .4750 | .4756 | .4761 | .4767 |
| **2.0** | .4772 | .4778 | .4783 | .4788 | .4793 | .4798 | .4803 | .4808 | .4812 | .4817 |
| **2.1** | .4821 | .4826 | .4830 | .4834 | .4838 | .4842 | .4846 | .4850 | .4854 | .4857 |
| **2.2** | .4861 | .4864 | .4868 | .4871 | .4875 | .4878 | .4881 | .4884 | .4887 | .4890 |
| **2.3** | .4893 | .4896 | .4898 | .4901 | .4904 | .4906 | .4909 | .4911 | .4913 | .4916 |
| **2.4** | .4918 | .4920 | .4922 | .4925 | .4927 | .4929 | .4931 | .4932 | .4934 | .4936 |
| **2.5** | .4938 | .4940 | .4941 | .4943 | .4945 | .4946 | .4948 | .4949 | .4951 | .4952 |
| **2.6** | .4953 | .4955 | .4956 | .4957 | .4959 | .4960 | .4961 | .4962 | .4963 | .4964 |
| **2.7** | .4965 | .4966 | .4967 | .4968 | .4969 | .4970 | .4971 | .4972 | .4973 | .4974 |
| **2.8** | .4974 | .4975 | .4976 | .4977 | .4977 | .4978 | .4979 | .4979 | .4980 | .4981 |
| **2.9** | .4981 | .4982 | .4982 | .4983 | .4984 | .4984 | .4985 | .4985 | .4986 | .4986 |
| **3.0** | .4987 | .4987 | .4987 | .4988 | .4988 | .4989 | .4989 | .4989 | .4990 | .4990 |

**Table A.3** Studentized range table

$q_{0.05}(p, f)$

| f \ p | 2 | 3 | 4 | 5 | 6 | 7 | 8 | 9 | 10 | 11 | 12 | 13 | 14 | 15 | 16 | 17 | 18 | 19 | 20 |
|---|---|---|---|---|---|---|---|---|---|---|---|---|---|---|---|---|---|---|---|
| 1 | 18.10 | 26.70 | 32.80 | 37.20 | 40.50 | 43.10 | 45.40 | 47.30 | 49.10 | 50.60 | 51.90 | 53.20 | 54.30 | 55.40 | 56.30 | 57.20 | 58.00 | 58.80 | 59.60 |
| 2 | 6.09 | 8.28 | 9.80 | 10.89 | 11.73 | 12.43 | 13.03 | 13.54 | 13.99 | 14.39 | 14.75 | 15.08 | 15.38 | 15.65 | 15.91 | 16.14 | 16.36 | 16.57 | 16.77 |
| 3 | 4.50 | 5.88 | 6.83 | 7.51 | 8.04 | 8.47 | 8.85 | 9.18 | 9.46 | 9.72 | 9.95 | 10.16 | 10.35 | 10.52 | 10.69 | 10.84 | 10.98 | 11.12 | 11.24 |
| 4 | 3.93 | 5.00 | 5.76 | 6.31 | 6.73 | 7.06 | 7.35 | 7.60 | 7.83 | 8.03 | 8.21 | 8.37 | 8.52 | 8.67 | 8.80 | 8.92 | 9.03 | 9.14 | 9.24 |
| 5 | 3.64 | 4.60 | 5.22 | 5.67 | 6.03 | 6.33 | 6.58 | 6.80 | 6.99 | 7.17 | 7.32 | 7.47 | 7.60 | 7.72 | 7.83 | 7.93 | 8.03 | 8.12 | 8.21 |
| 6 | 3.46 | 4.34 | 4.90 | 5.31 | 5.63 | 5.89 | 6.12 | 6.32 | 6.49 | 6.65 | 6.79 | 6.92 | 7.04 | 7.14 | 7.24 | 7.34 | 7.43 | 7.51 | 7.59 |
| 7 | 3.34 | 4.16 | 4.68 | 5.06 | 5.35 | 5.59 | 5.80 | 5.99 | 6.15 | 6.29 | 6.42 | 6.54 | 6.65 | 6.75 | 6.84 | 6.93 | 7.01 | 7.08 | 7.16 |
| 8 | 3.26 | 4.04 | 4.53 | 4.89 | 5.17 | 5.40 | 5.60 | 5.77 | 5.92 | 6.05 | 6.18 | 6.29 | 6.39 | 6.48 | 6.57 | 6.65 | 6.73 | 6.80 | 6.87 |
| 9 | 3.20 | 3.95 | 4.42 | 4.76 | 5.02 | 5.24 | 5.43 | 5.60 | 5.74 | 5.87 | 5.98 | 6.09 | 6.19 | 6.28 | 6.36 | 6.44 | 6.51 | 6.58 | 6.65 |
| 10 | 3.15 | 3.88 | 4.33 | 4.66 | 4.91 | 5.12 | 5.30 | 5.46 | 5.60 | 5.72 | 5.83 | 5.93 | 6.03 | 6.12 | 6.20 | 6.27 | 6.34 | 6.41 | 6.47 |
| 11 | 3.11 | 3.82 | 4.26 | 4.58 | 4.82 | 5.03 | 5.20 | 5.35 | 5.49 | 5.61 | 5.71 | 5.81 | 5.90 | 5.98 | 6.06 | 6.14 | 6.20 | 6.27 | 6.33 |
| 12 | 3.08 | 3.77 | 4.20 | 4.51 | 4.75 | 4.95 | 5.12 | 5.27 | 5.40 | 5.51 | 5.61 | 5.71 | 5.80 | 5.88 | 5.95 | 6.02 | 6.09 | 6.15 | 6.21 |
| 13 | 3.06 | 3.73 | 4.15 | 4.46 | 4.69 | 4.88 | 5.05 | 5.19 | 5.32 | 5.43 | 5.53 | 5.63 | 5.71 | 5.79 | 5.86 | 5.93 | 6.00 | 6.06 | 6.11 |
| 14 | 3.03 | 3.70 | 4.11 | 4.41 | 4.64 | 4.83 | 4.99 | 5.13 | 5.25 | 5.36 | 5.46 | 5.56 | 5.64 | 5.72 | 5.79 | 5.86 | 5.92 | 5.98 | 6.03 |
| 15 | 3.01 | 3.67 | 4.08 | 4.37 | 4.59 | 4.78 | 4.94 | 5.08 | 5.20 | 5.31 | 5.40 | 5.49 | 5.57 | 5.65 | 5.72 | 5.79 | 5.85 | 5.91 | 5.96 |
| 16 | 3.00 | 3.65 | 4.05 | 4.34 | 4.56 | 4.74 | 4.90 | 5.03 | 5.15 | 5.26 | 5.35 | 5.44 | 5.52 | 5.59 | 5.66 | 5.73 | 5.79 | 5.84 | 5.90 |
| 17 | 2.98 | 3.62 | 4.02 | 4.31 | 4.52 | 4.70 | 4.86 | 4.99 | 5.11 | 5.21 | 5.31 | 5.39 | 5.47 | 5.55 | 5.61 | 5.68 | 5.74 | 5.79 | 5.84 |
| 18 | 2.97 | 3.61 | 4.00 | 4.28 | 4.49 | 4.67 | 4.83 | 4.96 | 5.07 | 5.17 | 5.27 | 5.35 | 5.43 | 5.50 | 5.57 | 5.63 | 5.69 | 5.74 | 5.79 |
| 19 | 2.96 | 3.59 | 3.98 | 4.26 | 4.47 | 4.64 | 4.79 | 4.92 | 5.04 | 5.14 | 5.23 | 5.32 | 5.39 | 5.46 | 5.53 | 5.59 | 5.65 | 5.70 | 5.75 |
| 20 | 2.95 | 3.58 | 3.96 | 4.24 | 4.45 | 4.62 | 4.77 | 4.90 | 5.01 | 5.11 | 5.20 | 5.28 | 5.36 | 5.43 | 5.50 | 5.56 | 5.61 | 5.66 | 5.71 |
| 24 | 2.92 | 3.53 | 3.90 | 4.17 | 4.37 | 4.54 | 4.68 | 4.81 | 4.92 | 5.01 | 5.10 | 5.18 | 5.25 | 5.32 | 5.38 | 5.44 | 5.50 | 5.55 | 5.59 |
| 30 | 2.89 | 3.48 | 3.84 | 4.11 | 4.30 | 4.46 | 4.60 | 4.72 | 4.83 | 4.92 | 5.00 | 5.08 | 5.15 | 5.21 | 5.27 | 5.33 | 5.38 | 5.43 | 5.48 |
| 40 | 2.86 | 3.44 | 3.79 | 4.04 | 4.23 | 4.39 | 4.52 | 4.63 | 4.74 | 4.82 | 4.90 | 4.98 | 5.05 | 5.11 | 5.17 | 5.22 | 5.27 | 5.32 | 5.36 |
| 60 | 2.83 | 3.40 | 3.74 | 3.98 | 4.16 | 4.31 | 4.44 | 4.55 | 4.65 | 4.73 | 4.81 | 4.88 | 4.94 | 5.00 | 5.06 | 5.11 | 5.15 | 5.20 | 5.24 |
| 120 | 2.80 | 3.36 | 3.69 | 3.92 | 4.10 | 4.24 | 4.36 | 4.47 | 4.56 | 4.64 | 4.71 | 4.78 | 4.84 | 4.90 | 4.95 | 5.00 | 5.04 | 5.09 | 5.13 |
| ∞ | 2.77 | 3.32 | 3.63 | 3.86 | 4.03 | 4.17 | 4.29 | 4.39 | 4.47 | 4.55 | 4.62 | 4.68 | 4.74 | 4.80 | 4.84 | 4.98 | 4.93 | 4.97 | 5.01 |

(Continued)

**Table A.3** Studentized range table

$$q_{0.01}(p,f)$$

| f \ p | 2 | 3 | 4 | 5 | 6 | 7 | 8 | 9 | 10 | 11 | 12 | 13 | 14 | 15 | 16 | 17 | 18 | 19 | 20 |
|---|---|---|---|---|---|---|---|---|---|---|---|---|---|---|---|---|---|---|---|
| 1 | 90.0 | 135.0 | 164.0 | 186.0 | 202.0 | 216.0 | 227.0 | 237.0 | 246.0 | 253.0 | 260.0 | 266.0 | 272.0 | 277.0 | 282.0 | 286.0 | 290.0 | 294.0 | 298.0 |
| 2 | 14.00 | 19.00 | 22.30 | 24.70 | 26.60 | 28.20 | 29.50 | 30.70 | 31.70 | 32.60 | 33.40 | 34.10 | 34.80 | 35.4 | 36.0 | 36.5 | 37.0 | 37.5 | 37.90 |
| 3 | 8.26 | 10.60 | 12.20 | 13.30 | 14.20 | 15.0 | 15.60 | 16.20 | 16.70 | 17.10 | 17.50 | 17.90 | 18.20 | 18.5 | 18.8 | 19.1 | 19.3 | 19.5 | 19.80 |
| 4 | 6.51 | 8.12 | 9.17 | 9.96 | 10.60 | 11.1 | 11.50 | 11.90 | 12.30 | 12.60 | 12.80 | 13.10 | 13.30 | 13.5 | 13.7 | 13.9 | 14.1 | 14.2 | 14.40 |
| 5 | 5.70 | 6.97 | 7.80 | 8.42 | 8.91 | 9.32 | 9.67 | 9.97 | 10.24 | 10.48 | 10.70 | 10.89 | 11.08 | 11.24 | 11.4 | 11.55 | 11.68 | 11.81 | 11.93 |
| 6 | 5.24 | 6.33 | 7.03 | 7.56 | 7.97 | 8.32 | 8.61 | 8.87 | 9.10 | 9.30 | 9.49 | 9.65 | 9.81 | 9.95 | 10.08 | 10.21 | 10.32 | 10.43 | 10.54 |
| 7 | 4.95 | 5.92 | 6.54 | 7.01 | 7.37 | 7.68 | 7.94 | 8.17 | 8.37 | 8.55 | 8.71 | 8.86 | 9.00 | 9.12 | 9.24 | 9.35 | 9.46 | 9.55 | 9.65 |
| 8 | 4.74 | 5.63 | 6.20 | 6.63 | 6.96 | 7.24 | 7.47 | 7.68 | 7.87 | 8.03 | 8.18 | 8.31 | 8.44 | 8.55 | 8.66 | 8.76 | 8.85 | 8.94 | 9.03 |
| 9 | 4.60 | 5.43 | 5.96 | 6.35 | 6.66 | 6.91 | 7.13 | 7.32 | 7.49 | 7.65 | 7.78 | 7.91 | 8.03 | 8.13 | 8.23 | 8.32 | 8.41 | 8.49 | 8.57 |
| 10 | 4.48 | 5.27 | 5.77 | 6.14 | 6.43 | 6.67 | 6.87 | 7.05 | 7.21 | 7.36 | 7.48 | 7.60 | 7.71 | 7.81 | 7.91 | 7.99 | 8.07 | 8.15 | 8.22 |
| 11 | 4.39 | 5.14 | 5.62 | 5.97 | 6.25 | 6.48 | 6.67 | 6.84 | 6.99 | 7.13 | 7.25 | 7.36 | 7.46 | 7.56 | 7.65 | 7.73 | 7.81 | 7.88 | 7.95 |
| 12 | 4.32 | 5.04 | 5.50 | 5.84 | 6.10 | 6.32 | 6.51 | 6.67 | 6.81 | 6.94 | 7.06 | 7.17 | 7.26 | 7.36 | 7.44 | 7.52 | 7.59 | 7.66 | 7.73 |
| 13 | 4.26 | 4.96 | 5.40 | 5.73 | 5.98 | 6.19 | 6.37 | 6.53 | 6.67 | 6.79 | 6.90 | 7.01 | 7.10 | 7.19 | 7.27 | 7.34 | 7.42 | 7.48 | 7.55 |
| 14 | 4.21 | 4.89 | 5.32 | 5.63 | 5.88 | 6.08 | 6.26 | 6.41 | 6.54 | 6.66 | 6.77 | 6.87 | 6.96 | 7.05 | 7.12 | 7.20 | 7.27 | 7.33 | 7.39 |
| 15 | 4.17 | 4.83 | 5.25 | 5.56 | 5.80 | 5.99 | 6.16 | 6.31 | 6.44 | 6.55 | 6.66 | 6.76 | 6.84 | 6.93 | 7.00 | 7.07 | 7.14 | 7.20 | 7.26 |
| 16 | 4.13 | 4.78 | 5.19 | 5.49 | 5.72 | 5.92 | 6.08 | 6.22 | 6.35 | 6.46 | 6.56 | 6.66 | 6.74 | 6.82 | 6.90 | 6.97 | 7.03 | 7.09 | 7.15 |
| 17 | 4.10 | 4.74 | 5.14 | 5.43 | 5.66 | 5.85 | 6.01 | 6.15 | 6.27 | 6.38 | 6.48 | 6.57 | 6.66 | 6.73 | 6.80 | 6.87 | 6.94 | 7.00 | 7.05 |
| 18 | 4.07 | 4.70 | 5.09 | 5.38 | 5.60 | 5.79 | 5.94 | 6.08 | 6.20 | 6.31 | 6.41 | 6.50 | 6.58 | 6.65 | 6.72 | 6.79 | 6.85 | 6.91 | 6.96 |
| 19 | 4.05 | 4.67 | 5.05 | 5.33 | 5.55 | 5.73 | 5.89 | 6.02 | 6.14 | 6.25 | 6.34 | 6.43 | 6.51 | 6.58 | 6.65 | 6.72 | 6.78 | 6.84 | 6.89 |
| 20 | 4.02 | 4.64 | 5.02 | 5.29 | 5.51 | 5.69 | 5.84 | 5.97 | 6.09 | 6.19 | 6.29 | 6.37 | 6.45 | 6.52 | 6.59 | 6.65 | 6.71 | 6.76 | 6.82 |
| 24 | 3.96 | 4.54 | 4.91 | 5.17 | 5.37 | 5.54 | 5.69 | 5.81 | 5.92 | 6.02 | 6.11 | 6.19 | 6.26 | 6.33 | 6.39 | 6.45 | 6.51 | 6.56 | 6.61 |
| 30 | 3.89 | 4.45 | 4.80 | 5.05 | 5.24 | 5.40 | 5.54 | 5.65 | 5.76 | 5.85 | 5.93 | 6.01 | 6.08 | 6.14 | 6.20 | 6.26 | 6.31 | 6.36 | 6.41 |
| 40 | 3.82 | 4.37 | 4.70 | 4.93 | 5.11 | 5.27 | 5.39 | 5.50 | 5.60 | 5.69 | 5.77 | 5.84 | 5.90 | 5.96 | 6.02 | 6.07 | 6.12 | 6.17 | 6.21 |
| 60 | 3.76 | 4.28 | 4.60 | 4.82 | 4.99 | 5.13 | 5.25 | 5.36 | 5.45 | 5.53 | 5.60 | 5.67 | 5.73 | 5.79 | 5.84 | 5.89 | 5.93 | 5.98 | 6.02 |
| 120 | 3.70 | 4.20 | 4.50 | 4.71 | 4.87 | 5.01 | 5.12 | 5.21 | 5.30 | 5.38 | 5.44 | 5.51 | 5.56 | 5.61 | 5.66 | 5.71 | 5.75 | 5.79 | 5.83 |
| ∞ | 3.64 | 4.12 | 4.40 | 4.60 | 4.76 | 4.88 | 4.99 | 5.08 | 5.16 | 5.23 | 5.29 | 5.35 | 5.40 | 5.45 | 5.49 | 5.54 | 5.57 | 5.61 | 5.65 |

**Table A.4** $F$ distribution tables

$$F_{0.25\,(V_1,\,V_2)}$$

| $\begin{smallmatrix}v_1\\v_2\end{smallmatrix}$ | 1 | 2 | 3 | 4 | 5 | 6 | 7 | 8 | 9 | 10 | 12 | 15 | 20 | 24 | 30 | 40 | 60 | 120 | ∞ |
|---|---|---|---|---|---|---|---|---|---|---|---|---|---|---|---|---|---|---|---|
| | | | | | | | **Degrees of freedom for the numerator ($v_1$)** | | | | | | | | | | | | |
| 1 | 5.83 | 7.50 | 8.20 | 8.58 | 8.82 | 8.98 | 9.10 | 9.19 | 9.26 | 9.32 | 9.41 | 9.49 | 9.58 | 9.63 | 9.67 | 9.71 | 9.76 | 9.80 | 9.85 |
| 2 | 2.57 | 3.00 | 3.15 | 3.23 | 3.28 | 3.31 | 3.34 | 3.35 | 3.37 | 3.38 | 3.39 | 3.41 | 3.43 | 3.43 | 3.44 | 3.45 | 3.46 | 3.47 | 3.48 |
| 3 | 2.02 | 2.28 | 2.36 | 2.39 | 2.41 | 2.42 | 2.43 | 2.44 | 2.44 | 2.44 | 2.45 | 2.46 | 2.46 | 2.46 | 2.47 | 2.47 | 2.47 | 2.47 | 2.47 |
| 4 | 1.81 | 2.00 | 2.05 | 2.06 | 2.07 | 2.08 | 2.08 | 2.08 | 2.08 | 2.08 | 2.08 | 2.08 | 2.08 | 2.08 | 2.08 | 2.08 | 2.08 | 2.08 | 2.08 |
| 5 | 1.69 | 1.85 | 1.88 | 1.89 | 1.89 | 1.89 | 1.89 | 1.89 | 1.89 | 1.89 | 1.89 | 1.89 | 1.88 | 1.88 | 1.88 | 1.88 | 1.87 | 1.87 | 1.87 |
| 6 | 1.62 | 1.76 | 1.78 | 1.79 | 1.79 | 1.78 | 1.78 | 1.78 | 1.77 | 1.77 | 1.77 | 1.76 | 1.76 | 1.75 | 1.75 | 1.75 | 1.74 | 1.74 | 1.74 |
| 7 | 1.57 | 1.70 | 1.72 | 1.72 | 1.71 | 1.71 | 1.70 | 1.70 | 1.70 | 1.69 | 1.68 | 1.68 | 1.67 | 1.67 | 1.66 | 1.66 | 1.65 | 1.65 | 1.65 |
| 8 | 1.54 | 1.66 | 1.67 | 1.66 | 1.66 | 1.65 | 1.64 | 1.64 | 1.63 | 1.63 | 1.62 | 1.62 | 1.61 | 1.60 | 1.60 | 1.59 | 1.59 | 1.58 | 1.58 |
| 9 | 1.51 | 1.62 | 1.63 | 1.63 | 1.62 | 1.61 | 1.60 | 1.60 | 1.59 | 1.59 | 1.58 | 1.57 | 1.56 | 1.56 | 1.55 | 1.54 | 1.54 | 1.53 | 1.53 |
| 10 | 1.49 | 1.60 | 1.60 | 1.59 | 1.59 | 1.58 | 1.57 | 1.56 | 1.56 | 1.55 | 1.54 | 1.53 | 1.52 | 1.52 | 1.51 | 1.51 | 1.50 | 1.49 | 1.48 |
| 11 | 1.47 | 1.58 | 1.58 | 1.57 | 1.56 | 1.55 | 1.54 | 1.53 | 1.53 | 1.52 | 1.51 | 1.50 | 1.49 | 1.49 | 1.48 | 1.47 | 1.47 | 1.46 | 1.45 |
| 12 | 1.46 | 1.56 | 1.56 | 1.55 | 1.54 | 1.53 | 1.52 | 1.51 | 1.51 | 1.50 | 1.49 | 1.48 | 1.47 | 1.46 | 1.45 | 1.45 | 1.44 | 1.43 | 1.42 |
| 13 | 1.45 | 1.55 | 1.55 | 1.53 | 1.52 | 1.51 | 1.50 | 1.49 | 1.49 | 1.48 | 1.47 | 1.46 | 1.45 | 1.44 | 1.43 | 1.42 | 1.42 | 1.41 | 1.40 |
| 14 | 1.44 | 1.53 | 1.53 | 1.52 | 1.51 | 1.50 | 1.49 | 1.48 | 1.47 | 1.46 | 1.45 | 1.44 | 1.43 | 1.42 | 1.41 | 1.41 | 1.40 | 1.39 | 1.38 |
| 15 | 1.43 | 1.52 | 1.52 | 1.51 | 1.49 | 1.48 | 1.47 | 1.46 | 1.46 | 1.45 | 1.44 | 1.43 | 1.41 | 1.41 | 1.40 | 1.39 | 1.38 | 1.37 | 1.36 |
| 16 | 1.42 | 1.51 | 1.51 | 1.50 | 1.48 | 1.47 | 1.46 | 1.45 | 1.44 | 1.44 | 1.43 | 1.41 | 1.40 | 1.39 | 1.38 | 1.37 | 1.36 | 1.35 | 1.34 |
| 17 | 1.42 | 1.51 | 1.50 | 1.49 | 1.47 | 1.46 | 1.45 | 1.44 | 1.43 | 1.43 | 1.41 | 1.40 | 1.39 | 1.38 | 1.37 | 1.36 | 1.35 | 1.34 | 1.33 |
| 18 | 1.41 | 1.50 | 1.49 | 1.48 | 1.46 | 1.45 | 1.44 | 1.43 | 1.42 | 1.42 | 1.40 | 1.39 | 1.38 | 1.37 | 1.36 | 1.35 | 1.34 | 1.33 | 1.32 |
| 19 | 1.41 | 1.49 | 1.49 | 1.47 | 1.46 | 1.44 | 1.43 | 1.42 | 1.41 | 1.41 | 1.40 | 1.38 | 1.37 | 1.36 | 1.35 | 1.34 | 1.33 | 1.32 | 1.30 |
| 20 | 1.40 | 1.49 | 1.48 | 1.47 | 1.45 | 1.44 | 1.43 | 1.42 | 1.41 | 1.40 | 1.39 | 1.37 | 1.36 | 1.35 | 1.34 | 1.33 | 1.32 | 1.31 | 1.29 |
| 21 | 1.40 | 1.48 | 1.48 | 1.46 | 1.44 | 1.43 | 1.42 | 1.41 | 1.40 | 1.39 | 1.38 | 1.37 | 1.35 | 1.34 | 1.33 | 1.32 | 1.31 | 1.30 | 1.28 |
| 22 | 1.40 | 1.48 | 1.47 | 1.45 | 1.44 | 1.42 | 1.41 | 1.40 | 1.39 | 1.39 | 1.37 | 1.36 | 1.34 | 1.33 | 1.32 | 1.31 | 1.30 | 1.29 | 1.28 |
| 23 | 1.39 | 1.47 | 1.47 | 1.45 | 1.43 | 1.42 | 1.41 | 1.40 | 1.39 | 1.38 | 1.37 | 1.35 | 1.34 | 1.33 | 1.32 | 1.31 | 1.30 | 1.28 | 1.27 |
| 24 | 1.39 | 1.47 | 1.46 | 1.44 | 1.43 | 1.41 | 1.40 | 1.39 | 1.38 | 1.38 | 1.36 | 1.35 | 1.33 | 1.32 | 1.31 | 1.30 | 1.29 | 1.28 | 1.26 |
| 25 | 1.39 | 1.47 | 1.46 | 1.44 | 1.42 | 1.41 | 1.40 | 1.39 | 1.38 | 1.37 | 1.36 | 1.34 | 1.33 | 1.32 | 1.31 | 1.29 | 1.28 | 1.27 | 1.25 |
| 26 | 1.38 | 1.46 | 1.45 | 1.44 | 1.42 | 1.41 | 1.39 | 1.38 | 1.37 | 1.37 | 1.35 | 1.34 | 1.32 | 1.31 | 1.30 | 1.29 | 1.28 | 1.26 | 1.25 |
| 27 | 1.38 | 1.46 | 1.45 | 1.43 | 1.42 | 1.40 | 1.39 | 1.38 | 1.37 | 1.36 | 1.35 | 1.33 | 1.32 | 1.31 | 1.30 | 1.28 | 1.27 | 1.26 | 1.24 |
| 28 | 1.38 | 1.46 | 1.45 | 1.43 | 1.41 | 1.40 | 1.39 | 1.38 | 1.37 | 1.36 | 1.34 | 1.33 | 1.31 | 1.30 | 1.29 | 1.28 | 1.27 | 1.25 | 1.24 |
| 29 | 1.38 | 1.45 | 1.45 | 1.43 | 1.41 | 1.40 | 1.38 | 1.37 | 1.36 | 1.35 | 1.34 | 1.32 | 1.31 | 1.30 | 1.29 | 1.27 | 1.26 | 1.25 | 1.23 |
| 30 | 1.38 | 1.45 | 1.44 | 1.42 | 1.41 | 1.39 | 1.38 | 1.37 | 1.36 | 1.35 | 1.34 | 1.32 | 1.30 | 1.29 | 1.28 | 1.27 | 1.26 | 1.24 | 1.23 |
| 40 | 1.36 | 1.44 | 1.42 | 1.40 | 1.39 | 1.37 | 1.36 | 1.35 | 1.34 | 1.33 | 1.31 | 1.30 | 1.28 | 1.26 | 1.25 | 1.24 | 1.22 | 1.21 | 1.19 |
| 60 | 1.35 | 1.42 | 1.41 | 1.38 | 1.37 | 1.35 | 1.33 | 1.32 | 1.31 | 1.30 | 1.29 | 1.27 | 1.25 | 1.24 | 1.22 | 1.21 | 1.19 | 1.17 | 1.15 |
| 120 | 1.34 | 1.40 | 1.39 | 1.37 | 1.35 | 1.33 | 1.31 | 1.30 | 1.29 | 1.28 | 1.26 | 1.24 | 1.22 | 1.21 | 1.19 | 1.18 | 1.16 | 1.13 | 1.10 |
| ∞ | 1.32 | 1.39 | 1.37 | 1.35 | 1.33 | 1.31 | 1.29 | 1.28 | 1.27 | 1.25 | 1.24 | 1.22 | 1.19 | 1.18 | 1.16 | 1.14 | 1.12 | 1.08 | 1.00 |

Degrees freedom denominator ($v_2$)

*(Continued)*

**Table A.4** *F* Distribution tables

| $F_{0.10\,(V1,\,V2)}$ | | | | | | | | | | | | | | | | | | |
|---|---|---|---|---|---|---|---|---|---|---|---|---|---|---|---|---|---|---|
| **Degrees of freedom for the numerator ($v_1$)** | | | | | | | | | | | | | | | | | | |
| $v_1$ \ $v_2$ | 1 | 2 | 3 | 4 | 5 | 6 | 7 | 8 | 9 | 10 | 12 | 15 | 20 | 24 | 30 | 40 | 60 | 120 | ∞ |
| 1 | 39.86 | 49.50 | 53.59 | 55.83 | 57.24 | 58.20 | 58.91 | 59.44 | 59.86 | 60.19 | 60.71 | 61.22 | 61.74 | 62.00 | 62.26 | 62.53 | 62.79 | 63.06 | 63.33 |
| 2 | 8.53 | 9.00 | 9.16 | 9.24 | 9.29 | 9.33 | 9.35 | 9.37 | 9.38 | 9.39 | 9.41 | 9.42 | 9.44 | 9.45 | 9.46 | 9.47 | 9.47 | 9.48 | 9.49 |
| 3 | 5.54 | 5.46 | 5.39 | 5.34 | 5.31 | 5.28 | 5.27 | 5.25 | 5.24 | 5.23 | 5.22 | 5.20 | 5.18 | 5.18 | 5.17 | 5.16 | 5.15 | 5.14 | 5.13 |
| 4 | 4.54 | 4.32 | 4.19 | 4.11 | 4.05 | 4.01 | 3.98 | 3.95 | 3.94 | 3.92 | 3.90 | 3.87 | 3.84 | 3.83 | 3.82 | 3.80 | 3.79 | 3.78 | 3.76 |
| 5 | 4.06 | 3.78 | 3.62 | 3.52 | 3.45 | 3.40 | 3.37 | 3.34 | 3.32 | 3.30 | 3.27 | 3.24 | 3.21 | 3.19 | 3.17 | 3.16 | 3.14 | 3.12 | 3.10 |
| 6 | 3.78 | 3.46 | 3.29 | 3.18 | 3.11 | 3.05 | 3.01 | 2.98 | 2.96 | 2.94 | 2.90 | 2.87 | 2.84 | 2.82 | 2.80 | 2.78 | 2.76 | 2.74 | 2.72 |
| 7 | 3.59 | 3.26 | 3.07 | 2.96 | 2.88 | 2.83 | 2.78 | 2.75 | 2.72 | 2.70 | 2.67 | 2.63 | 2.59 | 2.58 | 2.56 | 2.54 | 2.51 | 2.49 | 2.47 |
| 8 | 3.46 | 3.11 | 2.92 | 2.81 | 2.73 | 2.67 | 2.62 | 2.59 | 2.56 | 2.54 | 2.50 | 2.46 | 2.42 | 2.40 | 2.38 | 2.36 | 2.34 | 2.32 | 2.29 |
| 9 | 3.36 | 3.01 | 2.81 | 2.69 | 2.61 | 2.55 | 2.51 | 2.47 | 2.44 | 2.42 | 2.38 | 2.34 | 2.30 | 2.28 | 2.25 | 2.23 | 2.21 | 2.18 | 2.16 |
| 10 | 3.29 | 2.92 | 2.73 | 2.61 | 2.52 | 2.46 | 2.41 | 2.38 | 2.35 | 2.32 | 2.28 | 2.24 | 2.20 | 2.18 | 2.16 | 2.13 | 2.11 | 2.08 | 2.06 |
| 11 | 3.23 | 2.86 | 2.66 | 2.54 | 2.45 | 2.39 | 2.34 | 2.30 | 2.27 | 2.25 | 2.21 | 2.17 | 2.12 | 2.10 | 2.08 | 2.05 | 2.03 | 2.00 | 1.97 |
| 12 | 3.18 | 2.81 | 2.61 | 2.48 | 2.39 | 2.33 | 2.28 | 2.24 | 2.21 | 2.19 | 2.15 | 2.10 | 2.06 | 2.04 | 2.01 | 1.99 | 1.96 | 1.93 | 1.90 |
| 13 | 3.14 | 2.76 | 2.56 | 2.43 | 2.35 | 2.28 | 2.23 | 2.20 | 2.16 | 2.14 | 2.10 | 2.05 | 2.01 | 1.98 | 1.96 | 1.93 | 1.90 | 1.88 | 1.85 |
| 14 | 3.10 | 2.73 | 2.52 | 2.39 | 2.31 | 2.24 | 2.19 | 2.15 | 2.12 | 2.10 | 2.05 | 2.01 | 1.96 | 1.94 | 1.91 | 1.89 | 1.86 | 1.83 | 1.80 |
| 15 | 3.07 | 2.70 | 2.49 | 2.36 | 2.27 | 2.21 | 2.16 | 2.12 | 2.09 | 2.06 | 2.02 | 1.97 | 1.92 | 1.90 | 1.87 | 1.85 | 1.82 | 1.79 | 1.76 |
| 16 | 3.05 | 2.67 | 2.46 | 2.33 | 2.24 | 2.18 | 2.13 | 2.09 | 2.06 | 2.03 | 1.99 | 1.94 | 1.89 | 1.87 | 1.84 | 1.81 | 1.78 | 1.75 | 1.72 |
| 17 | 3.03 | 2.64 | 2.44 | 2.31 | 2.22 | 2.15 | 2.10 | 2.06 | 2.03 | 2.00 | 1.96 | 1.91 | 1.86 | 1.84 | 1.81 | 1.78 | 1.75 | 1.72 | 1.69 |
| 18 | 3.01 | 2.62 | 2.42 | 2.29 | 2.20 | 2.13 | 2.08 | 2.04 | 2.00 | 1.98 | 1.93 | 1.89 | 1.84 | 1.81 | 1.78 | 1.75 | 1.72 | 1.69 | 1.66 |
| 19 | 2.99 | 2.61 | 2.40 | 2.27 | 2.18 | 2.11 | 2.06 | 2.02 | 1.98 | 1.96 | 1.91 | 1.86 | 1.81 | 1.79 | 1.76 | 1.73 | 1.70 | 1.67 | 1.63 |
| 20 | 2.97 | 2.59 | 2.38 | 2.25 | 2.16 | 2.09 | 2.04 | 2.00 | 1.96 | 1.94 | 1.89 | 1.84 | 1.79 | 1.77 | 1.74 | 1.71 | 1.68 | 1.64 | 1.61 |
| 21 | 2.96 | 2.57 | 2.36 | 2.23 | 2.14 | 2.08 | 2.02 | 1.98 | 1.95 | 1.92 | 1.87 | 1.83 | 1.78 | 1.75 | 1.72 | 1.69 | 1.66 | 1.62 | 1.59 |
| 22 | 2.95 | 2.56 | 2.35 | 2.22 | 2.13 | 2.06 | 2.01 | 1.97 | 1.93 | 1.90 | 1.86 | 1.81 | 1.76 | 1.73 | 1.70 | 1.67 | 1.64 | 1.60 | 1.57 |
| 23 | 2.94 | 2.55 | 2.34 | 2.21 | 2.11 | 2.05 | 1.99 | 1.96 | 1.92 | 1.89 | 1.84 | 1.80 | 1.74 | 1.72 | 1.69 | 1.66 | 1.62 | 1.59 | 1.55 |
| 24 | 2.93 | 2.54 | 2.33 | 2.19 | 2.10 | 2.04 | 1.98 | 1.94 | 1.91 | 1.88 | 1.83 | 1.78 | 1.73 | 1.70 | 1.67 | 1.64 | 1.61 | 1.57 | 1.53 |
| 25 | 2.92 | 2.53 | 2.32 | 2.18 | 2.09 | 2.02 | 1.97 | 1.93 | 1.89 | 1.87 | 1.82 | 1.77 | 1.72 | 1.69 | 1.66 | 1.63 | 1.59 | 1.56 | 1.52 |
| 26 | 2.91 | 2.52 | 2.31 | 2.17 | 2.08 | 2.01 | 1.96 | 1.92 | 1.88 | 1.86 | 1.81 | 1.76 | 1.71 | 1.68 | 1.65 | 1.61 | 1.58 | 1.54 | 1.50 |
| 27 | 2.90 | 2.51 | 2.30 | 2.17 | 2.07 | 2.00 | 1.95 | 1.91 | 1.87 | 1.85 | 1.80 | 1.75 | 1.70 | 1.67 | 1.64 | 1.60 | 1.57 | 1.53 | 1.49 |
| 28 | 2.89 | 2.50 | 2.29 | 2.16 | 2.06 | 2.00 | 1.94 | 1.90 | 1.87 | 1.84 | 1.79 | 1.74 | 1.69 | 1.66 | 1.63 | 1.59 | 1.56 | 1.52 | 1.48 |
| 29 | 2.89 | 2.50 | 2.28 | 2.15 | 2.06 | 1.99 | 1.93 | 1.89 | 1.86 | 1.83 | 1.78 | 1.73 | 1.68 | 1.65 | 1.62 | 1.58 | 1.55 | 1.51 | 1.47 |
| 30 | 2.88 | 2.49 | 2.28 | 2.14 | 2.03 | 1.98 | 1.93 | 1.88 | 1.85 | 1.82 | 1.77 | 1.72 | 1.67 | 1.64 | 1.61 | 1.57 | 1.54 | 1.50 | 1.46 |
| 40 | 2.84 | 2.44 | 2.23 | 2.09 | 2.00 | 1.93 | 1.87 | 1.83 | 1.79 | 1.76 | 1.71 | 1.66 | 1.61 | 1.57 | 1.54 | 1.51 | 1.47 | 1.42 | 1.38 |
| 60 | 2.79 | 2.39 | 2.18 | 2.04 | 1.95 | 1.87 | 1.82 | 1.77 | 1.74 | 1.71 | 1.66 | 1.60 | 1.54 | 1.51 | 1.48 | 1.44 | 1.40 | 1.35 | 1.29 |
| 120 | 2.75 | 2.35 | 2.13 | 1.99 | 1.90 | 1.82 | 1.77 | 1.72 | 1.68 | 1.65 | 1.60 | 1.55 | 1.48 | 1.45 | 1.41 | 1.37 | 1.32 | 1.26 | 1.19 |
| ∞ | 2.71 | 2.30 | 2.08 | 1.94 | 1.85 | 1.77 | 1.72 | 1.67 | 1.63 | 1.60 | 1.55 | 1.49 | 1.42 | 1.38 | 1.34 | 1.30 | 1.24 | 1.17 | 1.00 |

Degrees freedom denominator ($v_2$)

(*Continued*)

**Table A.4** $F$ distribution tables

| | $F_{0.05\,(V1,\,V2)}$ | | | | | | | | | | | | | | | | | | |
|---|---|---|---|---|---|---|---|---|---|---|---|---|---|---|---|---|---|---|---|
| | Degrees of freedom for the numerator ($v_1$) | | | | | | | | | | | | | | | | | | |
| $v_2$ \\ $v_1$ | 1 | 2 | 3 | 4 | 5 | 6 | 7 | 8 | 9 | 10 | 12 | 15 | 20 | 24 | 30 | 40 | 60 | 120 | $\infty$ |
| 1 | 161.40 | 199.50 | 215.70 | 224.60 | 230.20 | 234.00 | 236.80 | 238.90 | 240.50 | 241.90 | 243.90 | 245.90 | 248.00 | 249.10 | 250.10 | 251.10 | 252.20 | 253.30 | 254.30 |
| 2 | 18.51 | 19.00 | 19.16 | 19.25 | 19.30 | 19.33 | 19.35 | 19.37 | 19.38 | 19.40 | 19.41 | 19.43 | 19.45 | 19.45 | 19.46 | 19.47 | 19.48 | 19.49 | 19.50 |
| 3 | 10.13 | 9.55 | 9.28 | 9.12 | 9.01 | 8.94 | 8.89 | 8.85 | 8.81 | 8.79 | 8.74 | 8.70 | 8.66 | 8.64 | 8.62 | 8.59 | 8.57 | 8.55 | 8.53 |
| 4 | 7.71 | 6.94 | 6.59 | 6.39 | 6.26 | 6.16 | 6.09 | 6.04 | 6.00 | 5.96 | 5.91 | 5.86 | 5.80 | 5.77 | 5.75 | 5.72 | 5.69 | 5.66 | 5.63 |
| 5 | 6.61 | 5.79 | 5.41 | 5.19 | 5.05 | 4.95 | 4.88 | 4.82 | 4.77 | 4.74 | 4.68 | 4.62 | 4.56 | 4.53 | 4.50 | 4.46 | 4.43 | 4.40 | 4.36 |
| 6 | 5.99 | 5.14 | 4.76 | 4.53 | 4.39 | 4.28 | 4.21 | 4.15 | 4.10 | 4.06 | 4.00 | 3.94 | 3.87 | 3.84 | 3.81 | 3.77 | 3.74 | 3.70 | 3.67 |
| 7 | 5.59 | 4.74 | 4.35 | 4.12 | 3.97 | 3.87 | 3.79 | 3.73 | 3.68 | 3.64 | 3.57 | 3.51 | 3.44 | 3.41 | 3.38 | 3.34 | 3.30 | 3.27 | 3.23 |
| 8 | 5.32 | 4.46 | 4.07 | 3.84 | 3.69 | 3.58 | 3.50 | 3.44 | 3.39 | 3.35 | 3.28 | 3.22 | 3.15 | 3.12 | 3.08 | 3.04 | 3.01 | 2.97 | 2.93 |
| 9 | 5.12 | 4.26 | 3.86 | 3.63 | 3.48 | 3.37 | 3.29 | 3.23 | 3.18 | 3.14 | 3.07 | 3.01 | 2.94 | 2.90 | 2.86 | 2.83 | 2.79 | 2.75 | 2.71 |
| 10 | 4.96 | 4.10 | 3.71 | 3.48 | 3.33 | 3.22 | 3.14 | 3.07 | 3.02 | 2.98 | 2.91 | 2.85 | 2.77 | 2.74 | 2.70 | 2.66 | 2.62 | 2.58 | 2.54 |
| 11 | 4.84 | 3.98 | 3.59 | 3.36 | 3.20 | 3.09 | 3.01 | 2.95 | 2.90 | 2.85 | 2.79 | 2.72 | 2.65 | 2.61 | 2.57 | 2.53 | 2.49 | 2.45 | 2.40 |
| 12 | 4.75 | 3.89 | 3.49 | 3.26 | 3.11 | 3.00 | 2.91 | 2.85 | 2.80 | 2.75 | 2.69 | 2.62 | 2.54 | 2.51 | 2.47 | 2.43 | 2.38 | 2.34 | 2.30 |
| 13 | 4.67 | 3.81 | 3.41 | 3.18 | 3.03 | 2.92 | 2.83 | 2.77 | 2.71 | 2.67 | 2.60 | 2.53 | 2.46 | 2.42 | 2.38 | 2.34 | 2.30 | 2.25 | 2.21 |
| 14 | 4.60 | 3.74 | 3.34 | 3.11 | 2.96 | 2.85 | 2.76 | 2.70 | 2.65 | 2.60 | 2.53 | 2.46 | 2.39 | 2.35 | 2.31 | 2.27 | 2.22 | 2.18 | 2.13 |
| 15 | 4.54 | 3.68 | 3.29 | 3.06 | 2.90 | 2.79 | 2.71 | 2.64 | 2.59 | 2.54 | 2.48 | 2.40 | 2.33 | 2.29 | 2.25 | 2.20 | 2.16 | 2.11 | 2.07 |
| 16 | 4.49 | 3.63 | 3.24 | 3.01 | 2.85 | 2.74 | 2.66 | 2.59 | 2.54 | 2.49 | 2.42 | 2.35 | 2.28 | 2.24 | 2.19 | 2.15 | 2.11 | 2.06 | 2.01 |
| 17 | 4.45 | 3.59 | 3.20 | 2.96 | 2.81 | 2.70 | 2.61 | 2.55 | 2.49 | 2.45 | 2.38 | 2.31 | 2.23 | 2.19 | 2.15 | 2.10 | 2.06 | 2.01 | 1.96 |
| 18 | 4.41 | 3.55 | 3.16 | 2.93 | 2.77 | 2.66 | 2.58 | 2.51 | 2.46 | 2.41 | 2.34 | 2.27 | 2.19 | 2.15 | 2.11 | 2.06 | 2.02 | 1.97 | 1.92 |
| 19 | 4.38 | 3.52 | 3.13 | 2.90 | 2.74 | 2.63 | 2.54 | 2.48 | 2.42 | 2.38 | 2.31 | 2.23 | 2.16 | 2.11 | 2.07 | 2.03 | 1.98 | 1.93 | 1.88 |
| 20 | 4.35 | 3.49 | 3.10 | 2.87 | 2.71 | 2.60 | 2.51 | 2.45 | 2.39 | 2.35 | 2.28 | 2.20 | 2.12 | 2.08 | 2.04 | 1.99 | 1.95 | 1.90 | 1.84 |
| 21 | 4.32 | 3.47 | 3.07 | 2.84 | 2.68 | 2.57 | 2.49 | 2.42 | 2.37 | 2.32 | 2.25 | 2.18 | 2.10 | 2.05 | 2.01 | 1.96 | 1.92 | 1.87 | 1.81 |
| 22 | 4.30 | 3.44 | 3.05 | 2.82 | 2.66 | 2.55 | 2.46 | 2.40 | 2.34 | 2.30 | 2.23 | 2.15 | 2.07 | 2.03 | 1.98 | 1.94 | 1.89 | 1.84 | 1.78 |
| 23 | 4.28 | 3.42 | 3.03 | 2.80 | 2.64 | 2.53 | 2.44 | 2.37 | 2.32 | 2.27 | 2.20 | 2.13 | 2.05 | 2.01 | 1.96 | 1.91 | 1.86 | 1.81 | 1.76 |
| 24 | 4.26 | 3.40 | 3.01 | 2.78 | 2.62 | 2.51 | 2.42 | 2.36 | 2.30 | 2.25 | 2.18 | 2.11 | 2.03 | 1.98 | 1.94 | 1.89 | 1.84 | 1.79 | 1.73 |
| 25 | 4.24 | 3.39 | 2.99 | 2.76 | 2.60 | 2.49 | 2.40 | 2.34 | 2.28 | 2.24 | 2.16 | 2.09 | 2.01 | 1.96 | 1.92 | 1.87 | 1.82 | 1.77 | 1.71 |
| 26 | 4.23 | 3.37 | 2.98 | 2.74 | 2.59 | 2.47 | 2.39 | 2.32 | 2.27 | 2.22 | 2.15 | 2.07 | 1.99 | 1.95 | 1.90 | 1.85 | 1.80 | 1.75 | 1.69 |
| 27 | 4.21 | 3.35 | 2.96 | 2.73 | 2.57 | 2.46 | 2.37 | 2.31 | 2.25 | 2.20 | 2.13 | 2.06 | 1.97 | 1.93 | 1.88 | 1.84 | 1.79 | 1.73 | 1.67 |
| 28 | 4.20 | 3.34 | 2.95 | 2.71 | 2.56 | 2.45 | 2.36 | 2.29 | 2.24 | 2.19 | 2.12 | 2.04 | 1.96 | 1.91 | 1.87 | 1.82 | 1.77 | 1.71 | 1.65 |
| 29 | 4.18 | 3.33 | 2.93 | 2.70 | 2.55 | 2.43 | 2.35 | 2.28 | 2.22 | 2.18 | 2.10 | 2.03 | 1.94 | 1.90 | 1.85 | 1.81 | 1.75 | 1.70 | 1.64 |
| 30 | 4.17 | 3.32 | 2.92 | 2.69 | 2.53 | 2.42 | 2.33 | 2.27 | 2.21 | 2.16 | 2.09 | 2.01 | 1.93 | 1.89 | 1.84 | 1.79 | 1.74 | 1.68 | 1.62 |
| 40 | 4.08 | 3.23 | 2.84 | 2.61 | 2.45 | 2.34 | 2.25 | 2.18 | 2.12 | 2.08 | 2.00 | 1.92 | 1.84 | 1.79 | 1.74 | 1.69 | 1.64 | 1.58 | 1.51 |
| 60 | 4.00 | 3.15 | 2.76 | 2.53 | 2.37 | 2.25 | 2.17 | 2.10 | 2.04 | 1.99 | 1.92 | 1.84 | 1.75 | 1.70 | 1.65 | 1.59 | 1.53 | 1.47 | 1.39 |
| 120 | 3.92 | 3.07 | 2.68 | 2.45 | 2.29 | 2.17 | 2.09 | 2.02 | 1.96 | 1.91 | 1.83 | 1.75 | 1.66 | 1.61 | 1.55 | 1.55 | 1.43 | 1.35 | 1.25 |
| $\infty$ | 3.84 | 3.00 | 2.60 | 2.37 | 2.21 | 2.10 | 2.01 | 1.94 | 1.88 | 1.83 | 1.75 | 1.67 | 1.57 | 1.52 | 1.46 | 1.39 | 1.32 | 1.22 | 1.00 |

Degrees freedom denominator ($v_2$)

*(Continued)*

**Table A.4** *F* distribution tables

$$F_{0.025\,(v_1,\,v_2)}$$

**Degrees of freedom for the numerator ($v_1$)**

| $v_2$ \ $v_1$ | 1 | 2 | 3 | 4 | 5 | 6 | 7 | 8 | 9 | 10 | 12 | 15 | 20 | 24 | 30 | 40 | 60 | 120 | ∞ |
|---|---|---|---|---|---|---|---|---|---|---|---|---|---|---|---|---|---|---|---|
| 1 | 647.80 | 799.50 | 864.20 | 899.60 | 921.80 | 937.10 | 948.20 | 956.70 | 963.30 | 968.60 | 976.70 | 984.90 | 993.10 | 997.20 | 1001.0 | 1006.0 | 1010.0 | 1014.0 | 1018.0 |
| 2 | 38.51 | 39.00 | 39.17 | 39.25 | 39.30 | 39.33 | 39.36 | 39.37 | 39.39 | 39.40 | 39.41 | 39.43 | 39.45 | 39.46 | 39.46 | 39.47 | 39.48 | 39.49 | 39.50 |
| 3 | 17.44 | 16.04 | 15.44 | 15.10 | 14.88 | 14.73 | 14.62 | 14.54 | 14.47 | 14.42 | 14.34 | 14.25 | 14.17 | 14.12 | 14.08 | 14.04 | 13.99 | 13.95 | 13.90 |
| 4 | 12.22 | 10.65 | 9.98 | 9.60 | 9.36 | 9.20 | 9.07 | 8.98 | 8.90 | 8.84 | 8.75 | 8.66 | 8.56 | 8.51 | 8.46 | 8.41 | 8.36 | 8.31 | 8.26 |
| 5 | 10.01 | 8.43 | 7.76 | 7.39 | 7.15 | 6.98 | 6.85 | 6.76 | 6.68 | 6.62 | 6.52 | 6.43 | 6.33 | 6.28 | 6.23 | 6.18 | 6.12 | 6.07 | 6.02 |
| 6 | 8.81 | 7.26 | 6.60 | 6.23 | 5.99 | 5.82 | 5.70 | 5.60 | 5.52 | 5.46 | 5.37 | 5.27 | 5.17 | 5.12 | 5.07 | 5.01 | 4.96 | 4.90 | 4.85 |
| 7 | 8.07 | 6.54 | 5.89 | 5.52 | 5.29 | 5.12 | 4.99 | 4.90 | 4.82 | 4.76 | 4.67 | 4.57 | 4.47 | 4.42 | 4.36 | 4.31 | 4.25 | 4.20 | 4.14 |
| 8 | 7.57 | 6.06 | 5.42 | 5.05 | 4.82 | 4.65 | 4.53 | 4.43 | 4.36 | 4.30 | 4.20 | 4.10 | 4.00 | 3.95 | 3.89 | 3.84 | 3.78 | 3.73 | 3.67 |
| 9 | 7.21 | 5.71 | 5.08 | 4.72 | 4.48 | 4.32 | 4.20 | 4.10 | 4.03 | 3.96 | 3.87 | 3.77 | 3.67 | 3.61 | 3.56 | 3.51 | 3.45 | 3.39 | 3.33 |
| 10 | 6.94 | 5.46 | 4.83 | 4.47 | 4.24 | 4.07 | 3.95 | 3.85 | 3.78 | 3.72 | 3.62 | 3.52 | 3.42 | 3.37 | 3.31 | 3.26 | 3.20 | 3.14 | 3.08 |
| 11 | 6.72 | 5.26 | 4.63 | 4.28 | 4.04 | 3.88 | 3.76 | 3.66 | 3.59 | 3.53 | 3.43 | 3.33 | 3.23 | 3.17 | 3.12 | 3.06 | 3.00 | 2.94 | 2.88 |
| 12 | 6.55 | 5.10 | 4.47 | 4.12 | 3.89 | 3.73 | 3.61 | 3.51 | 3.44 | 3.37 | 3.28 | 3.18 | 3.07 | 3.02 | 2.96 | 2.91 | 2.85 | 2.79 | 2.72 |
| 13 | 6.41 | 4.97 | 4.35 | 4.00 | 3.77 | 3.60 | 3.48 | 3.39 | 3.31 | 3.25 | 3.15 | 3.05 | 2.95 | 2.89 | 2.84 | 2.78 | 2.72 | 2.66 | 2.60 |
| 14 | 6.30 | 4.86 | 4.24 | 3.89 | 3.66 | 3.50 | 3.38 | 3.29 | 3.21 | 3.15 | 3.05 | 2.95 | 2.84 | 2.79 | 2.73 | 2.67 | 2.61 | 2.55 | 2.49 |
| 15 | 6.20 | 4.77 | 4.15 | 3.80 | 3.58 | 3.41 | 3.29 | 3.20 | 3.12 | 3.06 | 2.96 | 2.86 | 2.76 | 2.70 | 2.64 | 2.59 | 2.52 | 2.46 | 2.40 |
| 16 | 6.12 | 4.69 | 4.08 | 3.73 | 3.50 | 3.34 | 3.22 | 3.12 | 3.05 | 2.99 | 2.89 | 2.79 | 2.68 | 2.63 | 2.57 | 2.51 | 2.45 | 2.38 | 2.32 |
| 17 | 6.04 | 4.62 | 4.01 | 3.66 | 3.44 | 3.28 | 3.16 | 3.06 | 2.98 | 2.92 | 2.82 | 2.72 | 2.62 | 2.56 | 2.50 | 2.44 | 2.38 | 2.32 | 2.25 |
| 18 | 5.98 | 4.56 | 3.95 | 3.61 | 3.38 | 3.22 | 3.10 | 3.01 | 2.93 | 2.87 | 2.77 | 2.67 | 2.56 | 2.50 | 2.44 | 2.38 | 2.32 | 2.26 | 2.19 |
| 19 | 5.92 | 4.51 | 3.90 | 3.56 | 3.33 | 3.17 | 3.05 | 2.96 | 2.88 | 2.82 | 2.72 | 2.62 | 2.51 | 2.45 | 2.39 | 2.33 | 2.27 | 2.20 | 2.13 |
| 20 | 5.87 | 4.46 | 3.86 | 3.51 | 3.29 | 3.13 | 3.01 | 2.91 | 2.84 | 2.77 | 2.68 | 2.57 | 2.46 | 2.41 | 2.35 | 2.29 | 2.22 | 2.16 | 2.09 |
| 21 | 5.83 | 4.42 | 3.82 | 3.48 | 3.25 | 3.09 | 2.97 | 2.87 | 2.80 | 2.73 | 2.64 | 2.53 | 2.42 | 2.37 | 2.31 | 2.25 | 2.18 | 2.11 | 2.04 |
| 22 | 5.79 | 4.38 | 3.78 | 3.44 | 3.22 | 3.05 | 2.93 | 2.84 | 2.76 | 2.70 | 2.60 | 2.50 | 2.39 | 2.33 | 2.27 | 2.21 | 2.14 | 2.08 | 2.00 |
| 23 | 5.75 | 4.35 | 3.75 | 3.41 | 3.18 | 3.02 | 2.90 | 2.81 | 2.73 | 2.67 | 2.57 | 2.47 | 2.36 | 2.30 | 2.24 | 2.18 | 2.11 | 2.04 | 1.97 |
| 24 | 5.72 | 4.32 | 3.72 | 3.38 | 3.15 | 2.99 | 2.87 | 2.78 | 2.70 | 2.64 | 2.54 | 2.44 | 2.33 | 2.27 | 2.21 | 2.15 | 2.08 | 2.01 | 1.94 |
| 25 | 5.69 | 4.29 | 3.69 | 3.35 | 3.13 | 2.97 | 2.85 | 2.75 | 2.68 | 2.61 | 2.51 | 2.41 | 2.30 | 2.24 | 2.18 | 2.12 | 2.05 | 1.98 | 1.91 |
| 26 | 5.66 | 4.27 | 3.67 | 3.33 | 3.10 | 2.94 | 2.82 | 2.73 | 2.65 | 2.59 | 2.49 | 2.39 | 2.28 | 2.22 | 2.16 | 2.09 | 2.03 | 1.95 | 1.88 |
| 27 | 5.63 | 4.24 | 3.65 | 3.31 | 3.08 | 2.92 | 2.80 | 2.71 | 2.63 | 2.57 | 2.47 | 2.36 | 2.25 | 2.19 | 2.13 | 2.07 | 2.00 | 1.93 | 1.85 |
| 28 | 5.61 | 4.22 | 3.63 | 3.29 | 3.06 | 2.90 | 2.78 | 2.69 | 2.61 | 2.55 | 2.45 | 2.34 | 2.23 | 2.17 | 2.11 | 2.05 | 1.98 | 1.91 | 1.83 |
| 29 | 5.59 | 4.20 | 1.61 | 3.27 | 3.04 | 2.88 | 2.76 | 2.67 | 2.59 | 2.53 | 2.43 | 2.32 | 2.21 | 2.15 | 2.09 | 2.03 | 1.96 | 1.89 | 1.81 |
| 30 | 5.57 | 4.18 | 3.59 | 3.25 | 3.03 | 2.87 | 2.75 | 2.65 | 2.57 | 2.51 | 2.41 | 2.31 | 2.20 | 2.14 | 2.07 | 2.01 | 1.94 | 1.87 | 1.79 |
| 40 | 5.42 | 4.05 | 3.46 | 3.13 | 2.90 | 2.74 | 2.62 | 2.53 | 2.45 | 2.39 | 2.29 | 2.18 | 2.07 | 2.01 | 1.94 | 1.88 | 1.80 | 1.72 | 1.64 |
| 60 | 5.29 | 3.93 | 3.34 | 3.01 | 2.79 | 2.63 | 2.51 | 2.41 | 2.33 | 2.27 | 2.17 | 2.06 | 1.94 | 1.88 | 1.82 | 1.74 | 1.67 | 1.58 | 1.48 |
| 120 | 5.15 | 3.80 | 3.23 | 2.89 | 2.67 | 2.52 | 2.39 | 2.30 | 2.22 | 2.16 | 2.05 | 1.94 | 1.82 | 1.76 | 1.69 | 1.61 | 1.53 | 1.43 | 1.31 |
| ∞ | 5.02 | 3.69 | 3.12 | 2.79 | 2.57 | 2.41 | 2.29 | 2.19 | 2.11 | 2.05 | 1.94 | 1.83 | 1.71 | 1.64 | 1.57 | 1.48 | 1.39 | 1.27 | 1.00 |

Degrees freedom denominator ($v_2$)

(*Continued*)

**Table A.4** $F$ distribution tables

| | | | | | | | | | | $F_{0.01\,(V1,\,V2)}$ | | | | | | | | | |
|---|---|---|---|---|---|---|---|---|---|---|---|---|---|---|---|---|---|---|---|
| | | | | | | | Degrees of freedom for the numerator $(v_1)$ | | | | | | | | | | | | |
| $v_2$ \ $v_1$ | 1 | 2 | 3 | 4 | 5 | 6 | 7 | 8 | 9 | 10 | 12 | 15 | 20 | 24 | 30 | 40 | 60 | 120 | ∞ |
| 1 | 4052.0 | 4999.5 | 5403.0 | 5625.0 | 5764.0 | 5859.0 | 5928.0 | 5982.0 | 6022.0 | 6056.0 | 6106.0 | 6157.0 | 6209.0 | 6235.0 | 6261.0 | 6287.0 | 6313.0 | 6339.0 | 6366.0 |
| 2 | 98.50 | 99.00 | 99.17 | 99.25 | 99.30 | 99.33 | 99.36 | 99.37 | 99.39 | 99.40 | 99.42 | 99.43 | 99.45 | 99.46 | 99.47 | 99.47 | 99.48 | 99.49 | 99.50 |
| 3 | 34.12 | 30.82 | 29.46 | 28.71 | 28.24 | 27.91 | 27.67 | 27.49 | 27.35 | 27.23 | 27.05 | 26.87 | 26.69 | 26.60 | 26.50 | 26.41 | 26.32 | 26.22 | 26.13 |
| 4 | 21.20 | 18.00 | 16.69 | 15.98 | 15.52 | 15.21 | 14.98 | 14.80 | 14.66 | 14.55 | 14.37 | 14.20 | 14.02 | 13.93 | 13.84 | 13.75 | 13.65 | 13.56 | 13.46 |
| 5 | 16.26 | 13.27 | 12.06 | 11.39 | 10.97 | 10.67 | 10.46 | 10.29 | 10.16 | 10.05 | 9.89 | 9.72 | 9.55 | 9.47 | 9.38 | 9.29 | 9.20 | 9.11 | 9.02 |
| 6 | 13.75 | 10.92 | 9.78 | 9.15 | 8.75 | 8.47 | 8.26 | 8.10 | 7.98 | 7.87 | 7.72 | 7.56 | 7.40 | 7.31 | 7.23 | 7.14 | 7.06 | 6.97 | 6.88 |
| 7 | 12.25 | 9.55 | 8.45 | 7.85 | 7.46 | 7.19 | 6.99 | 6.84 | 6.72 | 6.62 | 6.47 | 6.31 | 6.16 | 6.07 | 5.99 | 5.91 | 5.82 | 5.74 | 5.65 |
| 8 | 11.26 | 8.65 | 7.59 | 7.01 | 6.63 | 6.37 | 6.18 | 6.03 | 5.91 | 5.81 | 5.67 | 5.52 | 5.36 | 5.28 | 5.20 | 5.12 | 5.03 | 4.95 | 4.86 |
| 9 | 10.56 | 8.02 | 6.99 | 6.42 | 6.06 | 5.80 | 5.61 | 5.47 | 5.35 | 5.26 | 5.11 | 4.96 | 4.81 | 4.73 | 4.65 | 4.57 | 4.48 | 4.40 | 4.31 |
| 10 | 10.04 | 7.56 | 6.55 | 5.99 | 5.64 | 5.39 | 5.20 | 5.06 | 4.94 | 4.85 | 4.71 | 4.56 | 4.41 | 4.33 | 4.25 | 4.17 | 4.08 | 4.00 | 3.91 |
| 11 | 9.65 | 7.21 | 6.22 | 5.67 | 5.32 | 5.07 | 4.89 | 4.74 | 4.63 | 4.54 | 4.40 | 4.25 | 4.10 | 4.02 | 3.94 | 3.86 | 3.78 | 3.69 | 3.60 |
| 12 | 9.33 | 6.93 | 5.95 | 5.41 | 5.06 | 4.82 | 4.64 | 4.50 | 4.39 | 4.30 | 4.16 | 4.01 | 3.86 | 3.78 | 3.70 | 3.62 | 3.54 | 3.45 | 3.36 |
| 13 | 9.07 | 6.70 | 5.74 | 5.21 | 4.86 | 4.62 | 4.44 | 4.30 | 4.19 | 4.10 | 3.96 | 3.82 | 3.66 | 3.59 | 3.51 | 3.43 | 3.34 | 3.25 | 3.17 |
| 14 | 8.86 | 6.51 | 5.56 | 5.04 | 4.69 | 4.46 | 4.28 | 4.14 | 4.03 | 3.94 | 3.80 | 3.66 | 3.51 | 3.43 | 3.35 | 3.27 | 3.18 | 3.09 | 3.00 |
| 15 | 8.68 | 6.36 | 5.42 | 4.89 | 4.36 | 4.32 | 4.14 | 4.00 | 3.89 | 3.80 | 3.67 | 3.52 | 3.37 | 3.29 | 3.21 | 3.13 | 3.05 | 2.96 | 2.87 |
| 16 | 8.53 | 6.23 | 5.29 | 4.77 | 4.44 | 4.20 | 4.03 | 3.89 | 3.78 | 3.69 | 3.55 | 3.41 | 3.26 | 3.18 | 3.10 | 3.02 | 2.93 | 2.84 | 2.75 |
| 17 | 8.40 | 6.11 | 5.18 | 4.67 | 4.34 | 4.10 | 3.93 | 3.79 | 3.68 | 3.59 | 3.46 | 3.31 | 3.16 | 3.08 | 3.00 | 2.92 | 2.83 | 2.75 | 2.65 |
| 18 | 8.29 | 6.01 | 5.09 | 4.58 | 4.25 | 4.01 | 3.84 | 3.71 | 3.60 | 3.51 | 3.37 | 3.23 | 3.08 | 3.00 | 2.92 | 2.84 | 2.75 | 2.66 | 2.57 |
| 19 | 8.18 | 5.93 | 5.01 | 4.50 | 4.17 | 3.94 | 3.77 | 3.63 | 3.52 | 3.43 | 3.30 | 3.15 | 3.00 | 2.92 | 2.84 | 2.76 | 2.67 | 2.58 | 2.49 |
| 20 | 8.10 | 5.85 | 4.94 | 4.43 | 4.10 | 3.87 | 3.70 | 3.56 | 3.46 | 3.37 | 3.23 | 3.09 | 2.94 | 2.86 | 2.78 | 2.69 | 2.61 | 2.52 | 2.42 |
| 21 | 8.02 | 5.78 | 4.87 | 4.37 | 4.04 | 3.81 | 3.64 | 3.51 | 3.40 | 3.31 | 3.17 | 3.03 | 2.88 | 2.80 | 2.72 | 2.64 | 2.55 | 2.46 | 2.36 |
| 22 | 7.95 | 5.72 | 4.82 | 4.31 | 3.99 | 3.76 | 3.59 | 3.45 | 3.35 | 3.26 | 3.12 | 2.98 | 2.83 | 2.75 | 2.67 | 2.58 | 2.50 | 2.40 | 2.31 |
| 23 | 7.88 | 5.66 | 4.76 | 4.26 | 3.94 | 3.71 | 3.54 | 3.41 | 3.30 | 3.21 | 3.07 | 2.93 | 2.78 | 2.70 | 2.62 | 2.54 | 2.45 | 2.35 | 2.26 |
| 24 | 7.82 | 5.61 | 4.72 | 4.22 | 3.90 | 3.67 | 3.50 | 3.36 | 3.26 | 3.17 | 3.03 | 2.89 | 2.74 | 2.66 | 2.58 | 2.49 | 2.40 | 2.31 | 2.21 |
| 25 | 7.77 | 5.57 | 4.68 | 4.18 | 3.85 | 3.63 | 3.46 | 3.32 | 3.22 | 3.13 | 2.99 | 2.85 | 2.70 | 2.62 | 2.54 | 2.45 | 2.36 | 2.27 | 2.17 |
| 26 | 7.72 | 5.53 | 4.64 | 4.14 | 3.82 | 3.59 | 3.42 | 3.29 | 3.18 | 3.09 | 2.96 | 2.81 | 2.66 | 2.58 | 2.50 | 2.42 | 2.33 | 2.23 | 2.13 |
| 27 | 7.68 | 5.49 | 4.60 | 4.11 | 3.78 | 3.56 | 3.39 | 3.26 | 3.15 | 3.06 | 2.93 | 2.78 | 2.63 | 2.55 | 2.47 | 2.38 | 2.29 | 2.20 | 2.10 |
| 28 | 7.64 | 5.45 | 4.57 | 4.07 | 3.75 | 3.53 | 3.36 | 3.23 | 3.12 | 3.03 | 2.90 | 2.75 | 2.60 | 2.52 | 2.44 | 2.35 | 2.26 | 2.17 | 2.06 |
| 29 | 7.60 | 5.42 | 4.54 | 4.04 | 3.73 | 3.50 | 3.33 | 3.20 | 3.09 | 3.00 | 2.87 | 2.73 | 2.57 | 2.49 | 2.41 | 2.33 | 2.23 | 2.14 | 2.03 |
| 30 | 7.56 | 5.39 | 4.51 | 4.02 | 3.70 | 3.47 | 3.30 | 3.17 | 3.07 | 2.98 | 2.84 | 2.70 | 2.55 | 2.47 | 2.39 | 2.30 | 2.21 | 2.11 | 2.01 |
| 40 | 7.31 | 5.18 | 4.31 | 3.83 | 3.51 | 3.29 | 3.12 | 2.99 | 2.89 | 2.80 | 2.66 | 2.52 | 2.37 | 2.29 | 2.20 | 2.11 | 2.02 | 1.92 | 1.80 |
| 60 | 7.08 | 4.98 | 4.13 | 3.65 | 3.34 | 3.12 | 2.95 | 2.82 | 2.72 | 2.63 | 2.50 | 2.35 | 2.20 | 2.12 | 2.03 | 1.94 | 1.84 | 1.73 | 1.60 |
| 120 | 6.85 | 4.79 | 3.95 | 3.48 | 3.17 | 2.96 | 2.79 | 2.66 | 2.56 | 2.47 | 2.34 | 2.19 | 2.03 | 1.95 | 1.86 | 1.76 | 1.66 | 1.53 | 1.38 |
| ∞ | 6.63 | 4.61 | 3.78 | 3.32 | 3.02 | 2.80 | 2.64 | 2.51 | 2.41 | 2.32 | 2.18 | 2.04 | 1.88 | 1.79 | 1.70 | 1.59 | 1.47 | 1.32 | 1.00 |

Degrees freedom denominator $(v_2)$

**Table A.5** Chi square table

| df | $1-\alpha =$ $\chi^2_{0.005}$ | $\chi^2_{0.025}$ | $\chi^2_{0.05}$ | $\alpha = 0.10$ $\chi^2_{0.90}$ | $0.05$ $\chi^2_{0.95}$ | $0.025$ $\chi^2_{0.975}$ | $0.01$ $\chi^2_{0.99}$ | $0.005$ $\chi^2_{0.995}$ |
|---|---|---|---|---|---|---|---|---|
| 1 | 0.0000393 | 0.000982 | 0.00393 | 2.706 | 3.841 | 5.024 | 6.635 | 7.879 |
| 2 | 0.0100 | 0.0506 | 0.103 | 4.605 | 5.991 | 7.378 | 9.210 | 10.597 |
| 3 | 0.0717 | 0.216 | 0.352 | 6.251 | 7.815 | 9.348 | 11.345 | 12.838 |
| 4 | 0.207 | 0.484 | 0.711 | 7.779 | 9.488 | 11.143 | 13.277 | 14.860 |
| 5 | 0.412 | 0.831 | 1.145 | 9.236 | 11.070 | 12.832 | 15.086 | 16.750 |
| 6 | 0.676 | 1.237 | 1.635 | 10.645 | 12.592 | 14.449 | 16.812 | 18.548 |
| 7 | 0.989 | 1.690 | 2.167 | 12.017 | 14.067 | 16.013 | 18.475 | 20.278 |
| 8 | 1.344 | 2.180 | 2.733 | 13.362 | 15.507 | 17.535 | 20.090 | 21.955 |
| 9 | 1.735 | 2.700 | 3.325 | 14.684 | 16.919 | 19.023 | 21.666 | 23.589 |
| 10 | 2.156 | 3.247 | 3.940 | 15.987 | 18.307 | 20.483 | 23.209 | 25.188 |
| 11 | 2.603 | 3.816 | 4.575 | 17.275 | 19.675 | 21.920 | 24.725 | 26.757 |
| 12 | 3.074 | 4.404 | 5.226 | 18.549 | 21.026 | 23.336 | 26.217 | 28.300 |
| 13 | 3.565 | 5.009 | 5.892 | 19.812 | 22.362 | 24.736 | 27.688 | 29.819 |
| 14 | 4.075 | 5.629 | 6.571 | 21.064 | 23.685 | 26.119 | 29.141 | 31.319 |
| 15 | 4.601 | 6.262 | 7.261 | 22.307 | 24.996 | 27.488 | 30.578 | 32.801 |
| 16 | 5.142 | 6.908 | 7.962 | 23.542 | 26.296 | 28.845 | 32.000 | 34.267 |
| 17 | 5.697 | 7.564 | 8.672 | 24.769 | 27.587 | 30.191 | 33.409 | 35.718 |
| 18 | 6.265 | 8.231 | 9.390 | 25.989 | 28.869 | 31.526 | 34.805 | 37.156 |
| 19 | 6.844 | 8.907 | 10.117 | 27.204 | 30.144 | 32.852 | 36.191 | 38.582 |
| 20 | 7.434 | 9.591 | 10.851 | 28.412 | 31.410 | 34.170 | 37.566 | 39.997 |
| 21 | 8.034 | 10.283 | 11.591 | 29.615 | 32.671 | 35.479 | 38.932 | 41.401 |
| 22 | 8.643 | 10.982 | 12.338 | 30.813 | 33.924 | 36.781 | 40.289 | 42.796 |
| 23 | 9.260 | 11.688 | 13.091 | 32.007 | 35.172 | 38.076 | 41.638 | 44.181 |
| 24 | 9.886 | 12.401 | 13.848 | 33.196 | 36.415 | 39.364 | 42.980 | 45.558 |
| 25 | 10.520 | 13.120 | 14.611 | 34.382 | 37.652 | 40.646 | 44.314 | 46.928 |
| 26 | 11.160 | 13.844 | 15.379 | 35.563 | 38.885 | 41.923 | 45.642 | 48.290 |
| 27 | 11.808 | 14.573 | 16.151 | 36.741 | 40.113 | 43.194 | 46.963 | 49.645 |
| 28 | 12.461 | 15.308 | 16.928 | 37.916 | 41.337 | 44.461 | 48.278 | 50.993 |
| 29 | 13.121 | 16.047 | 17.708 | 39.087 | 42.557 | 45.722 | 49.588 | 52.336 |
| 30 | 13.787 | 16.791 | 18.493 | 40.256 | 43.773 | 46.979 | 50.892 | 53.672 |
| 35 | 17.192 | 20.569 | 22.465 | 46.059 | 49.802 | 53.203 | 57.342 | 60.275 |
| 40 | 20.707 | 24.433 | 26.509 | 51.805 | 55.758 | 59.342 | 63.691 | 66.766 |
| 45 | 24.311 | 28.366 | 30.612 | 57.505 | 61.656 | 65.410 | 69.957 | 73.166 |
| 50 | 27.991 | 32.357 | 34.764 | 63.167 | 67.505 | 71.420 | 76.154 | 79.490 |
| 60 | 35.535 | 40.482 | 43.188 | 74.397 | 79.082 | 83.298 | 88.379 | 91.952 |
| 70 | 43.275 | 48.758 | 51.739 | 85.527 | 90.531 | 95.023 | 100.425 | 104.215 |
| 80 | 51.172 | 57.153 | 60.391 | 96.578 | 101.879 | 106.629 | 112.329 | 116.321 |
| 90 | 59.196 | 65.647 | 69.126 | 107.565 | 113.145 | 118.136 | 124.116 | 128.299 |
| 100 | 67.328 | 74.222 | 77.929 | 118.498 | 124.342 | 129.561 | 135.807 | 140.169 |

**Table A.6** Quantiles of the Mann-Whitney test statistic

| $n_1$ | $\alpha$ | $n_2=2$ | 3 | 4 | 5 | 6 | 7 | 8 | 9 | 10 | 11 | 12 | 13 | 14 | 15 | 16 | 17 | 18 | 19 | 20 |
|---|---|---|---|---|---|---|---|---|---|---|---|---|---|---|---|---|---|---|---|---|
| 2 | .001 | 0 | 0 | 0 | 0 | 0 | 0 | 0 | 0 | 0 | 0 | 0 | 0 | 0 | 0 | 0 | 0 | 0 | 0 | 0 |
|   | .005 | 0 | 0 | 0 | 0 | 0 | 0 | 0 | 0 | 0 | 0 | 0 | 0 | 0 | 0 | 0 | 0 | 0 | 1 | 1 |
|   | .01  | 0 | 0 | 0 | 0 | 0 | 0 | 1 | 1 | 1 | 1 | 0 | 1 | 1 | 1 | 1 | 1 | 1 | 2 | 2 |
|   | .025 | 0 | 0 | 0 | 1 | 1 | 1 | 2 | 2 | 2 | 2 | 2 | 2 | 2 | 2 | 2 | 3 | 3 | 3 | 3 |
|   | .05  | 0 | 0 | 1 | 1 | 2 | 2 | 3 | 3 | 4 | 4 | 5 | 5 | 5 | 6 | 6 | 7 | 7 | 8 | 8 |
|   | .10  | 0 | 1 | 1 | 2 | 2 | 2 | 3 | 3 | 4 | 4 | 5 | 5 | 6 | 6 | 7 | 7 | 8 | 8 | 8 |
| 3 | .001 | 0 | 0 | 0 | 0 | 0 | 0 | 0 | 0 | 0 | 0 | 0 | 0 | 0 | 0 | 0 | 0 | 0 | 0 | 1 |
|   | .005 | 0 | 0 | 0 | 0 | 0 | 1 | 1 | 1 | 1 | 1 | 2 | 2 | 2 | 2 | 2 | 2 | 3 | 3 | 4 |
|   | .01  | 0 | 0 | 0 | 1 | 2 | 2 | 3 | 3 | 4 | 4 | 5 | 5 | 6 | 6 | 6 | 7 | 7 | 8 | 8 |
|   | .025 | 0 | 0 | 1 | 2 | 3 | 3 | 4 | 5 | 5 | 6 | 6 | 7 | 8 | 8 | 9 | 10 | 10 | 11 | 12 |
|   | .05  | 0 | 0 | 2 | 3 | 3 | 4 | 5 | 6 | 7 | 8 | 9 | 9 | 11 | 11 | 12 | 13 | 14 | 15 | 16 |
|   | .10  | 0 | 1 | 2 | 3 | 4 | 5 | 6 | 7 | 8 | 8 | 9 | 10 | 11 | 12 | 13 | 14 | 15 | 15 | 16 |
| 4 | .001 | 0 | 0 | 0 | 0 | 0 | 0 | 0 | 0 | 0 | 0 | 0 | 1 | 1 | 1 | 1 | 1 | 1 | 1 | 1 |
|   | .005 | 0 | 0 | 0 | 0 | 1 | 1 | 2 | 2 | 3 | 3 | 4 | 4 | 5 | 5 | 6 | 6 | 7 | 8 | 9 |
|   | .01  | 0 | 0 | 0 | 1 | 2 | 2 | 3 | 4 | 4 | 5 | 6 | 6 | 7 | 9 | 8 | 9 | 10 | 11 | 11 |
|   | .025 | 0 | 1 | 1 | 2 | 3 | 4 | 5 | 5 | 6 | 7 | 8 | 9 | 10 | 11 | 12 | 12 | 13 | 14 | 15 |
|   | .05  | 0 | 1 | 2 | 3 | 4 | 5 | 6 | 7 | 8 | 9 | 10 | 11 | 12 | 13 | 15 | 16 | 17 | 18 | 19 |
|   | .10  | 1 | 2 | 4 | 5 | 6 | 7 | 8 | 10 | 11 | 12 | 13 | 14 | 16 | 17 | 18 | 19 | 21 | 22 | 23 |
| 5 | .001 | 0 | 0 | 0 | 0 | 0 | 0 | 1 | 2 | 2 | 3 | 3 | 4 | 4 | 5 | 6 | 6 | 7 | 8 | 8 |
|   | .005 | 0 | 1 | 1 | 2 | 3 | 4 | 3 | 4 | 5 | 6 | 7 | 8 | 8 | 9 | 10 | 11 | 12 | 13 | 14 |
|   | .01  | 0 | 1 | 2 | 3 | 4 | 6 | 5 | 6 | 7 | 8 | 9 | 10 | 11 | 12 | 13 | 14 | 15 | 16 | 17 |
|   | .025 | 0 | 2 | 3 | 5 | 6 | 7 | 7 | 8 | 9 | 10 | 12 | 13 | 14 | 15 | 16 | 18 | 19 | 20 | 21 |
|   | .05  | 1 | 3 | 5 | 6 | 8 | 9 | 9 | 10 | 12 | 13 | 14 | 16 | 17 | 19 | 20 | 21 | 23 | 24 | 26 |
|   | .10  | 2 | 3 | 5 | 6 | 8 | 9 | 11 | 13 | 14 | 16 | 18 | 19 | 21 | 23 | 24 | 26 | 28 | 29 | 31 |
|   | .001 | 0 | 0 | 0 | 0 | 0 | 0 | 2 | 3 | 4 | 5 | 5 | 6 | 7 | 8 | 9 | 10 | 11 | 12 | 13 |
|   | .005 | 0 | 0 | 1 | 2 | 3 | 4 | 5 | 6 | 7 | 8 | 10 | 11 | 12 | 13 | 14 | 16 | 17 | 18 | 19 |

*(Continued)*

**Table A.6** Quantiles of the Mann-Whitney test statistic

| $n_1$ | $\alpha$ | $n_2=2$ | 3 | 4 | 5 | 6 | 7 | 8 | 9 | 10 | 11 | 12 | 13 | 14 | 15 | 16 | 17 | 18 | 19 | 20 |
|---|---|---|---|---|---|---|---|---|---|---|---|---|---|---|---|---|---|---|---|---|
| 6 | .01 | 0 | 0 | 2 | 3 | 4 | 5 | 7 | 8 | 9 | 10 | 12 | 13 | 14 | 16 | 17 | 19 | 20 | 21 | 23 |
|   | .025 | 0 | 2 | 3 | 4 | 6 | 7 | 9 | 11 | 12 | 14 | 15 | 17 | 18 | 20 | 22 | 23 | 25 | 26 | 28 |
|   | .05 | 1 | 3 | 4 | 6 | 8 | 9 | 11 | 13 | 15 | 17 | 18 | 20 | 22 | 24 | 26 | 27 | 29 | 31 | 33 |
|   | .10 | 2 | 4 | 6 | 8 | 10 | 12 | 14 | 16 | 18 | 20 | 22 | 24 | 26 | 28 | 30 | 32 | 35 | 37 | 39 |
|   | .001 | 0 | 0 | 0 | 0 | 1 | 2 | 3 | 4 | 6 | 7 | 8 | 9 | 10 | 11 | 12 | 14 | 15 | 16 | 17 |
|   | .005 | 0 | 0 | 1 | 2 | 4 | 5 | 7 | 8 | 10 | 11 | 13 | 14 | 16 | 17 | 19 | 20 | 22 | 23 | 25 |
| 7 | .01 | 0 | 1 | 2 | 4 | 5 | 7 | 8 | 10 | 12 | 13 | 15 | 17 | 18 | 20 | 22 | 24 | 25 | 27 | 29 |
|   | .025 | 0 | 3 | 4 | 6 | 7 | 9 | 11 | 13 | 15 | 17 | 19 | 21 | 23 | 25 | 27 | 29 | 31 | 33 | 35 |
|   | .05 | 1 | 4 | 5 | 7 | 9 | 12 | 14 | 16 | 18 | 20 | 22 | 25 | 27 | 29 | 31 | 34 | 36 | 38 | 40 |
|   | .10 | 2 | 5 | 7 | 9 | 12 | 14 | 17 | 19 | 22 | 24 | 27 | 29 | 32 | 34 | 37 | 39 | 42 | 44 | 47 |
|   | .001 | 0 | 0 | 0 | 1 | 2 | 3 | 5 | 6 | 7 | 9 | 10 | 12 | 13 | 15 | 16 | 18 | 19 | 21 | 22 |
|   | .005 | 0 | 1 | 2 | 3 | 5 | 7 | 8 | 10 | 12 | 14 | 16 | 18 | 19 | 21 | 23 | 25 | 27 | 29 | 31 |
| 8 | .01 | 0 | 1 | 3 | 5 | 7 | 8 | 10 | 12 | 14 | 16 | 18 | 21 | 23 | 25 | 27 | 29 | 31 | 33 | 35 |
|   | .025 | 1 | 3 | 5 | 7 | 9 | 11 | 14 | 16 | 18 | 20 | 23 | 25 | 27 | 30 | 32 | 35 | 37 | 39 | 42 |
|   | .05 | 2 | 5 | 6 | 9 | 11 | 14 | 16 | 19 | 21 | 24 | 27 | 29 | 32 | 34 | 37 | 40 | 42 | 45 | 48 |
|   | .10 | 3 | 6 | 8 | 11 | 14 | 17 | 20 | 23 | 25 | 28 | 31 | 34 | 37 | 40 | 43 | 46 | 49 | 52 | 55 |
|   | .001 | 0 | 0 | 0 | 2 | 3 | 4 | 6 | 8 | 9 | 11 | 13 | 15 | 16 | 18 | 20 | 22 | 24 | 26 | 27 |
|   | .005 | 0 | 1 | 2 | 4 | 6 | 8 | 10 | 12 | 14 | 17 | 19 | 21 | 23 | 25 | 28 | 30 | 32 | 34 | 37 |
| 9 | .01 | 0 | 2 | 4 | 6 | 8 | 10 | 12 | 15 | 17 | 19 | 22 | 24 | 27 | 29 | 32 | 34 | 37 | 39 | 41 |
|   | .025 | 1 | 4 | 5 | 8 | 11 | 13 | 16 | 18 | 21 | 24 | 27 | 29 | 32 | 35 | 38 | 40 | 43 | 46 | 49 |
|   | .05 | 2 | 5 | 7 | 10 | 13 | 16 | 19 | 22 | 25 | 28 | 31 | 34 | 37 | 40 | 43 | 46 | 49 | 52 | 55 |
|   | .10 | 4 | 7 | 10 | 13 | 16 | 19 | 23 | 26 | 29 | 32 | 36 | 39 | 42 | 46 | 49 | 53 | 56 | 59 | 63 |
|   | .001 | 0 | 0 | 1 | 2 | 4 | 6 | 7 | 9 | 11 | 13 | 15 | 18 | 20 | 22 | 24 | 26 | 28 | 30 | 33 |
|   | .005 | 0 | 1 | 3 | 5 | 7 | 10 | 12 | 14 | 17 | 19 | 22 | 25 | 27 | 30 | 32 | 35 | 38 | 40 | 43 |
| 10 | .01 | 0 | 2 | 4 | 7 | 9 | 12 | 14 | 17 | 20 | 23 | 25 | 28 | 31 | 34 | 37 | 39 | 42 | 45 | 48 |
|   | .025 | 1 | 4 | 6 | 9 | 12 | 15 | 18 | 21 | 24 | 27 | 30 | 34 | 37 | 40 | 43 | 46 | 49 | 53 | 56 |
|   | .05 | 2 | 5 | 8 | 12 | 15 | 18 | 21 | 25 | 28 | 32 | 35 | 38 | 42 | 45 | 49 | 52 | 56 | 59 | 63 |
|   | .10 | 4 | 7 | 11 | 14 | 18 | 22 | 25 | 29 | 33 | 37 | 40 | 44 | 48 | 52 | 55 | 59 | 63 | 67 | 71 |
|   | .001 | 0 | 0 | 1 | 3 | 5 | 7 | 9 | 11 | 13 | 16 | 18 | 21 | 23 | 25 | 28 | 30 | 33 | 35 | 38 |
|   | .005 | 0 | 1 | 3 | 6 | 8 | 11 | 14 | 17 | 19 | 22 | 25 | 28 | 31 | 34 | 37 | 40 | 43 | 46 | 49 |

*(Continued)*

**Table A.6** Quantiles of the Mann-Whitney test statistic

| $n_1$ | $\alpha$ | $n_2=2$ | 3 | 4 | 5 | 6 | 7 | 8 | 9 | 10 | 11 | 12 | 13 | 14 | 15 | 16 | 17 | 18 | 19 | 20 |
|---|---|---|---|---|---|---|---|---|---|---|---|---|---|---|---|---|---|---|---|---|
| 11 | .01 | 0 | 2 | 5 | 8 | 10 | 13 | 16 | 19 | 23 | 26 | 29 | 32 | 35 | 38 | 42 | 45 | 48 | 51 | 54 |
|  | .025 | 1 | 4 | 7 | 10 | 14 | 17 | 20 | 24 | 27 | 31 | 34 | 38 | 41 | 45 | 48 | 52 | 56 | 59 | 63 |
|  | .05 | 2 | 6 | 9 | 13 | 17 | 20 | 24 | 28 | 32 | 35 | 39 | 43 | 47 | 51 | 55 | 58 | 62 | 66 | 70 |
|  | .10 | 4 | 8 | 12 | 16 | 20 | 24 | 28 | 32 | 37 | 41 | 45 | 49 | 53 | 58 | 62 | 66 | 70 | 74 | 79 |
|  | .001 | 0 | 0 | 1 | 3 | 5 | 8 | 10 | 13 | 15 | 18 | 21 | 24 | 26 | 29 | 32 | 35 | 38 | 41 | 43 |
|  | .005 | 0 | 2 | 4 | 7 | 10 | 13 | 16 | 19 | 22 | 25 | 28 | 32 | 35 | 38 | 42 | 45 | 48 | 52 | 55 |
| 12 | .01 | 0 | 3 | 6 | 9 | 12 | 15 | 18 | 22 | 25 | 29 | 32 | 36 | 39 | 43 | 47 | 50 | 54 | 57 | 61 |
|  | .025 | 2 | 5 | 8 | 12 | 15 | 19 | 23 | 27 | 30 | 34 | 38 | 42 | 46 | 50 | 54 | 58 | 62 | 66 | 70 |
|  | .05 | 3 | 6 | 10 | 14 | 18 | 22 | 27 | 31 | 35 | 39 | 43 | 48 | 52 | 56 | 61 | 65 | 69 | 73 | 78 |
|  | .10 | 5 | 9 | 13 | 18 | 22 | 27 | 31 | 36 | 40 | 45 | 50 | 54 | 59 | 64 | 68 | 73 | 78 | 82 | 87 |
|  | .001 | 0 | 0 | 2 | 4 | 6 | 9 | 12 | 15 | 18 | 21 | 24 | 27 | 30 | 33 | 36 | 39 | 43 | 46 | 49 |
|  | .005 | 0 | 2 | 4 | 8 | 11 | 14 | 18 | 21 | 25 | 28 | 32 | 35 | 39 | 43 | 46 | 50 | 54 | 58 | 61 |
| 13 | .01 | 0 | 3 | 6 | 10 | 13 | 17 | 21 | 24 | 28 | 32 | 36 | 40 | 44 | 48 | 52 | 56 | 60 | 64 | 68 |
|  | .025 | 2 | 5 | 9 | 13 | 17 | 21 | 25 | 29 | 34 | 38 | 42 | 46 | 51 | 55 | 60 | 64 | 68 | 73 | 77 |
|  | .05 | 3 | 7 | 11 | 16 | 20 | 25 | 29 | 34 | 38 | 43 | 48 | 52 | 57 | 62 | 66 | 71 | 76 | 81 | 85 |
|  | .10 | 5 | 10 | 14 | 19 | 24 | 29 | 34 | 39 | 44 | 49 | 54 | 59 | 64 | 69 | 75 | 80 | 85 | 90 | 95 |
|  | .001 | 0 | 0 | 2 | 4 | 7 | 10 | 13 | 16 | 20 | 23 | 26 | 30 | 33 | 37 | 40 | 44 | 47 | 51 | 55 |
|  | .005 | 1 | 2 | 5 | 8 | 12 | 16 | 19 | 23 | 27 | 31 | 35 | 39 | 43 | 47 | 51 | 55 | 59 | 64 | 68 |
| 14 | .01 | 0 | 3 | 7 | 11 | 14 | 18 | 23 | 27 | 31 | 35 | 39 | 44 | 48 | 52 | 57 | 61 | 66 | 70 | 74 |
|  | .025 | 2 | 6 | 10 | 14 | 18 | 23 | 27 | 32 | 37 | 41 | 46 | 51 | 56 | 60 | 65 | 70 | 75 | 79 | 84 |
|  | .05 | 4 | 8 | 12 | 17 | 22 | 27 | 32 | 37 | 42 | 47 | 52 | 57 | 62 | 67 | 72 | 78 | 83 | 88 | 93 |
|  | .10 | 5 | 11 | 16 | 21 | 26 | 32 | 37 | 42 | 48 | 53 | 59 | 64 | 70 | 75 | 81 | 86 | 92 | 98 | 103 |
|  | .001 | 0 | 0 | 2 | 5 | 8 | 11 | 15 | 18 | 22 | 25 | 29 | 33 | 37 | 41 | 44 | 48 | 52 | 56 | 60 |
|  | .005 | 1 | 3 | 6 | 9 | 13 | 17 | 21 | 25 | 30 | 34 | 38 | 43 | 47 | 52 | 56 | 61 | 65 | 70 | 74 |
| 15 | .01 | 1 | 4 | 8 | 12 | 16 | 20 | 25 | 29 | 34 | 38 | 43 | 48 | 52 | 57 | 62 | 67 | 71 | 76 | 81 |
|  | .025 | 2 | 6 | 11 | 15 | 20 | 25 | 30 | 35 | 40 | 45 | 50 | 55 | 60 | 65 | 71 | 76 | 81 | 86 | 91 |
|  | .05 | 4 | 8 | 13 | 19 | 24 | 29 | 34 | 40 | 45 | 51 | 56 | 62 | 67 | 73 | 78 | 84 | 89 | 95 | 101 |
|  | .10 | 6 | 11 | 17 | 23 | 28 | 34 | 40 | 46 | 52 | 58 | 64 | 69 | 75 | 81 | 87 | 93 | 99 | 105 | 111 |
|  | .001 | 0 | 0 | 3 | 6 | 9 | 12 | 16 | 20 | 24 | 28 | 32 | 36 | 40 | 44 | 49 | 53 | 57 | 61 | 66 |
|  | .005 | 0 | 3 | 6 | 10 | 14 | 19 | 23 | 28 | 32 | 37 | 42 | 46 | 51 | 56 | 61 | 66 | 71 | 75 | 80 |

(Continued)

**Table A.6** Quantiles of the Mann-Whitney test statistic

| $n_1$ | $\alpha$ | $n_2=2$ | 3 | 4 | 5 | 6 | 7 | 8 | 9 | 10 | 11 | 12 | 13 | 14 | 15 | 16 | 17 | 18 | 19 | 20 |
|---|---|---|---|---|---|---|---|---|---|---|---|---|---|---|---|---|---|---|---|---|
| 16 | .01 | 1 | 4 | 8 | 13 | 17 | 22 | 27 | 32 | 37 | 42 | 47 | 52 | 57 | 62 | 67 | 72 | 77 | 83 | 88 |
|  | .025 | 2 | 7 | 12 | 16 | 22 | 27 | 32 | 38 | 43 | 48 | 54 | 60 | 65 | 71 | 76 | 82 | 87 | 93 | 99 |
|  | .05 | 4 | 9 | 15 | 20 | 26 | 31 | 37 | 43 | 49 | 55 | 61 | 66 | 72 | 78 | 84 | 90 | 96 | 102 | 108 |
|  | .10 | 6 | 12 | 18 | 24 | 30 | 37 | 43 | 49 | 55 | 62 | 68 | 75 | 81 | 87 | 94 | 100 | 107 | 113 | 120 |
|  | .001 | 0 | 1 | 3 | 6 | 10 | 14 | 18 | 22 | 26 | 30 | 35 | 39 | 44 | 48 | 53 | 58 | 62 | 67 | 71 |
|  | .005 | 0 | 3 | 7 | 11 | 16 | 20 | 25 | 30 | 35 | 40 | 45 | 50 | 55 | 61 | 66 | 71 | 76 | 82 | 87 |
| 17 | .01 | 1 | 5 | 9 | 14 | 19 | 24 | 29 | 34 | 39 | 45 | 50 | 56 | 61 | 67 | 72 | 78 | 83 | 89 | 94 |
|  | .025 | 3 | 7 | 12 | 18 | 23 | 29 | 35 | 40 | 46 | 52 | 58 | 64 | 70 | 76 | 82 | 88 | 94 | 100 | 106 |
|  | .05 | 4 | 10 | 16 | 21 | 27 | 34 | 40 | 46 | 52 | 58 | 65 | 71 | 78 | 84 | 90 | 97 | 103 | 110 | 116 |
|  | .10 | 7 | 13 | 19 | 26 | 32 | 39 | 46 | 53 | 59 | 66 | 73 | 80 | 86 | 93 | 100 | 107 | 114 | 121 | 128 |
|  | .001 | 0 | 1 | 4 | 7 | 11 | 15 | 19 | 24 | 28 | 33 | 38 | 43 | 47 | 52 | 57 | 62 | 67 | 72 | 77 |
|  | .005 | 0 | 3 | 7 | 12 | 17 | 22 | 27 | 32 | 38 | 43 | 48 | 54 | 59 | 65 | 71 | 76 | 82 | 88 | 93 |
| 18 | .01 | 1 | 5 | 10 | 15 | 20 | 25 | 31 | 37 | 42 | 48 | 54 | 60 | 66 | 71 | 77 | 83 | 89 | 95 | 101 |
|  | .025 | 3 | 8 | 13 | 19 | 25 | 31 | 37 | 43 | 50 | 56 | 62 | 68 | 75 | 81 | 87 | 94 | 100 | 107 | 113 |
|  | .05 | 5 | 11 | 17 | 23 | 29 | 36 | 42 | 49 | 56 | 62 | 69 | 76 | 83 | 89 | 96 | 103 | 110 | 117 | 124 |
|  | .10 | 7 | 14 | 21 | 28 | 35 | 42 | 49 | 56 | 63 | 70 | 78 | 85 | 92 | 99 | 107 | 114 | 121 | 129 | 136 |
|  | .001 | 0 | 1 | 4 | 8 | 12 | 16 | 21 | 26 | 30 | 35 | 41 | 46 | 51 | 56 | 61 | 67 | 72 | 78 | 83 |
|  | .005 | 1 | 4 | 8 | 13 | 18 | 23 | 29 | 34 | 40 | 46 | 52 | 58 | 64 | 70 | 75 | 82 | 88 | 94 | 100 |
| 19 | .01 | 2 | 5 | 10 | 16 | 21 | 27 | 33 | 39 | 45 | 51 | 57 | 64 | 70 | 77 | 83 | 89 | 96 | 102 | 108 |
|  | .025 | 3 | 9 | 14 | 20 | 26 | 33 | 39 | 46 | 53 | 59 | 66 | 73 | 79 | 86 | 93 | 100 | 107 | 114 | 120 |
|  | .05 | 5 | 11 | 18 | 24 | 31 | 38 | 45 | 52 | 59 | 66 | 74 | 81 | 88 | 95 | 102 | 110 | 117 | 124 | 131 |
|  | .10 | 8 | 15 | 22 | 29 | 37 | 44 | 52 | 59 | 67 | 74 | 82 | 90 | 98 | 105 | 113 | 121 | 129 | 136 | 144 |
|  | .001 | 0 | 1 | 4 | 8 | 13 | 17 | 22 | 27 | 33 | 38 | 43 | 49 | 55 | 60 | 66 | 71 | 77 | 83 | 89 |
|  | .005 | 1 | 4 | 9 | 14 | 19 | 25 | 31 | 37 | 43 | 49 | 55 | 61 | 68 | 74 | 80 | 87 | 93 | 100 | 106 |
| 20 | .01 | 2 | 6 | 11 | 17 | 23 | 28 | 35 | 41 | 48 | 54 | 61 | 68 | 74 | 81 | 88 | 94 | 101 | 108 | 115 |
|  | .025 | 3 | 9 | 15 | 21 | 28 | 35 | 42 | 49 | 56 | 63 | 70 | 77 | 84 | 91 | 99 | 106 | 113 | 120 | 128 |
|  | .05 | 5 | 12 | 19 | 26 | 33 | 40 | 48 | 55 | 63 | 70 | 78 | 85 | 93 | 101 | 108 | 116 | 124 | 131 | 139 |
|  | .10 | 8 | 16 | 23 | 31 | 39 | 47 | 55 | 63 | 71 | 79 | 87 | 95 | 103 | 111 | 120 | 128 | 136 | 144 | 152 |

**Table A.7** Binomial probability distribution

$$P(t \mid n, P) = \binom{n}{t} p^t q^{n-t}$$

| $n = 1$ | | | | | | | | | |
|---|---|---|---|---|---|---|---|---|---|

| $\dfrac{p}{t}$ | 0.01 | 0.02 | 0.03 | 0.04 | 0.05 | 0.06 | 0.07 | 0.08 | 0.09 | 0.10 |
|---|---|---|---|---|---|---|---|---|---|---|
| 0 | 0.9900 | 0.9800 | 0.9700 | 0.9600 | 0.9500 | 0.9400 | 0.9300 | 0.9200 | 0.9100 | 0.9000 |
| 1 | .0100 | .0200 | .0300 | .0400 | .0500 | .0600 | .0700 | .0800 | 0.0900 | 0.1000 |

| | 0.11 | 0.12 | 0.13 | 0.14 | 0.15 | 0.16 | 0.17 | 0.18 | 0.19 | 0.20 |
|---|---|---|---|---|---|---|---|---|---|---|
| 0 | 0.8900 | 0.8800 | 0.8700 | 0.8600 | 0.8500 | 0.8400 | 0.8300 | 0.8200 | 0.8100 | 0.8000 |
| 1 | 0.1100 | 0.1200 | 0.1300 | 0.1400 | 0.1500 | 0.1600 | .1700 | 0.1800 | 0.1900 | 0.2000 |

| | 0.21 | 0.22 | 0.23 | 0.24 | 0.25 | 0.26 | 0.27 | 0.28 | 0.29 | 0.30 |
|---|---|---|---|---|---|---|---|---|---|---|
| 0 | 0.7900 | 0.7800 | 0.7700 | 0.7600 | 0.7500 | 0.7400 | .7300 | 0.7200 | 0.7100 | 0.7000 |
| 1 | 0.2100 | 0.2200 | 0.2300 | 0.2400 | 0.2500 | 0.2600 | .2700 | 0.2800 | 0.2900 | 0.3000 |

| | 0.31 | 0.32 | 0.33 | 0.34 | 0.35 | 0.36 | 0.37 | 0.38 | 0.39 | 0.40 |
|---|---|---|---|---|---|---|---|---|---|---|
| 0 | 0.6900 | 0.6800 | 0.6700 | 0.6600 | .6500 | 0.6400 | 0.6300 | 0.6200 | 0.6100 | 0.6000 |
| 1 | 0.3100 | 0.3200 | 0.3300 | 0.3400 | .3500 | 0.3600 | 0.3700 | 0.3800 | 0.3900 | 0.4000 |

| | 0.41 | 0.42 | 0.43 | 0.44 | 0.45 | 0.46 | 0.47 | 0.48 | 0.49 | 0.50 |
|---|---|---|---|---|---|---|---|---|---|---|
| 0 | 0.5900 | 0.5800 | 0.5700 | 0.5600 | 0.5500 | 0.5400 | 0.5300 | 0.5200 | 0.5100 | 0.5000 |
| 1 | 0.4100 | 0.4200 | 0.4300 | 0.4400 | 0.4500 | 0.4600 | 0.4700 | 0.4800 | 0.4900 | 0.5000 |

| $n = 2$ | | | | | | | | | |
|---|---|---|---|---|---|---|---|---|---|

| $\dfrac{p}{t}$ | 0.01 | 0.02 | 0.03 | 0.04 | 0.05 | 0.06 | 0.07 | 0.08 | 0.09 | 0.10 |
|---|---|---|---|---|---|---|---|---|---|---|
| 0 | 0.9801 | 0.9604 | 0.9409 | 0.9216 | 0.9025 | 0.8836 | 0.8649 | 0.8464 | 0.8281 | 0.8100 |
| 1 | 0.0198 | 0.0392 | 0.0582 | 0.0768 | 0.0950 | 0.1128 | 0.1302 | 0.1472 | 0.1638 | 0.1800 |
| 2 | 0.0001 | 0.0004 | 0.0009 | 0.0016 | 0.0025 | 0.0036 | 0.0049 | 0.0064 | 0.0081 | 0.0100 |

| | 0.11 | 0.12 | 0.13 | 0.14 | 0.15 | 0.16 | 0.17 | 0.18 | 0.19 | 0.20 |
|---|---|---|---|---|---|---|---|---|---|---|
| 0 | 0.7921 | 0.7744 | 0.7569 | 0.7396 | 0.7225 | 0.7056 | 0.6889 | 0.6724 | 0.6561 | 0.6400 |
| 1 | 0.1958 | 0.2112 | 0.2262 | 0.2408 | 0.2550 | 0.2688 | 0.2822 | 0.2952 | 0.3078 | 0.3200 |
| 2 | 0.0121 | 0.0144 | 0.0169 | 0.0196 | 0.0225 | 0.0256 | 0.0289 | 0.0324 | 0.0361 | 0.0400 |

| | 0.21 | 0.22 | 0.23 | 0.24 | 0.25 | 0.26 | 0.27 | 0.28 | 0.29 | 0.30 |
|---|---|---|---|---|---|---|---|---|---|---|
| 0 | 0.6241 | 0.6084 | 0.5929 | 0.5776 | 0.5625 | 0.5476 | 0.5329 | 0.5184 | 0.5041 | 0.4900 |
| 1 | 0.3318 | 0.3432 | 0.3542 | 0.3648 | 0.3750 | 0.3848 | 0.3942 | 0.4032 | 0.4118 | 0.4200 |
| 2 | 0.0441 | 0.0484 | 0.0529 | 0.0576 | 0.0625 | 0.0676 | 0.0729 | 0.0784 | 0.0841 | 0.0900 |

(*Continued*)

**Table A.7** Binomial probability distribution

|   | 0.31 | 0.32 | 0.33 | 0.34 | 0.35 | 0.36 | 0.37 | 0.38 | 0.39 | 0.40 |
|---|------|------|------|------|------|------|------|------|------|------|
| 0 | 0.4761 | 0.4624 | 0.4489 | 0.4356 | 0.4225 | 0.4096 | 0.3969 | 0.3844 | 0.3721 | 0.3600 |
| 1 | 0.4278 | 0.4352 | 0.4422 | 0.4488 | 0.4550 | 0.4608 | 0.4662 | 0.4712 | 0.4758 | 0.4800 |
| 2 | 0.0961 | 0.1024 | 0.1089 | 0.1156 | 0.1225 | 0.1296 | 0.1369 | 0.1444 | 0.1521 | 0.1600 |

|   | 0.41 | 0.42 | 0.43 | 0.44 | 0.45 | 0.46 | 0.47 | 0.48 | 0.49 | 0.50 |
|---|------|------|------|------|------|------|------|------|------|------|
| 0 | 0.3481 | 0.3364 | 0.3249 | 0.3136 | 0.3025 | 0.2916 | 0.2809 | 0.2704 | 0.2601 | 0.2500 |
| 1 | 0.4838 | 0.4872 | 0.4902 | 0.4928 | 0.4950 | 0.4968 | 0.4982 | 0.4992 | 0.4998 | 0.5000 |
| 2 | 0.1681 | 0.1764 | 0.1849 | 0.1936 | 0.2025 | 0.2116 | 0.2209 | 0.2304 | 0.2401 | 0.2500 |

$n = 3$

| $t$ \ $p$ | 0.01 | 0.02 | 0.03 | 0.04 | 0.05 | 0.06 | 0.07 | 0.08 | 0.09 | 0.10 |
|---|------|------|------|------|------|------|------|------|------|------|
| 0 | 0.9704 | 0.9412 | 0.9127 | 0.8847 | 0.8574 | 0.8306 | 0.8044 | 0.7787 | 0.7536 | 0.7290 |
| 1 | 0.0294 | 0.0576 | 0.0847 | 0.1106 | 0.1354 | 0.1590 | 0.1816 | 0.2031 | 0.2236 | 0.2430 |
| 2 | 0.0003 | 0.0012 | 0.0026 | 0.0046 | 0.0071 | 0.0102 | 0.0137 | 0.0177 | 0.0221 | 0.0270 |
| 3 | 0.0000 | 0.0000 | 0.0000 | 0.0001 | 0.0001 | 0.0002 | 0.0003 | 0.0005 | 0.0007 | 0.0010 |

|   | 0.11 | 0.12 | 0.13 | 0.14 | 0.15 | 0.16 | 0.17 | 0.18 | 0.19 | 0.20 |
|---|------|------|------|------|------|------|------|------|------|------|
| 0 | 0.7050 | 0.6815 | 0.6585 | 0.6361 | 0.6141 | 0.5927 | 0.5718 | 0.5514 | 0.5314 | 0.5120 |
| 1 | 0.2614 | 0.2788 | 0.2952 | 0.3106 | 0.3251 | 0.3387 | 0.3513 | 0.3631 | 0.3740 | 0.3840 |
| 2 | 0.0323 | 0.0380 | 0.0441 | 0.0506 | 0.0574 | 0.0645 | 0.0720 | 0.0797 | 0.0877 | 0.0960 |
| 3 | 0.0013 | 0.0017 | 0.0022 | 0.0027 | 0.0034 | 0.0041 | 0.0049 | 0.0058 | 0.0069 | 0.0080 |

|   | 0.21 | 0.22 | 0.23 | 0.24 | 0.25 | 0.26 | 0.27 | 0.28 | 0.29 | 0.30 |
|---|------|------|------|------|------|------|------|------|------|------|
| 0 | 0.4930 | 0.4746 | 0.4565 | 0.4390 | 0.4219 | 0.4052 | 0.3890 | 0.3732 | 0.3579 | 0.3430 |
| 1 | 0.3932 | 0.4015 | 0.4091 | 0.4159 | 0.4219 | 0.4271 | 0.4316 | 0.4355 | 0.4386 | 0.4410 |
| 2 | 0.1045 | 0.1133 | 0.1222 | 0.1313 | 0.1406 | 0.1501 | 0.1597 | 0.1693 | 0.1791 | 0.1890 |
| 3 | 0.0093 | 0.0106 | 0.0122 | 0.0138 | 0.0156 | 0.0176 | 0.0197 | 0.0220 | 0.0244 | 0.0270 |

|   | 0.31 | 0.32 | 0.33 | 0.34 | 0.35 | 0.36 | 0.37 | 0.38 | 0.39 | 0.40 |
|---|------|------|------|------|------|------|------|------|------|------|
| 0 | 0.3285 | 0.3144 | 0.3008 | 0.2875 | 0.2746 | 0.2621 | 0.2500 | 0.2383 | 0.2270 | 0.2160 |
| 1 | 0.4428 | 0.4439 | 0.4444 | 0.4443 | 0.4436 | 0.4424 | 0.4406 | 0.4382 | 0.4354 | 0.4320 |
| 2 | 0.1989 | 0.2089 | 0.2189 | 0.2289 | 0.2389 | 0.2488 | 0.2587 | 0.2686 | 0.2783 | 0.2880 |
| 3 | 0.0298 | 0.0328 | 0.0359 | 0.0393 | 0.0429 | 0.0467 | 0.0507 | 0.0549 | 0.0593 | 0.0640 |

|   | 0.41 | 0.42 | 0.43 | 0.44 | 0.45 | 0.46 | 0.47 | 0.48 | 0.49 | 0.50 |
|---|------|------|------|------|------|------|------|------|------|------|
| 0 | 0.2054 | 0.1951 | 0.1852 | 0.1756 | 0.1664 | 0.1575 | 0.1489 | 0.1406 | 0.1327 | 0.1250 |
| 1 | 0.4282 | 0.4239 | 0.4191 | 0.4140 | 0.4084 | 0.4024 | 0.3961 | 0.3894 | 0.3823 | 0.3750 |
| 2 | 0.2975 | 0.3069 | 0.3162 | 0.3252 | 0.3341 | 0.3428 | 0.3512 | 0.3594 | 0.3674 | 0.3750 |
| 3 | 0.0689 | 0.0741 | 0.0795 | 0.0852 | 0.0911 | 0.0973 | 0.1038 | 0.1106 | 0.1176 | 0.1250 |

(*Continued*)

**Table A.7** Binomial probability distribution

| | | | | | $n=4$ | | | | | |
|---|---|---|---|---|---|---|---|---|---|---|
| $p$ / $t$ | 0.01 | 0.02 | 0.03 | 0.04 | 0.05 | 0.06 | 0.07 | 0.08 | 0.09 | 0.10 |
| 0 | 0.9606 | 0.9224 | 0.8853 | 0.8493 | 0.8145 | 0.7807 | 0.7481 | 0.7164 | 0.6857 | 0.6561 |
| 1 | 0.0388 | 0.0753 | 0.1095 | 0.1416 | 0.1715 | 0.1993 | 0.2252 | 0.2492 | 0.2713 | 0.2916 |
| 2 | 0.0006 | 0.0023 | 0.0051 | 0.0088 | 0.0135 | 0.0191 | 0.0254 | 0.0325 | 0.0402 | 0.0486 |
| 3 | 0.0000 | 0.0000 | 0.0001 | 0.0002 | 0.0005 | 0.0008 | 0.0013 | 0.0019 | 0.0027 | 0.0036 |
| 4 | 0.0000 | 0.0000 | 0.0000 | 0.0000 | 0.0000 | 0.0000 | 0.0000 | 0.0000 | 0.0001 | 0.0001 |
| | 0.11 | 0.12 | 0.13 | 0.14 | 0.15 | 0.16 | 0.17 | 0.18 | 0.19 | 0.20 |
| 0 | 0.6274 | 0.5997 | 0.5729 | 0.5470 | 0.5220 | 0.4979 | 0.4746 | 0.4521 | 0.4305 | 0.4096 |
| 1 | 0.3102 | 0.3271 | 0.3424 | 0.3562 | 0.3685 | 0.3793 | 0.3888 | 0.3970 | 0.4039 | 0.4096 |
| 2 | 0.0575 | 0.0669 | 0.0767 | 0.0870 | 0.0975 | 0.1084 | 0.1195 | 0.1307 | 0.1421 | 0.1536 |
| 3 | 0.0047 | 0.0061 | 0.0076 | 0.0094 | 0.0115 | 0.0138 | 0.0163 | 0.0191 | 0.0222 | 0.0256 |
| 4 | 0.0001 | 0.0002 | 0.0003 | 0.0004 | 0.0005 | 0.0007 | 0.0008 | 0.0010 | 0.0013 | 0.0016 |
| | 0.21 | 0.22 | 0.23 | 0.24 | 0.25 | 0.26 | 0.27 | 0.28 | 0.29 | 0.30 |
| 0 | 0.3895 | 0.3702 | 0.3515 | 0.3336 | 0.3164 | 0.2999 | 0.2840 | 0.2687 | 0.2541 | 0.2401 |
| 1 | 0.4142 | 0.4176 | 0.4200 | 0.4214 | 0.4219 | 0.4214 | 0.4201 | 0.4180 | 0.4152 | 0.4116 |
| 2 | 0.1651 | 0.1767 | 0.1882 | 0.1996 | 0.2109 | 0.2221 | 0.2331 | 0.2439 | 0.2544 | 0.2646 |
| 3 | 0.0293 | 0.0332 | 0.0375 | 0.0420 | 0.0469 | 0.0520 | 0.0575 | 0.0632 | 0.0693 | 0.0756 |
| 4 | 0.0019 | 0.0023 | 0.0028 | 0.0033 | 0.0039 | 0.0046 | 0.0053 | 0.0061 | 0.0071 | 0.0081 |
| | 0.31 | 0.32 | 0.33 | 0.34 | 0.35 | 0.36 | 0.37 | 0.38 | 0.39 | 0.40 |
| 0 | 0.2267 | 0.2138 | 0.2015 | 0.1897 | 0.1785 | 0.1678 | 0.1575 | 0.1478 | 0.1385 | 0.1296 |
| 1 | 0.4074 | 0.4025 | 0.3970 | 0.3910 | 0.3845 | 0.3775 | 0.3701 | 0.3623 | 0.3541 | 0.3456 |
| 2 | 0.2745 | 0.2841 | 0.2933 | 0.3021 | 0.3105 | 0.3185 | 0.3260 | 0.3330 | 0.3396 | 0.3456 |
| 3 | 0.0822 | 0.0891 | 0.0963 | 0.1038 | 0.1115 | 0.1194 | 0.1276 | 0.1361 | 0.1447 | 0.1536 |
| 4 | 0.0092 | 0.0105 | 0.0119 | 0.0134 | 0.0150 | 0.0168 | 0.0187 | 0.0209 | 0.0231 | 0.0256 |
| | 0.41 | 0.42 | 0.43 | 0.44 | 0.45 | 0.46 | 0.47 | 0.48 | 0.49 | 0.50 |
| 0 | 0.1212 | 0.1132 | 0.1056 | 0.0983 | 0.0915 | 0.0850 | 0.0789 | 0.0731 | 0.0677 | 0.0625 |
| 1 | 0.3368 | 0.3278 | 0.3185 | 0.3091 | 0.2995 | 0.2897 | 0.2799 | 0.2700 | 0.2600 | 0.2500 |
| 2 | 0.3511 | 0.3560 | 0.3604 | 0.3643 | 0.3675 | 0.3702 | 0.3723 | 0.3738 | 0.3747 | 0.3750 |
| 3 | 0.1627 | 0.1719 | 0.1813 | 0.1908 | 0.2005 | 0.2102 | 0.2201 | 0.2300 | 0.2400 | 0.2500 |
| 4 | 0.0283 | 0.0311 | 0.0342 | 0.0375 | 0.0410 | 0.0448 | 0.0488 | 0.0531 | 0.0576 | 0.0625 |
| | | | | | $n=5$ | | | | | |
| $p$ / $t$ | 0.01 | 0.02 | 0.03 | 0.04 | 0.05 | 0.06 | 0.07 | 0.08 | 0.09 | 0.10 |
| 0 | 0.9510 | 0.9039 | 0.8587 | 0.8154 | 0.7738 | 0.7339 | 0.6957 | 0.6591 | 0.6240 | 0.5905 |
| 1 | 0.0480 | 0.0922 | 0.1328 | 0.1699 | 0.2036 | 0.2342 | 0.2618 | 0.2866 | 0.3086 | 0.3280 |
| 2 | 0.0010 | 0.0038 | 0.0082 | 0.0142 | 0.0214 | 0.0299 | 0.0394 | 0.0498 | 0.0610 | 0.0729 |
| 3 | 0.0000 | 0.0001 | 0.0003 | 0.0006 | 0.0011 | 0.0019 | 0.0030 | 0.0043 | 0.0060 | 0.0081 |
| 4 | 0.0000 | 0.0000 | 0.0000 | 0.0000 | 0.0000 | 0.0001 | 0.0001 | 0.0002 | 0.0003 | 0.0004 |

(*Continued*)

**Table A.7** Binomial probability distribution

|   | 0.11 | 0.12 | 0.13 | 0.14 | 0.15 | 0.16 | 0.17 | 0.18 | 0.19 | 0.20 |
|---|------|------|------|------|------|------|------|------|------|------|
| 0 | 0.5584 | 0.5277 | 0.4984 | 0.4704 | 0.4437 | 0.4182 | 0.3939 | 0.3707 | 0.3487 | 0.3277 |
| 1 | 0.3451 | 0.3598 | 0.3724 | 0.3829 | 0.3915 | 0.3983 | 0.4034 | 0.4069 | 0.4089 | 0.4096 |
| 2 | 0.0853 | 0.0981 | 0.1113 | 0.1247 | 0.1382 | 0.1517 | 0.1652 | 0.1786 | 0.1919 | 0.2048 |
| 3 | 0.0105 | 0.0134 | 0.0166 | 0.0203 | 0.0244 | 0.0289 | 0.0338 | 0.0392 | 0.0450 | 0.0512 |
| 4 | 0.0007 | 0.0009 | 0.0012 | 0.0017 | 0.0022 | 0.0028 | 0.0035 | 0.0043 | 0.0053 | 0.0064 |
| 5 | 0.0000 | 0.0000 | 0.0000 | 0.0001 | 0.0001 | 0.0001 | 0.0001 | 0.0002 | 0.0002 | 0.0003 |

|   | 0.21 | 0.22 | 0.23 | 0.24 | 0.25 | 0.26 | 0.27 | 0.28 | 0.29 | 0.30 |
|---|------|------|------|------|------|------|------|------|------|------|
| 0 | 0.3077 | 0.2887 | 0.2707 | 0.2536 | 0.2373 | 0.2219 | 0.2073 | 0.1935 | 0.1804 | 0.1681 |
| 1 | 0.4090 | 0.4072 | 0.4043 | 0.4003 | 0.3955 | 0.3898 | 0.3834 | 0.3762 | 0.3685 | 0.3602 |
| 2 | 0.2174 | 0.2297 | 0.2415 | 0.2529 | 0.2637 | 0.2739 | 0.2836 | 0.2926 | 0.3010 | 0.3087 |
| 3 | 0.0578 | 0.0648 | 0.0721 | 0.0798 | 0.0879 | 0.0962 | 0.1049 | 0.1138 | 0.1229 | 0.1323 |
| 4 | 0.0077 | 0.0091 | 0.0108 | 0.0126 | 0.0146 | 0.0169 | 0.0194 | 0.0221 | 0.0251 | 0.0284 |
| 5 | 0.0004 | 0.0005 | 0.0006 | 0.0008 | 0.0010 | 0.0012 | 0.0014 | 0.0017 | 0.0021 | 0.0024 |

|   | 0.31 | 0.32 | 0.33 | 0.34 | 0.35 | 0.36 | 0.37 | 0.38 | 0.39 | 0.40 |
|---|------|------|------|------|------|------|------|------|------|------|
| 0 | 0.1564 | 0.1454 | 0.1350 | 0.1252 | 0.1160 | 0.1074 | 0.0992 | 0.0916 | 0.0845 | 0.0778 |
| 1 | 0.3513 | 0.3421 | 0.3325 | 0.3226 | 0.3124 | 0.3020 | 0.2914 | 0.2808 | 0.2700 | 0.2592 |
| 2 | 0.3157 | 0.3220 | 0.3275 | 0.3323 | 0.3364 | 0.3397 | 0.3423 | 0.3441 | 0.3452 | 0.3456 |
| 3 | 0.1418 | 0.1515 | 0.1613 | 0.1712 | 0.1811 | 0.1911 | 0.2010 | 0.2109 | 0.2207 | 0.2304 |
| 4 | 0.0319 | 0.0357 | 0.0397 | 0.0441 | 0.0488 | 0.0537 | 0.0590 | 0.0646 | 0.0706 | 0.0768 |
| 5 | 0.0029 | 0.0034 | 0.0039 | 0.0045 | 0.0053 | 0.0060 | 0.0069 | 0.0079 | 0.0090 | 0.0102 |

|   | 0.41 | 0.42 | 0.43 | 0.44 | 0.45 | 0.46 | 0.47 | 0.48 | 0.49 | 0.50 |
|---|------|------|------|------|------|------|------|------|------|------|
| 0 | 0.0715 | 0.0656 | 0.0602 | 0.0551 | 0.0503 | 0.0459 | 0.0418 | 0.0380 | 0.0345 | 0.0312 |
| 1 | 0.2484 | 0.2376 | 0.2270 | 0.2164 | 0.2059 | 0.1956 | 0.1854 | 0.1755 | 0.1657 | 0.1562 |
| 2 | 0.3452 | 0.3442 | 0.3424 | 0.3400 | 0.3369 | 0.3332 | 0.3289 | 0.3240 | 0.3185 | 0.3125 |
| 3 | 0.2399 | 0.2492 | 0.2583 | 0.2671 | 0.2757 | 0.2838 | 0.2916 | 0.2990 | 0.3060 | 0.3125 |
| 4 | 0.0834 | 0.0902 | 0.0974 | 0.1049 | 0.1128 | 0.1209 | 0.1293 | 0.1380 | 0.1470 | 0.1562 |
| 5 | 0.0116 | 0.0131 | 0.0147 | 0.0165 | 0.0185 | 0.0206 | 0.0229 | 0.0255 | 0.0282 | 0.0312 |

$n = 6$

| $p$ / $t$ | 0.01 | 0.02 | 0.03 | 0.04 | 0.05 | 0.06 | 0.07 | 0.08 | 0.09 | 0.10 |
|---|------|------|------|------|------|------|------|------|------|------|
| 0 | 0.9415 | 0.8858 | 0.8330 | 0.7828 | 0.7351 | 0.6899 | 0.6470 | 0.6064 | 0.5679 | 0.5314 |
| 1 | 0.0571 | 0.1085 | 0.1546 | 0.1957 | 0.2321 | 0.2642 | 0.2922 | 0.3164 | 0.3370 | 0.3543 |
| 2 | 0.0014 | 0.0055 | 0.0120 | 0.0204 | 0.0305 | 0.0422 | 0.0550 | 0.0688 | 0.0833 | 0.0984 |
| 3 | 0.0000 | 0.0002 | 0.0005 | 0.0011 | 0.0021 | 0.0036 | 0.0055 | 0.0080 | 0.0110 | 0.0146 |
| 4 | 0.0000 | 0.0000 | 0.0000 | 0.0000 | 0.0001 | 0.0002 | 0.0003 | 0.0005 | 0.0008 | 0.0012 |
| 5 | 0.0000 | 0.0000 | 0.0000 | 0.0000 | 0.0000 | 0.0000 | 0.0000 | 0.0000 | 0.0000 | 0.0001 |

(*Continued*)

**Table A.7** Binomial probability distribution

| | 0.11 | 0.12 | 0.13 | 0.14 | 0.15 | 0.16 | 0.17 | 0.18 | 0.19 | 0.20 |
|---|---|---|---|---|---|---|---|---|---|---|
| 0 | 0.4970 | 0.4644 | 0.4336 | 0.4046 | 0.3771 | 0.3513 | 0.3269 | 0.3040 | 0.2824 | 0.2621 |
| 1 | 0.3685 | 0.3800 | 0.3888 | 0.3952 | 0.3993 | 0.4015 | 0.4018 | 0.4004 | 0.3975 | 0.3932 |
| 2 | 0.1139 | 0.1295 | 0.1452 | 0.1608 | 0.1762 | 0.1912 | 0.2057 | 0.2197 | 0.2331 | 0.2458 |
| 3 | 0.0188 | 0.0236 | 0.0289 | 0.0349 | 0.0415 | 0.0486 | 0.0562 | 0.0643 | 0.0729 | 0.0819 |
| 4 | 0.0017 | 0.0024 | 0.0032 | 0.0043 | 0.0055 | 0.0069 | 0.0086 | 0.0106 | 0.0128 | 0.0154 |
| 5 | 0.0001 | 0.0001 | 0.0002 | 0.0003 | 0.0004 | 0.0005 | 0.0007 | 0.0009 | 0.0012 | 0.0015 |
| 6 | 0.0000 | 0.0000 | 0.0000 | 0.0000 | 0.0000 | 0.0000 | 0.0000 | 0.0000 | 0.0000 | 0.0001 |

| | 0.21 | 0.22 | 0.23 | 0.24 | 0.25 | 0.26 | 0.27 | 0.28 | 0.29 | 0.30 |
|---|---|---|---|---|---|---|---|---|---|---|
| 0 | 0.2431 | 0.2552 | 0.2084 | 0.1927 | 0.1780 | 0.1642 | 0.1513 | 0.1393 | 0.1281 | 0.1176 |
| 1 | 0.3877 | 0.3811 | 0.3735 | 0.3651 | 0.3560 | 0.3462 | 0.3358 | 0.3251 | 0.3139 | 0.3025 |
| 2 | 0.2577 | 0.2687 | 0.2789 | 0.2882 | 0.2966 | 0.3041 | 0.3105 | 0.3160 | 0.3206 | 0.3241 |
| 3 | 0.0913 | 0.1011 | 0.1111 | 0.1214 | 0.1318 | 0.1424 | 0.1531 | 0.1639 | 0.1746 | 0.1852 |
| 4 | 0.0182 | 0.0214 | 0.0249 | 0.0287 | 0.0330 | 0.0375 | 0.0425 | 0.0478 | 0.0535 | 0.0595 |
| 5 | 0.0019 | 0.0024 | 0.0030 | 0.0036 | 0.0044 | 0.0053 | 0.0063 | 0.0074 | 0.0087 | 0.0102 |
| 6 | 0.0001 | 0.0001 | 0.0001 | 0.0002 | 0.0002 | 0.0003 | 0.0004 | 0.0005 | 0.0006 | 0.0007 |

| | 0.31 | 0.32 | 0.33 | 0.34 | 0.35 | 0.36 | 0.37 | 0.38 | 0.39 | 0.40 |
|---|---|---|---|---|---|---|---|---|---|---|
| 0 | 0.1079 | 0.0989 | 0.0905 | 0.0827 | 0.0754 | 0.0687 | 0.0625 | 0.0568 | 0.0515 | 0.0467 |
| 1 | 0.2909 | 0.2792 | 0.2673 | 0.2555 | 0.2437 | 0.2319 | 0.2203 | 0.2089 | 0.1976 | 0.1866 |
| 2 | 0.3267 | 0.3284 | 0.3292 | 0.3290 | 0.3280 | 0.3261 | 0.3235 | 0.3201 | 0.3159 | 0.3110 |
| 3 | 0.1957 | 0.2061 | 0.2162 | 0.2260 | 0.2355 | 0.2446 | 0.2533 | 0.2616 | 0.2693 | 0.2765 |
| 4 | 0.0660 | 0.0727 | 0.0799 | 0.0873 | 0.0951 | 0.1032 | 0.1116 | 0.1202 | 0.1291 | 0.1382 |
| 5 | 0.0119 | 0.0137 | 0.0157 | 0.0180 | 0.0205 | 0.0232 | 0.0262 | 0.0295 | 0.0330 | 0.0369 |
| 6 | 0.0009 | 0.0011 | 0.0013 | 0.0015 | 0.0018 | 0.0022 | 0.0026 | 0.0030 | 0.0035 | 0.0041 |

| | 0.41 | 0.42 | 0.43 | 0.44 | 0.45 | 0.46 | 0.47 | 0.48 | 0.49 | 0.50 |
|---|---|---|---|---|---|---|---|---|---|---|
| 0 | 0.0422 | 0.0381 | 0.0343 | 0.0308 | 0.0277 | 0.0248 | 0.0222 | 0.0198 | 0.0176 | 0.0156 |
| 1 | 0.1759 | 0.1654 | 0.1552 | 0.1454 | 0.1359 | 0.1267 | 0.1179 | 0.1095 | 0.1014 | 0.0938 |
| 2 | 0.3055 | 0.2994 | 0.2928 | 0.2856 | 0.2780 | 0.2699 | 0.2615 | 0.2527 | 0.2436 | 0.2344 |
| 3 | 0.2831 | 0.2891 | 0.2945 | 0.2992 | 0.3032 | 0.3065 | 0.3091 | 0.3110 | 0.3121 | 0.3125 |
| 4 | 0.1475 | 0.1570 | 0.1666 | 0.1763 | 0.1861 | 0.1958 | 0.2056 | 0.2153 | 0.2249 | 0.2344 |
| 5 | 0.0410 | 0.0455 | 0.0503 | 0.0554 | 0.0609 | 0.0667 | 0.0729 | 0.0795 | 0.0864 | 0.0938 |
| 6 | 0.0048 | 0.0055 | 0.0063 | 0.0073 | 0.0083 | 0.0095 | 0.0108 | 0.0122 | 0.0138 | 0.0156 |

$n = 7$

| $p$ / $t$ | 0.01 | 0.02 | 0.03 | 0.04 | 0.05 | 0.06 | 0.07 | 0.08 | 0.09 | 0.10 |
|---|---|---|---|---|---|---|---|---|---|---|
| 0 | 0.9321 | 0.8681 | 0.8080 | 0.7514 | 0.6983 | 0.6485 | 0.6017 | 0.5578 | 0.5168 | 0.4783 |
| 1 | 0.0659 | 0.1240 | 0.1749 | 0.2192 | 0.2573 | 0.2897 | 0.3170 | 0.3396 | 0.3578 | 0.3720 |
| 2 | 0.0020 | 0.0076 | 0.0162 | 0.0274 | 0.0406 | 0.0555 | 0.0716 | 0.0886 | 0.1061 | 0.1240 |
| 3 | 0.0000 | 0.0003 | 0.0008 | 0.0019 | 0.0036 | 0.0059 | 0.0090 | 0.0128 | 0.0175 | 0.0230 |
| 4 | 0.0000 | 0.0000 | 0.0000 | 0.0001 | 0.0002 | 0.0004 | 0.0007 | 0.0011 | 0.0017 | 0.0026 |
| 5 | 0.0000 | 0.0000 | 0.0000 | 0.0000 | 0.0000 | 0.0000 | 0.0000 | 0.0001 | 0.0001 | 0.0002 |

(*Continued*)

**Table A.7** Binomial probability distribution

| | 0.11 | 0.12 | 0.13 | 0.14 | 0.15 | 0.16 | 0.17 | 0.18 | 0.19 | 0.20 |
|---|---|---|---|---|---|---|---|---|---|---|
| 0 | 0.4423 | 0.4087 | 0.3773 | 0.3479 | 0.3206 | 0.2951 | 0.2714 | 0.2493 | 0.2288 | 0.2097 |
| 1 | 0.3827 | 0.3901 | 0.3946 | 0.3965 | 0.3960 | 0.3935 | 0.3891 | 0.3830 | 0.3756 | 0.3670 |
| 2 | 0.1419 | 0.1596 | 0.1769 | 0.1936 | 0.2097 | 0.2248 | 0.2391 | 0.2523 | 0.2643 | 0.2753 |
| 3 | 0.0292 | 0.0363 | 0.0441 | 0.0525 | 0.0617 | 0.0714 | 0.0816 | 0.0923 | 0.1033 | 0.1147 |
| 4 | 0.0036 | 0.0049 | 0.0066 | 0.0086 | 0.0109 | 0.0136 | 0.0167 | 0.0203 | 0.0242 | 0.0287 |
| 5 | 0.0003 | 0.0004 | 0.0006 | 0.0008 | 0.0012 | 0.0016 | 0.0021 | 0.0027 | 0.0034 | 0.0043 |
| 6 | 0.0000 | 0.0000 | 0.0000 | 0.0000 | 0.0001 | 0.0001 | 0.0001 | 0.0002 | 0.0003 | 0.0004 |

| | 0.21 | 0.22 | 0.23 | 0.24 | 0.25 | 0.26 | 0.27 | 0.28 | 0.29 | 0.30 |
|---|---|---|---|---|---|---|---|---|---|---|
| 0 | 0.1920 | 0.1757 | 0.1605 | 0.1465 | 0.1335 | 0.1215 | 0.1105 | 0.1003 | 0.0910 | 0.0824 |
| 1 | 0.3573 | 0.3468 | 0.3356 | 0.3237 | 0.3115 | 0.2989 | 0.2860 | 0.2731 | 0.2600 | 0.2471 |
| 2 | 0.2850 | 0.2935 | 0.3007 | 0.3067 | 0.3115 | 0.3150 | 0.3174 | 0.3186 | 0.3186 | 0.3177 |
| 3 | 0.1263 | 0.1379 | 0.1497 | 0.1614 | 0.1730 | 0.1845 | 0.1956 | 0.2065 | 0.2169 | 0.2269 |
| 4 | 0.0336 | 0.0389 | 0.0447 | 0.0510 | 0.0577 | 0.0648 | 0.0724 | 0.0803 | 0.0886 | 0.0972 |
| 5 | 0.0054 | 0.0066 | 0.0080 | 0.0097 | 0.0115 | 0.0137 | 0.0161 | 0.0187 | 0.0217 | 0.0250 |
| 6 | 0.0005 | 0.0006 | 0.0008 | 0.0010 | 0.0013 | 0.0016 | 0.0020 | 0.0024 | 0.0030 | 0.0036 |
| 7 | 0.0000 | 0.0000 | 0.0000 | 0.0000 | 0.0001 | 0.0001 | 0.0001 | 0.0001 | 0.0002 | 0.0002 |

| | 0.31 | 0.32 | 0.33 | 0.34 | 0.35 | 0.36 | 0.37 | 0.38 | 0.39 | 0.40 |
|---|---|---|---|---|---|---|---|---|---|---|
| 0 | 0.0745 | 0.0672 | 0.0606 | 0.0546 | 0.0490 | 0.0440 | 0.0394 | 0.0352 | 0.0314 | 0.0280 |
| 1 | 0.2342 | 0.2215 | 0.2090 | 0.1967 | 0.1848 | 0.1732 | 0.1619 | 0.1511 | 0.1407 | 0.1306 |
| 2 | 0.3156 | 0.3127 | 0.3088 | 0.3040 | 0.2985 | 0.2922 | 0.2853 | 0.2778 | 0.2698 | 0.2613 |
| 3 | 0.2363 | 0.2452 | 0.2535 | 0.2610 | 0.2679 | 0.2740 | 0.2793 | 0.2838 | 0.2875 | 0.2903 |
| 4 | 0.1062 | 0.1154 | 0.1248 | 0.1345 | 0.1442 | 0.1541 | 0.1640 | 0.1739 | 0.1838 | 0.1935 |
| 5 | 0.0286 | 0.0326 | 0.0369 | 0.0416 | 0.0466 | 0.0520 | 0.0578 | 0.0640 | 0.0705 | 0.0774 |
| 6 | 0.0043 | 0.0051 | 0.0061 | 0.0071 | 0.0084 | 0.0098 | 0.0113 | 0.0131 | 0.0150 | 0.0172 |
| 7 | 0.0003 | 0.0003 | 0.0004 | 0.0005 | 0.0006 | 0.0008 | 0.0009 | 0.0011 | 0.0014 | 0.0016 |

| | 0.41 | 0.42 | 0.43 | 0.44 | 0.45 | 0.46 | 0.47 | 0.48 | 0.49 | 0.50 |
|---|---|---|---|---|---|---|---|---|---|---|
| 0 | 0.0249 | 0.0221 | 0.0195 | 0.0173 | 0.0152 | 0.0134 | 0.0117 | 0.0103 | 0.0090 | 0.0078 |
| 1 | 0.1211 | 0.1119 | 0.1032 | 0.0950 | 0.0872 | 0.0798 | 0.0729 | 0.0664 | 0.0604 | 0.0547 |
| 2 | 0.2524 | 0.2431 | 0.2336 | 0.2239 | 0.2140 | 0.2040 | 0.1940 | 0.1840 | 0.1740 | 0.1641 |
| 3 | 0.2923 | 0.2934 | 0.2937 | 0.2932 | 0.2918 | 0.2897 | 0.2867 | 0.2830 | 0.2786 | 0.2734 |
| 4 | 0.2031 | 0.2125 | 0.2216 | 0.2304 | 0.2388 | 0.2468 | 0.2543 | 0.2612 | 0.2676 | 0.2734 |
| 5 | 0.0847 | 0.0923 | 0.1003 | 0.1086 | 0.1172 | 0.1261 | 0.1353 | 0.1447 | 0.1543 | 0.1641 |
| 6 | 0.0196 | 0.0223 | 0.0252 | 0.0284 | 0.0320 | 0.0358 | 0.0400 | 0.0445 | 0.0494 | 0.0547 |
| 7 | 0.0019 | 0.0023 | 0.0027 | 0.0032 | 0.0037 | 0.0044 | 0.0051 | 0.0059 | 0.0068 | 0.0078 |

$n = 8$

| $t$ $\backslash$ $p$ | 0.01 | 0.02 | 0.03 | 0.04 | 0.05 | 0.06 | 0.07 | 0.08 | 0.09 | 0.10 |
|---|---|---|---|---|---|---|---|---|---|---|
| 0 | 0.9227 | 0.8508 | 0.7837 | 0.7214 | 0.6634 | 0.6096 | 0.5596 | 0.5132 | 0.4703 | 0.4305 |
| 1 | 0.0746 | 0.1389 | 0.1939 | 0.2405 | 0.2793 | 0.3113 | 0.3370 | 0.3570 | 0.3721 | 0.3826 |
| 2 | 0.0026 | 0.0099 | 0.0210 | 0.0351 | 0.0515 | 0.0695 | 0.0888 | 0.1087 | 0.1288 | 0.1488 |

(*Continued*)

**Table A.7** Binomial probability distribution

| 3 | 0.0001 | 0.0004 | 0.0013 | 0.0029 | 0.0054 | 0.0089 | 0.0134 | 0.0189 | 0.0255 | 0.0331 |
| 4 | 0.0000 | 0.0000 | 0.0001 | 0.0002 | 0.0004 | 0.0007 | 0.0013 | 0.0021 | 0.0031 | 0.0046 |
| 5 | 0.0000 | 0.0000 | 0.0000 | 0.0000 | 0.0000 | 0.0000 | 0.0001 | 0.0001 | 0.0002 | 0.0004 |

| | 0.11 | 0.12 | 0.13 | 0.14 | 0.15 | 0.16 | 0.17 | 0.18 | 0.19 | 0.20 |
|---|---|---|---|---|---|---|---|---|---|---|
| 0 | 0.3937 | 0.3596 | 0.3282 | 0.2992 | 0.2725 | 0.2479 | 0.2252 | 0.2044 | 0.1853 | 0.1678 |
| 1 | 0.3892 | 0.3923 | 0.3923 | 0.3897 | 0.3847 | 0.3777 | 0.3691 | 0.3590 | 0.3477 | 0.3355 |
| 2 | 0.1684 | 0.1872 | 0.2052 | 0.2220 | 0.2376 | 0.2518 | 0.2646 | 0.2758 | 0.2855 | 0.2936 |
| 3 | 0.0416 | 0.0511 | 0.0613 | 0.0723 | 0.0839 | 0.0959 | 0.1084 | 0.1211 | 0.1339 | 0.1468 |
| 4 | 0.0064 | 0.0087 | 0.0115 | 0.0147 | 0.0185 | 0.0228 | 0.0277 | 0.0332 | 0.0393 | 0.0459 |
| 5 | 0.0006 | 0.0009 | 0.0014 | 0.0019 | 0.0026 | 0.0035 | 0.0045 | 0.0058 | 0.0074 | 0.0092 |
| 6 | 0.0000 | 0.0001 | 0.0001 | 0.0002 | 0.0002 | 0.0003 | 0.0005 | 0.0006 | 0.0009 | 0.0011 |
| 7 | 0.0000 | 0.0000 | 0.0000 | 0.0000 | 0.0000 | 0.0000 | 0.0000 | 0.0000 | 0.0001 | 0.0001 |

| | 0.21 | 0.22 | 0.23 | 0.24 | 0.25 | 0.26 | 0.27 | 0.28 | 0.29 | 0.30 |
|---|---|---|---|---|---|---|---|---|---|---|
| 0 | 0.1517 | 0.1370 | 0.1236 | 0.1113 | 0.1001 | 0.0899 | 0.0806 | 0.0722 | 0.0646 | 0.0576 |
| 1 | 0.3226 | 0.3092 | 0.2953 | 0.2812 | 0.2670 | 0.2527 | 0.2386 | 0.2247 | 0.2110 | 0.1977 |
| 2 | 0.3002 | 0.3052 | 0.3087 | 0.3108 | 0.3115 | 0.3108 | 0.3089 | 0.3058 | 0.3017 | 0.2965 |
| 3 | 0.1596 | 0.1722 | 0.1844 | 0.1963 | 0.2076 | 0.2184 | 0.2285 | 0.2379 | 0.2464 | 0.2541 |
| 4 | 0.0530 | 0.0607 | 0.0689 | 0.0775 | 0.0865 | 0.0959 | 0.1056 | 0.1156 | 0.1258 | 0.1361 |
| 5 | 0.0113 | 0.0137 | 0.0165 | 0.0196 | 0.0231 | 0.0270 | 0.0313 | 0.0360 | 0.0411 | 0.0467 |
| 6 | 0.0015 | 0.0019 | 0.0025 | 0.0031 | 0.0038 | 0.0047 | 0.0058 | 0.0070 | 0.0084 | 0.0100 |
| 7 | 0.0001 | 0.0002 | 0.0002 | 0.0003 | 0.0004 | 0.0005 | 0.0006 | 0.0008 | 0.0010 | 0.0012 |
| 8 | 0.0000 | 0.0000 | 0.0000 | 0.0000 | 0.0000 | 0.0000 | 0.0000 | 0.0000 | 0.0001 | 0.0001 |

| | 0.31 | 0.32 | 0.33 | 0.34 | 0.35 | 0.36 | 0.37 | 0.38 | 0.39 | 0.40 |
|---|---|---|---|---|---|---|---|---|---|---|
| 0 | 0.0514 | 0.0457 | 0.0406 | 0.0360 | 0.0319 | 0.0281 | 0.0248 | 0.0218 | 0.0192 | 0.0168 |
| 1 | 0.1847 | 0.1721 | 0.1600 | 0.1484 | 0.1373 | 0.1267 | 0.1166 | 0.1071 | 0.0981 | 0.0896 |
| 2 | 0.2904 | 0.2835 | 0.2758 | 0.2675 | 0.2587 | 0.2494 | 0.2397 | 0.2297 | 0.2194 | 0.2090 |
| 3 | 0.2609 | 0.2668 | 0.2717 | 0.2756 | 0.2786 | 0.2805 | 0.2815 | 0.2815 | 0.2806 | 0.2787 |
| 4 | 0.1465 | 0.1569 | 0.1673 | 0.1775 | 0.1875 | 0.1973 | 0.2067 | 0.2157 | 0.2242 | 0.2322 |
| 5 | 0.0527 | 0.0591 | 0.0659 | 0.0732 | 0.0808 | 0.0888 | 0.0971 | 0.1058 | 0.1147 | 0.1239 |
| 6 | 0.0118 | 0.0139 | 0.0162 | 0.0188 | 0.0217 | 0.0250 | 0.0285 | 0.0324 | 0.0367 | 0.0413 |
| 7 | 0.0015 | 0.0019 | 0.0023 | 0.0028 | 0.0033 | 0.0040 | 0.0048 | 0.0057 | 0.0067 | 0.0079 |
| 8 | 0.0001 | 0.0001 | 0.0001 | 0.0002 | 0.0002 | 0.0003 | 0.0004 | 0.0004 | 0.0005 | 0.0007 |

| | 0.41 | 0.42 | 0.43 | 0.44 | 0.45 | 0.46 | 0.47 | 0.48 | 0.49 | 0.50 |
|---|---|---|---|---|---|---|---|---|---|---|
| 0 | 0.0147 | 0.0128 | 0.0111 | 0.0097 | 0.0084 | 0.0072 | 0.0062 | 0.0053 | 0.0046 | 0.0039 |
| 1 | 0.0816 | 0.0742 | 0.0672 | 0.0608 | 0.0548 | 0.0493 | 0.0442 | 0.0395 | 0.0352 | 0.0312 |
| 2 | 0.1985 | 0.1880 | 0.1776 | 0.1672 | 0.1569 | 0.1469 | 0.1371 | 0.1275 | 0.1183 | 0.1094 |
| 3 | 0.2759 | 0.2723 | 0.2679 | 0.2627 | 0.2568 | 0.2503 | 0.2431 | 0.2355 | 0.2273 | 0.2188 |
| 4 | 0.2397 | 0.2465 | 0.2526 | 0.2580 | 0.2627 | 0.2665 | 0.2695 | 0.2717 | 0.2730 | 0.2734 |
| 5 | 0.1332 | 0.1428 | 0.1525 | 0.1622 | 0.1719 | 0.1816 | 0.1912 | 0.2006 | 0.2098 | 0.2188 |

(*Continued*)

**Table A.7** Binomial probability distribution

| 6 | 0.0463 | 0.0517 | 0.0575 | 0.0637 | 0.0703 | 0.0774 | 0.0848 | 0.0926 | 0.1008 | 0.1094 |
| 7 | 0.0092 | 0.0107 | 0.0124 | 0.0143 | 0.0164 | 0.0188 | 0.0215 | 0.0244 | 0.0277 | 0.0312 |
| 8 | 0.0008 | 0.0010 | 0.0012 | 0.0014 | 0.0017 | 0.0020 | 0.0024 | 0.0028 | 0.0033 | 0.0039 |

$n = 9$

| $t$ \ $p$ | 0.01 | 0.02 | 0.03 | 0.04 | 0.05 | 0.06 | 0.07 | 0.08 | 0.09 | 0.10 |
|---|---|---|---|---|---|---|---|---|---|---|
| 0 | 0.9135 | 0.8337 | 0.7602 | 0.6925 | 0.6302 | 0.5730 | 0.5204 | 0.4722 | 0.4279 | 0.3874 |
| 1 | 0.0830 | 0.1531 | 0.2116 | 0.2597 | 0.2985 | 0.3292 | 0.3525 | 0.3695 | 0.3809 | 0.3874 |
| 2 | 0.0034 | 0.0125 | 0.0262 | 0.0433 | 0.0629 | 0.0840 | 0.1061 | 0.1285 | 0.1507 | 0.1722 |
| 3 | 0.0001 | 0.0006 | 0.0019 | 0.0042 | 0.0077 | 0.0125 | 0.0186 | 0.0261 | 0.0348 | 0.0446 |
| 4 | 0.0000 | 0.0000 | 0.0001 | 0.0003 | 0.0006 | 0.0012 | 0.0021 | 0.0034 | 0.0052 | 0.0074 |
| 5 | 0.0000 | 0.0000 | 0.0000 | 0.0000 | 0.0000 | 0.0001 | 0.0002 | 0.0003 | 0.0005 | 0.0008 |
| 6 | 0.0000 | 0.0000 | 0.0000 | 0.0000 | 0.0000 | 0.0000 | 0.0000 | 0.0000 | 0.0000 | 0.0001 |

| | 0.11 | 0.12 | 0.13 | 0.14 | 0.15 | 0.16 | 0.17 | 0.18 | 0.19 | 0.20 |
|---|---|---|---|---|---|---|---|---|---|---|
| 0 | 0.3504 | 0.3165 | 0.2855 | 0.2573 | 0.2316 | 0.2082 | 0.1869 | 0.1676 | 0.1501 | 0.1342 |
| 1 | 0.3897 | 0.3884 | 0.3840 | 0.3770 | 0.3679 | 0.3569 | 0.3446 | 0.3312 | 0.3169 | 0.3020 |
| 2 | 0.1927 | 0.2119 | 0.2295 | 0.2455 | 0.2597 | 0.2720 | 0.2823 | 0.2908 | 0.2973 | 0.3020 |
| 3 | 0.0556 | 0.0674 | 0.0800 | 0.0933 | 0.1069 | 0.1209 | 0.1349 | 0.1489 | 0.1627 | 0.1762 |
| 4 | 0.0103 | 0.0138 | 0.0179 | 0.0228 | 0.0283 | 0.0345 | 0.0415 | 0.0490 | 0.0573 | 0.0661 |
| 5 | 0.0013 | 0.0019 | 0.0027 | 0.0037 | 0.0050 | 0.0066 | 0.0085 | 0.0108 | 0.0134 | 0.0165 |
| 6 | 0.0001 | 0.0002 | 0.0003 | 0.0004 | 0.0006 | 0.0008 | 0.0012 | 0.0016 | 0.0021 | 0.0028 |
| 7 | 0.0000 | 0.0000 | 0.0000 | 0.0000 | 0.0000 | 0.0001 | 0.0001 | 0.0001 | 0.0002 | 0.0003 |

| | 0.21 | 0.22 | 0.23 | 0.24 | 0.25 | 0.26 | 0.27 | 0.28 | 0.29 | 0.30 |
|---|---|---|---|---|---|---|---|---|---|---|
| 0 | 0.1199 | 0.1069 | 0.0952 | 0.0846 | 0.0751 | 0.0665 | 0.0589 | 0.0520 | 0.0458 | 0.0404 |
| 1 | 0.2867 | 0.2713 | 0.2558 | 0.2404 | 0.2253 | 0.2104 | 0.1960 | 0.1820 | 0.1685 | 0.1556 |
| 2 | 0.3049 | 0.3061 | 0.3056 | 0.3037 | 0.3003 | 0.2957 | 0.2899 | 0.2831 | 0.2754 | 0.2668 |
| 3 | 0.1891 | 0.2014 | 0.2130 | 0.2238 | 0.2336 | 0.2424 | 0.2502 | 0.2569 | 0.2624 | 0.2668 |
| 4 | 0.0754 | 0.0852 | 0.0954 | 0.1060 | 0.1168 | 0.1278 | 0.1388 | 0.1499 | 0.1608 | 0.1715 |
| 5 | 0.0200 | 0.0240 | 0.0285 | 0.0335 | 0.0389 | 0.0449 | 0.0513 | 0.0583 | 0.0657 | 0.0735 |
| 6 | 0.0036 | 0.0045 | 0.0057 | 0.0070 | 0.0087 | 0.0105 | 0.0127 | 0.0151 | 0.0179 | 0.0210 |
| 7 | 0.0004 | 0.0005 | 0.0007 | 0.0010 | 0.0012 | 0.0016 | 0.0020 | 0.0025 | 0.0031 | 0.0039 |
| 8 | 0.0000 | 0.0000 | 0.0001 | 0.0001 | 0.0001 | 0.0001 | 0.0002 | 0.0002 | 0.0003 | 0.0004 |

| | 0.31 | 0.32 | 0.33 | 0.34 | 0.35 | 0.36 | 0.37 | 0.38 | 0.39 | 0.40 |
|---|---|---|---|---|---|---|---|---|---|---|
| 0 | 0.0355 | 0.0311 | 0.0272 | 0.0238 | 0.0207 | 0.0180 | 0.0156 | 0.0135 | 0.0117 | 0.0101 |
| 1 | 0.1433 | 0.1317 | 0.1206 | 0.1102 | 0.1004 | 0.0912 | 0.0826 | 0.0747 | 0.0673 | 0.0605 |
| 2 | 0.2576 | 0.2478 | 0.2376 | 0.2270 | 0.2162 | 0.2052 | 0.1941 | 0.1831 | 0.1721 | 0.1612 |
| 3 | 0.2701 | 0.2721 | 0.2731 | 0.2729 | 0.2716 | 0.2693 | 0.2660 | 0.2618 | 0.2567 | 0.2508 |
| 4 | 0.1820 | 0.1921 | 0.2017 | 0.2109 | 0.2194 | 0.2272 | 0.2344 | 0.2407 | 0.2462 | 0.2508 |
| 5 | 0.0818 | 0.0904 | 0.0994 | 0.1086 | 0.1181 | 0.1278 | 0.1376 | 0.1475 | 0.1574 | 0.1672 |
| 6 | 0.0245 | 0.0284 | 0.0326 | 0.0373 | 0.0424 | 0.0479 | 0.0539 | 0.0603 | 0.0671 | 0.0743 |

(*Continued*)

**Table A.7** Binomial probability distribution

| | | | | | | | | | |
|---|---|---|---|---|---|---|---|---|---|
| 7 | 0.0047 | 0.0057 | 0.0069 | 0.0082 | 0.0098 | 0.0116 | 0.0136 | 0.0158 | 0.0184 | 0.0212 |
| 8 | 0.0005 | 0.0007 | 0.0008 | 0.0011 | 0.0013 | 0.0016 | 0.0020 | 0.0024 | 0.0029 | 0.0035 |
| 9 | 0.0000 | 0.0000 | 0.0000 | 0.0001 | 0.0001 | 0.0001 | 0.0001 | 0.0002 | 0.0002 | 0.0003 |

| | 0.41 | 0.42 | 0.43 | 0.44 | 0.45 | 0.46 | 0.47 | 0.48 | 0.49 | 0.50 |
|---|---|---|---|---|---|---|---|---|---|---|
| 0 | 0.0087 | 0.0074 | 0.0064 | 0.0054 | 0.0046 | 0.0039 | 0.0033 | 0.0028 | 0.0023 | 0.0020 |
| 1 | 0.0542 | 0.0484 | 0.0431 | 0.0383 | 0.0339 | 0.0299 | 0.0263 | 0.0231 | 0.0202 | 0.0176 |
| 2 | 0.1506 | 0.1402 | 0.1301 | 0.1204 | 0.1110 | 0.1020 | 0.0934 | 0.0853 | 0.0776 | 0.0703 |
| 3 | 0.2442 | 0.2369 | 0.2291 | 0.2207 | 0.2119 | 0.2027 | 0.1933 | 0.1837 | 0.1739 | 0.1641 |
| 4 | 0.2545 | 0.2573 | 0.2592 | 0.2601 | 0.2600 | 0.2590 | 0.2571 | 0.2543 | 0.2506 | 0.2461 |
| 5 | 0.1769 | 0.1863 | 0.1955 | 0.2044 | 0.2128 | 0.2207 | 0.2280 | 0.2347 | 0.2408 | 0.2461 |
| 6 | 0.0819 | 0.0900 | 0.0983 | 0.1070 | 0.1160 | 0.1253 | 0.1348 | 0.1445 | 0.1542 | 0.1641 |
| 7 | 0.0244 | 0.0279 | 0.0318 | 0.0360 | 0.0407 | 0.0458 | 0.0512 | 0.0571 | 0.0635 | 0.0703 |
| 8 | 0.0042 | 0.0051 | 0.0060 | 0.0071 | 0.0083 | 0.0097 | 0.0114 | 0.0132 | 0.0153 | 0.0176 |
| 9 | 0.0003 | 0.0004 | 0.0005 | 0.0006 | 0.0008 | 0.0009 | 0.0011 | 0.0014 | 0.0016 | 0.0020 |

$n = 10$

| $p$ / $t$ | 0.01 | 0.02 | 0.03 | 0.04 | 0.05 | 0.06 | 0.07 | 0.08 | 0.09 | 0.10 |
|---|---|---|---|---|---|---|---|---|---|---|
| 0 | 0.9044 | 0.8171 | 0.7374 | 0.6648 | 0.5987 | 0.5386 | 0.4840 | 0.4344 | 0.3894 | 0.3487 |
| 1 | 0.0914 | 0.1667 | 0.2281 | 0.2770 | 0.3151 | 0.3438 | 0.3643 | 0.3777 | 0.3851 | 0.3874 |
| 2 | 0.0042 | 0.0153 | 0.0317 | 0.0519 | 0.0746 | 0.0988 | 0.1234 | 0.1478 | 0.1714 | 0.1937 |
| 3 | 0.0001 | 0.0008 | 0.0026 | 0.0058 | 0.0105 | 0.0168 | 0.0248 | 0.0343 | 0.0452 | 0.0574 |
| 4 | 0.0000 | 0.0000 | 0.0001 | 0.0004 | 0.0010 | 0.0019 | 0.0033 | 0.0052 | 0.0078 | 0.0112 |
| 5 | 0.0000 | 0.0000 | 0.0000 | 0.0000 | 0.0001 | 0.0001 | 0.0003 | 0.0005 | 0.0009 | 0.0015 |
| 6 | 0.0000 | 0.0000 | 0.0000 | 0.0000 | 0.0000 | 0.0000 | 0.0000 | 0.0000 | 0.0001 | 0.0001 |

| | 0.11 | 0.12 | 0.13 | 0.14 | 0.15 | 0.16 | 0.17 | 0.18 | 0.19 | 0.20 |
|---|---|---|---|---|---|---|---|---|---|---|
| 0 | 0.3118 | 0.2785 | 0.2484 | 0.2213 | 0.1969 | 0.1749 | 0.1552 | 0.1374 | 0.1216 | 0.1074 |
| 1 | 0.3854 | 0.3798 | 0.3712 | 0.3603 | 0.3474 | 0.3331 | 0.3178 | 0.3017 | 0.2852 | 0.2684 |
| 2 | 0.2143 | 0.2330 | 0.2496 | 0.2639 | 0.2759 | 0.2856 | 0.2929 | 0.2980 | 0.3010 | 0.3020 |
| 3 | 0.0706 | 0.0847 | 0.0995 | 0.1146 | 0.1298 | 0.1450 | 0.1600 | 0.1745 | 0.1883 | 0.2013 |
| 4 | 0.0153 | 0.0202 | 0.0260 | 0.0326 | 0.0401 | 0.0483 | 0.0573 | 0.0670 | 0.0773 | 0.0881 |
| 5 | 0.0023 | 0.0033 | 0.0047 | 0.0064 | 0.0085 | 0.0111 | 0.0141 | 0.0177 | 0.0218 | 0.0264 |
| 6 | 0.0002 | 0.0004 | 0.0006 | 0.0009 | 0.0012 | 0.0018 | 0.0024 | 0.0032 | 0.0043 | 0.0055 |
| 7 | 0.0000 | 0.0000 | 0.0000 | 0.0001 | 0.0001 | 0.0002 | 0.0003 | 0.0004 | 0.0006 | 0.0008 |
| 8 | 0.0000 | 0.0000 | 0.0000 | 0.0000 | 0.0000 | 0.0000 | 0.0000 | 0.0000 | 0.0001 | 0.0001 |

| | 0.21 | 0.22 | 0.23 | 0.24 | 0.25 | 0.26 | 0.27 | 0.28 | 0.29 | 0.30 |
|---|---|---|---|---|---|---|---|---|---|---|
| 0 | 0.0947 | 0.0834 | 0.0733 | 0.0643 | 0.0563 | 0.0492 | 0.0430 | 0.0374 | 0.0326 | 0.0282 |
| 1 | 0.2517 | 0.2351 | 0.2188 | 0.2030 | 0.1877 | 0.1730 | 0.1590 | 0.1456 | 0.1330 | 0.1211 |
| 2 | 0.3011 | 0.2984 | 0.2942 | 0.2885 | 0.2816 | 0.2735 | 0.2646 | 0.2548 | 0.2444 | 0.2335 |
| 3 | 0.2134 | 0.2244 | 0.2343 | 0.2429 | 0.2503 | 0.2563 | 0.2609 | 0.2642 | 0.2662 | 0.2668 |
| 4 | 0.0993 | 0.1108 | 0.1225 | 0.1343 | 0.1460 | 0.1576 | 0.1689 | 0.1798 | 0.1903 | 0.2001 |
| 5 | 0.0317 | 0.0375 | 0.0439 | 0.0509 | 0.0584 | 0.0664 | 0.0750 | 0.0839 | 0.0933 | 0.1029 |
| 6 | 0.0070 | 0.0088 | 0.0109 | 0.0134 | 0.0162 | 0.0195 | 0.0231 | 0.0272 | 0.0317 | 0.0368 |

(*Continued*)

**Table A.7** Binomial probability distribution

| | | | | | | | | | | |
|---|---|---|---|---|---|---|---|---|---|---|
| 7 | 0.0011 | 0.0014 | 0.0019 | 0.0024 | 0.0031 | 0.0039 | 0.0049 | 0.0060 | 0.0074 | 0.0090 |
| 8 | 0.0001 | 0.0002 | 0.0002 | 0.0003 | 0.0004 | 0.0005 | 0.0007 | 0.0009 | 0.0011 | 0.0014 |
| 9 | 0.0000 | 0.0000 | 0.0000 | 0.0000 | 0.0000 | 0.0000 | 0.0001 | 0.0001 | 0.0001 | 0.0001 |

| | 0.31 | 0.32 | 0.33 | 0.34 | 0.35 | 0.36 | 0.37 | 0.38 | 0.39 | 0.40 |
|---|---|---|---|---|---|---|---|---|---|---|
| 0 | 0.0245 | 0.0211 | 0.0182 | 0.0157 | 0.0135 | 0.0115 | 0.0098 | 0.0084 | 0.0071 | 0.0060 |
| 1 | 0.1099 | 0.0995 | 0.0898 | 0.0808 | 0.0725 | 0.0649 | 0.0578 | 0.0514 | 0.0456 | 0.0403 |
| 2 | 0.2222 | 0.2107 | 0.1990 | 0.0873 | 0.1757 | 0.1642 | 0.1529 | 0.1419 | 0.1312 | 0.1209 |
| 3 | 0.2662 | 0.2644 | 0.2614 | 0.2573 | 0.2522 | 0.2462 | 0.2394 | 0.2319 | 0.2237 | 0.2150 |
| 4 | 0.2093 | 0.2177 | 0.2253 | 0.2320 | 0.2377 | 0.2424 | 0.2461 | 0.2487 | 0.2503 | 0.2508 |
| 5 | 0.1128 | 0.1229 | 0.1332 | 0.1434 | 0.1536 | 0.1636 | 0.1734 | 0.1829 | 0.1920 | 0.2007 |
| 6 | 0.0422 | 0.0482 | 0.0547 | 0.0616 | 0.0689 | 0.0767 | 0.0849 | 0.0934 | 0.1023 | 0.1115 |
| 7 | 0.0108 | 0.0130 | 0.0154 | 0.0181 | 0.0212 | 0.0247 | 0.0285 | 0.0327 | 0.0374 | 0.0425 |
| 8 | 0.0018 | 0.0023 | 0.0028 | 0.0035 | 0.0043 | 0.0052 | 0.0063 | 0.0075 | 0.0090 | 0.0106 |
| 9 | 0.0002 | 0.0002 | 0.0003 | 0.0004 | 0.0005 | 0.0006 | 0.0008 | 0.0010 | 0.0013 | 0.0016 |
| 10 | 0.0000 | 0.0000 | 0.0000 | 0.0000 | 0.0000 | 0.0000 | 0.0000 | 0.0001 | 0.0001 | 0.0001 |

| | 0.41 | 0.42 | 0.43 | 0.44 | 0.45 | 0.46 | 0.47 | 0.48 | 0.49 | 0.50 |
|---|---|---|---|---|---|---|---|---|---|---|
| 0 | 0.0051 | 0.0043 | 0.0036 | 0.0030 | 0.0025 | 0.0021 | 0.0017 | 0.0014 | 0.0012 | 0.0010 |
| 1 | 0.0355 | 0.0312 | 0.0273 | 0.0238 | 0.0207 | 0.0180 | 0.0155 | 0.0133 | 0.0114 | 0.0098 |
| 2 | 0.1111 | 0.1017 | 0.0927 | 0.0843 | 0.0763 | 0.0688 | 0.0619 | 0.0554 | 0.0494 | 0.0439 |
| 3 | 0.2058 | 0.1963 | 0.1865 | 0.1765 | 0.1665 | 0.1564 | 0.1464 | 0.1364 | 0.1267 | 0.1172 |
| 4 | 0.2503 | 0.2488 | 0.2462 | 0.2427 | 0.2384 | 0.2331 | 0.2271 | 0.2204 | 0.2130 | 0.2051 |
| 5 | 0.2087 | 0.2162 | 0.2229 | 0.2289 | 0.2340 | 0.2383 | 0.2417 | 0.2441 | 0.2456 | 0.2461 |
| 6 | 0.1209 | 0.1304 | 0.1401 | 0.1499 | 0.1596 | 0.1692 | 0.1786 | 0.1878 | 0.1966 | 0.2051 |
| 7 | 0.0480 | 0.0540 | 0.0604 | 0.0673 | 0.0746 | 0.0824 | 0.0905 | 0.0991 | 0.1080 | 0.1172 |
| 8 | 0.0125 | 0.0147 | 0.0171 | 0.0198 | 0.0229 | 0.0263 | 0.0301 | 0.0343 | 0.0389 | 0.0439 |
| 9 | 0.0019 | 0.0024 | 0.0029 | 0.0035 | 0.0042 | 0.0050 | 0.0059 | 0.0070 | 0.0083 | 0.0098 |
| 10 | 0.0001 | 0.0002 | 0.0002 | 0.0003 | 0.0003 | 0.0004 | 0.0005 | 0.0006 | 0.0008 | 0.0010 |

$n = 11$

| $t$ \ $p$ | 0.01 | 0.02 | 0.03 | 0.04 | 0.05 | 0.06 | 0.07 | 0.08 | 0.09 | 0.10 |
|---|---|---|---|---|---|---|---|---|---|---|
| 0 | 0.8953 | 0.8007 | 0.7153 | 0.6382 | 0.5688 | 0.5063 | 0.4501 | 0.3996 | 0.3544 | 0.3138 |
| 1 | 0.0995 | 0.1798 | 0.2433 | 0.2925 | 0.3293 | 0.3555 | 0.3727 | 0.3823 | 0.3855 | 0.3835 |
| 2 | 0.0050 | 0.0183 | 0.0376 | 0.0609 | 0.0867 | 0.1135 | 0.1403 | 0.1662 | 0.1906 | 0.2131 |
| 3 | 0.0002 | 0.0011 | 0.0035 | 0.0076 | 0.0137 | 0.0217 | 0.0317 | 0.0434 | 0.0566 | 0.0710 |
| 4 | 0.0000 | 0.0000 | 0.0002 | 0.0006 | 0.0014 | 0.0028 | 0.0048 | 0.0075 | 0.0112 | 0.0158 |
| 5 | 0.0000 | 0.0000 | 0.0000 | 0.0000 | 0.0001 | 0.0002 | 0.0005 | 0.0009 | 0.0015 | 0.0025 |
| 6 | 0.0000 | 0.0000 | 0.0000 | 0.0000 | 0.0000 | 0.0000 | 0.0000 | 0.0001 | 0.0002 | 0.0003 |

(*Continued*)

**Table A.7** Binomial probability distribution

|   | 0.11 | 0.12 | 0.13 | 0.14 | 0.15 | 0.16 | 0.17 | 0.18 | 0.19 | 0.20 |
|---|------|------|------|------|------|------|------|------|------|------|
| 0 | 0.2775 | 0.2451 | 0.2161 | 0.1903 | 0.1673 | 0.1469 | 0.1288 | 0.1127 | 0.9085 | 0.0859 |
| 1 | 0.3773 | 0.3676 | 0.3552 | 0.3408 | 0.3248 | 0.3078 | 0.2901 | 0.2721 | 0.2541 | 0.2362 |
| 2 | 0.2332 | 0.2507 | 0.2654 | 0.2774 | 0.2866 | 0.2932 | 0.2971 | 0.2987 | 0.2980 | 0.2953 |
| 3 | 0.0865 | 0.1025 | 0.1190 | 0.1355 | 0.1517 | 0.1675 | 0.1826 | 0.1967 | 0.2097 | 0.2215 |
| 4 | 0.0214 | 0.0280 | 0.0356 | 0.0441 | 0.0536 | 0.0638 | 0.0748 | 0.0864 | 0.0984 | 0.1107 |
| 5 | 0.0037 | 0.0053 | 0.0074 | 0.0101 | 0.0132 | 0.0170 | 0.0214 | 0.0265 | 0.0323 | 0.0388 |
| 6 | 0.0005 | 0.0007 | 0.0011 | 0.0016 | 0.0023 | 0.0032 | 0.0044 | 0.0058 | 0.0076 | 0.0097 |
| 7 | 0.0000 | 0.0001 | 0.0001 | 0.0002 | 0.0003 | 0.0004 | 0.0006 | 0.0009 | 0.0013 | 0.0017 |
| 8 | 0.0000 | 0.0000 | 0.0000 | 0.0000 | 0.0000 | 0.0000 | 0.0001 | 0.0001 | 0.0001 | 0.0002 |

|   | 0.21 | 0.22 | 0.23 | 0.24 | 0.25 | 0.26 | 0.27 | 0.28 | 0.29 | 0.30 |
|---|------|------|------|------|------|------|------|------|------|------|
| 0 | 0.0748 | 0.0650 | 0.0564 | 0.0489 | 0.0422 | 0.0364 | 0.0314 | 0.0270 | 0.0231 | 0.0198 |
| 1 | 0.2187 | 0.2017 | 0.1854 | 0.1697 | 0.1549 | 0.1408 | 0.1276 | 0.1153 | 0.1038 | 0.0932 |
| 2 | 0.2907 | 0.2845 | 0.2768 | 0.2680 | 0.2581 | 0.2474 | 0.2360 | 0.2242 | 0.2121 | 0.1998 |
| 3 | 0.2318 | 0.2407 | 0.2481 | 0.2539 | 0.2581 | 0.2608 | 0.2619 | 0.2616 | 0.2599 | 0.2568 |
| 4 | 0.1232 | 0.1358 | 0.1482 | 0.1603 | 0.1721 | 0.1832 | 0.1937 | 0.2035 | 0.2123 | 0.2201 |
| 5 | 0.0459 | 0.0536 | 0.0620 | 0.0709 | 0.0803 | 0.0901 | 0.1003 | 0.1108 | 0.1214 | 0.1321 |
| 6 | 0.0122 | 0.0151 | 0.0185 | 0.0224 | 0.0268 | 0.0317 | 0.0371 | 0.0431 | 0.0496 | 0.0566 |
| 7 | 0.0023 | 0.0030 | 0.0039 | 0.0050 | 0.0064 | 0.0079 | 0.0098 | 0.0120 | 0.0145 | 0.0173 |
| 8 | 0.0003 | 0.0004 | 0.0006 | 0.0008 | 0.0011 | 0.0014 | 0.0018 | 0.0023 | 0.0030 | 0.0037 |
| 9 | 0.0000 | 0.0000 | 0.0001 | 0.0001 | 0.0001 | 0.0002 | 0.0002 | 0.0003 | 0.0004 | 0.0005 |

|   | 0.31 | 0.32 | 0.33 | 0.34 | 0.35 | 0.36 | 0.37 | 0.38 | 0.39 | 0.40 |
|---|------|------|------|------|------|------|------|------|------|------|
| 0 | 0.0169 | 0.0144 | 0.0122 | 0.0104 | 0.0088 | 0.0074 | 0.0062 | 0.0052 | 0.0044 | 0.0036 |
| 1 | 0.0834 | 0.0744 | 0.0662 | 0.0587 | 0.0518 | 0.0457 | 0.0401 | 0.0351 | 0.0306 | 0.0266 |
| 2 | 0.1874 | 0.1751 | 0.1630 | 0.1511 | 0.1395 | 0.1284 | 0.1177 | 0.1075 | 0.0978 | 0.0887 |
| 3 | 0.2526 | 0.2472 | 0.2408 | 0.2335 | 0.2254 | 0.2167 | 0.2074 | 0.1977 | 0.1876 | 0.1774 |
| 4 | 0.2269 | 0.2326 | 0.2372 | 0.2406 | 0.2428 | 0.2438 | 0.2436 | 0.2423 | 0.2399 | 0.2365 |
| 5 | 0.1427 | 0.1533 | 0.1636 | 0.1735 | 0.1830 | 0.1920 | 0.2003 | 0.2079 | 0.2148 | 0.2207 |
| 6 | 0.0641 | 0.0721 | 0.0806 | 0.0894 | 0.0985 | 0.1080 | 0.1176 | 0.1274 | 0.1373 | 0.1471 |
| 7 | 0.0206 | 0.0242 | 0.0283 | 0.0329 | 0.0379 | 0.0434 | 0.0494 | 0.0558 | 0.0627 | 0.0701 |
| 8 | 0.0046 | 0.0057 | 0.0070 | 0.0085 | 0.0102 | 0.0122 | 0.0145 | 0.0171 | 0.0200 | 0.0234 |
| 9 | 0.0007 | 0.0009 | 0.0011 | 0.0015 | 0.0018 | 0.0023 | 0.0028 | 0.0035 | 0.0043 | 0.0052 |
| 10 | 0.0001 | 0.0001 | 0.0001 | 0.0001 | 0.0002 | 0.0003 | 0.0003 | 0.0004 | 0.0005 | 0.0007 |

|   | 0.41 | 0.42 | 0.43 | 0.44 | 0.45 | 0.46 | 0.47 | 0.48 | 0.49 | 0.50 |
|---|------|------|------|------|------|------|------|------|------|------|
| 0 | 0.0030 | 0.0025 | 0.0021 | 0.0017 | 0.0014 | 0.0011 | 0.0009 | 0.0008 | 0.0006 | 0.0005 |
| 1 | 0.0231 | 0.0199 | 0.0171 | 0.0147 | 0.0125 | 0.0107 | 0.0090 | 0.0076 | 0.0064 | 0.0054 |
| 2 | 0.0801 | 0.0721 | 0.0646 | 0.0577 | 0.0513 | 0.0454 | 0.0401 | 0.0352 | 0.0308 | 0.0269 |
| 3 | 0.1670 | 0.1566 | 0.1462 | 0.1359 | 0.1259 | 0.1161 | 0.1067 | 0.0976 | 0.0888 | 0.0806 |
| 4 | 0.2321 | 0.2267 | 0.2206 | 0.2136 | 0.2060 | 0.1978 | 0.1892 | 0.1801 | 0.1707 | 0.1611 |
| 5 | 0.2258 | 0.2299 | 0.2329 | 0.2350 | 0.2360 | 0.2359 | 0.2348 | 0.2327 | 0.2296 | 0.2256 |

*(Continued)*

**Table A.7** Binomial probability distribution

| t | | | | | | | | | | |
|---|---|---|---|---|---|---|---|---|---|---|
| 6 | 0.1569 | 0.1664 | 0.1757 | 0.1846 | 0.1931 | 0.2010 | 0.2083 | 0.2148 | 0.2206 | 0.2256 |
| 7 | 0.0779 | 0.0861 | 0.0947 | 0.1036 | 0.1128 | 0.1223 | 0.1319 | 0.1416 | 0.1514 | 0.1611 |
| 8 | 0.0271 | 0.0312 | 0.0357 | 0.0407 | 0.0462 | 0.0521 | 0.0585 | 0.0654 | 0.0727 | 0.0806 |
| 9 | 0.0063 | 0.0075 | 0.0090 | 0.0107 | 0.0126 | 0.0148 | 0.0173 | 0.0201 | 0.0233 | 0.0269 |
| 10 | 0.0009 | 0.0011 | 0.0014 | 0.0017 | 0.0021 | 0.0025 | 0.0031 | 0.0037 | 0.0045 | 0.0054 |
| 11 | 0.0001 | 0.0001 | 0.0001 | 0.0001 | 0.0002 | 0.0002 | 0.0002 | 0.0003 | 0.0004 | 0.0005 |

$n = 12$

| $p$ / $t$ | 0.01 | 0.02 | 0.03 | 0.04 | 0.05 | 0.06 | 0.07 | 0.08 | 0.09 | 0.10 |
|---|---|---|---|---|---|---|---|---|---|---|
| 0 | 0.8864 | 0.7847 | 0.6938 | 0.6127 | 0.5404 | 0.4759 | 0.4186 | 0.3677 | 0.3225 | 0.2824 |
| 1 | 0.1074 | 0.1922 | 0.2575 | 0.3064 | 0.3413 | 0.3645 | 0.3781 | 0.3837 | 0.3827 | 0.3766 |
| 2 | 0.0060 | 0.0216 | 0.0438 | 0.0702 | 0.0988 | 0.1280 | 0.1565 | 0.1835 | 0.2082 | 0.2301 |
| 3 | 0.0002 | 0.0015 | 0.0045 | 0.0098 | 0.0173 | 0.0272 | 0.0393 | 0.0532 | 0.0686 | 0.0852 |
| 4 | 0.0000 | 0.0001 | 0.0003 | 0.0009 | 0.0021 | 0.0039 | 0.0067 | 0.0104 | 0.0153 | 0.0213 |
| 5 | 0.0000 | 0.0000 | 0.0000 | 0.0001 | 0.0002 | 0.0004 | 0.0008 | 0.0014 | 0.0024 | 0.0038 |
| 6 | 0.0000 | 0.0000 | 0.0000 | 0.0000 | 0.0000 | 0.0000 | 0.0001 | 0.0001 | 0.0003 | 0.0005 |

| | 0.11 | 0.12 | 0.13 | 0.14 | 0.15 | 0.16 | 0.17 | 0.18 | 0.19 | 0.20 |
|---|---|---|---|---|---|---|---|---|---|---|
| 0 | 0.2470 | 0.2157 | 0.1880 | 0.1637 | 0.1422 | 0.1234 | 0.1069 | 0.0924 | 0.0798 | 0.0687 |
| 1 | 0.3663 | 0.3529 | 0.3372 | 0.3197 | 0.3012 | 0.2821 | 0.2627 | 0.2434 | 0.2245 | 0.2062 |
| 2 | 0.2490 | 0.2647 | 0.2771 | 0.2863 | 0.2924 | 0.2955 | 0.2960 | 0.2939 | 0.2897 | 0.2835 |
| 3 | 0.1026 | 0.1203 | 0.1380 | 0.1553 | 0.1720 | 0.1876 | 0.2021 | 0.2151 | 0.2265 | 0.2362 |
| 4 | 0.0285 | 0.0369 | 0.0464 | 0.0569 | 0.0683 | 0.0804 | 0.0931 | 0.1062 | 0.1195 | 0.1329 |
| 5 | 0.0056 | 0.0081 | 0.0111 | 0.0148 | 0.0193 | 0.0245 | 0.0305 | 0.0373 | 0.0449 | 0.0532 |
| 6 | 0.0008 | 0.0013 | 0.0019 | 0.0028 | 0.0040 | 0.0054 | 0.0073 | 0.0096 | 0.0123 | 0.01505 |
| 7 | 0.0001 | 0.0001 | 0.0002 | 0.0004 | 0.0006 | 0.0009 | 0.0013 | 0.0018 | 0.0025 | 0.0033 |
| 8 | 0.0000 | 0.0000 | 0.0000 | 0.0000 | 0.0001 | 0.0001 | 0.0002 | 0.0002 | 0.0004 | 0.0005 |
| 9 | 0.0000 | 0.0000 | 0.0000 | 0.0000 | 0.0000 | 0.0000 | 0.0000 | 0.0000 | 0.0000 | 0.0001 |

| | 0.21 | 0.22 | 0.23 | 0.24 | 0.25 | 0.26 | 0.27 | 0.28 | 0.29 | 0.30 |
|---|---|---|---|---|---|---|---|---|---|---|
| 0 | 0.0591 | 0.0507 | 0.0434 | 0.0371 | 0.0317 | 0.0270 | 0.0229 | 0.0194 | 0.0164 | 0.0138 |
| 1 | 0.1885 | 0.1717 | 0.1557 | 0.1407 | 0.1267 | 0.1137 | 0.1016 | 0.0906 | 0.0804 | 0.0712 |
| 2 | 0.2756 | 0.2663 | 0.2558 | 0.2444 | 0.2323 | 0.2197 | 0.2068 | 0.1937 | 0.1807 | 0.1678 |
| 3 | 0.2442 | 0.2503 | 0.2547 | 0.2573 | 0.2581 | 0.2573 | 0.2549 | 0.2511 | 0.2460 | 0.2397 |
| 4 | 0.1460 | 0.1589 | 0.1712 | 0.1828 | 0.1936 | 0.2034 | 0.2122 | 0.2197 | 0.2261 | 0.2311 |
| 5 | 0.0621 | 0.0717 | 0.0818 | 0.0924 | 0.1032 | 0.1143 | 0.1255 | 0.1367 | 0.1477 | 0.1585 |
| 6 | 0.0193 | 0.0236 | 0.0285 | 0.0340 | 0.0401 | 0.0469 | 0.0542 | 0.0620 | 0.0704 | 0.0792 |
| 7 | 0.0044 | 0.0057 | 0.0073 | 0.0092 | 0.0115 | 0.0141 | 0.0172 | 0.0207 | 0.0246 | 0.0291 |
| 8 | 0.0007 | 0.0010 | 0.0014 | 0.0018 | 0.0024 | 0.0031 | 0.0040 | 0.0050 | 0.0063 | 0.0078 |
| 9 | 0.0001 | 0.0001 | 0.0002 | 0.0003 | 0.0004 | 0.0005 | 0.0007 | 0.0009 | 0.0011 | 0.0015 |
| 10 | 0.0000 | 0.0000 | 0.0000 | 0.0000 | 0.0000 | 0.0001 | 0.0001 | 0.0001 | 0.0001 | 0.0002 |

(*Continued*)

**Table A.7** Binomial probability distribution

| | 0.31 | 0.32 | 0.33 | 0.34 | 0.35 | 0.36 | 0.37 | 0.38 | 0.39 | 0.40 |
|---|---|---|---|---|---|---|---|---|---|---|
| 0 | 0.0016 | 0.0098 | 0.0082 | 0.0068 | 0.0057 | 0.0047 | 0.0039 | 0.0032 | 0.0027 | 0.0022 |
| 1 | 0.0628 | 0.0552 | 0.0484 | 0.0422 | 0.0368 | 0.0319 | 0.0276 | 0.0237 | 0.0204 | 0.0174 |
| 2 | 0.1552 | 0.1429 | 0.1310 | 0.1197 | 0.1088 | 0.0986 | 0.0890 | 0.0800 | 0.0716 | 0.0639 |
| 3 | 0.2324 | 0.2241 | 0.2151 | 0.2055 | 0.1954 | 0.1849 | 0.1742 | 0.1634 | 0.1526 | 0.1419 |
| 4 | 0.2349 | 0.2373 | 0.2384 | 0.2382 | 0.2367 | 0.2340 | 0.2302 | 0.2254 | 0.2195 | 0.2128 |
| 5 | 0.1688 | 0.1787 | 0.1879 | 0.1963 | 0.2039 | 0.2106 | 0.2163 | 0.2210 | 0.2246 | 0.2270 |
| 6 | 0.0885 | 0.0981 | 0.1079 | 0.1180 | 0.1281 | 0.1382 | 0.1482 | 0.1580 | 0.1675 | 0.1766 |
| 7 | 0.0341 | 0.0396 | 0.0456 | 0.0521 | 0.0591 | 0.0666 | 0.0746 | 0.0830 | 0.0918 | 0.1009 |
| 8 | 0.0096 | 0.0116 | 0.0140 | 0.0168 | 0.0199 | 0.0234 | 0.0274 | 0.0318 | 0.0367 | 0.0420 |
| 9 | 0.0019 | 0.0024 | 0.0031 | 0.0038 | 0.0048 | 0.0059 | 0.0071 | 0.0087 | 0.0104 | 0.0125 |
| 10 | 0.0003 | 0.0003 | 0.0005 | 0.0006 | 0.0008 | 0.0010 | 0.0013 | 0.0016 | 0.0020 | 0.0025 |
| 11 | 0.0000 | 0.0000 | 0.0000 | 0.0001 | 0.0001 | 0.0001 | 0.0001 | 0.0002 | 0.0002 | 0.0003 |

| | 0.41 | 0.42 | 0.43 | 0.44 | 0.45 | 0.46 | 0.47 | 0.48 | 0.49 | 0.50 |
|---|---|---|---|---|---|---|---|---|---|---|
| 0 | 0.0018 | 0.0014 | 0.0012 | 0.0010 | 0.0008 | 0.0006 | 0.0005 | 0.0004 | 0.0003 | 0.0002 |
| 1 | 0.0148 | 0.0126 | 0.0106 | 0.0090 | 0.0075 | 0.0063 | 0.0052 | 0.0043 | 0.0036 | 0.0029 |
| 2 | 0.0567 | 0.0502 | 0.0442 | 0.0388 | 0.0339 | 0.0294 | 0.0255 | 0.0220 | 0.0189 | 0.0161 |
| 3 | 0.1314 | 0.1211 | 0.1111 | 0.1015 | 0.0923 | 0.0836 | 0.0754 | 0.0676 | 0.0604 | 0.0537 |
| 4 | 0.2054 | 0.1973 | 0.1886 | 0.1794 | 0.1700 | 0.1602 | 0.1504 | 0.1405 | 0.1306 | 0.1208 |
| 5 | 0.2284 | 0.2285 | 0.2276 | 0.2256 | 0.2225 | 0.2184 | 0.2134 | 0.2075 | 0.2008 | 0.1934 |
| 6 | 0.1851 | 0.1931 | 0.2003 | 0.2068 | 0.2124 | 0.2171 | 0.2208 | 0.2234 | 0.2250 | 0.2256 |
| 7 | 0.1103 | 0.1198 | 0.1295 | 0.1393 | 0.1498 | 0.1585 | 0.1678 | 0.1768 | 0.1853 | 0.1934 |
| 8 | 0.0479 | 0.0542 | 0.0611 | 0.0684 | 0.0762 | 0.0844 | 0.0930 | 0.1020 | 0.1113 | 0.1208 |
| 9 | 0.0148 | 0.0175 | 0.0205 | 0.0239 | 0.0277 | 0.0319 | 0.0367 | 0.0418 | 0.0475 | 0.0537 |
| 10 | 0.0031 | 0.0038 | 0.0046 | 0.0056 | 0.0068 | 0.0082 | 0.0098 | 0.0116 | 0.0137 | 0.0161 |
| 11 | 0.0004 | 0.0005 | 0.0006 | 0.0008 | 0.0010 | 0.0013 | 0.0016 | 0.0019 | 0.0024 | 0.0029 |
| 12 | 0.0000 | .0000 | 0.0000 | 0.0001 | 0.0001 | 0.0001 | 0.0001 | 0.0001 | 0.0002 | 0.0002 |

|  |  |  |  |  | $n = 13$ |  |  |  |  |  |
|---|---|---|---|---|---|---|---|---|---|---|
| $t$ \ $p$ | 0.01 | 0.02 | 0.03 | 0.04 | 0.05 | 0.06 | 0.07 | 0.08 | 0.09 | 0.10 |
| 0 | 0.8775 | 0.7690 | 0.6730 | 0.5882 | 0.5133 | 0.4474 | 0.3893 | 0.3383 | 0.2935 | 0.2542 |
| 1 | 0.1152 | 0.2040 | 0.2706 | 0.3186 | 0.3512 | 0.3712 | 0.3809 | 0.3824 | 0.3773 | 0.3672 |
| 2 | 0.0070 | 0.0250 | 0.0502 | 0.0797 | 0.1109 | 0.1422 | 0.1720 | 0.1995 | 0.2239 | 0.2448 |
| 3 | 0.0003 | 0.0019 | 0.0057 | 0.0122 | 0.0214 | 0.0333 | 0.0475 | 0.0636 | 0.0812 | 0.0997 |
| 4 | 0.0000 | 0.0001 | 0.0004 | 0.0013 | 0.0028 | 0.0053 | 0.0089 | 0.0138 | 0.0201 | 0.0277 |
| 5 | 0.0000 | 0.0000 | 0.0000 | 0.0001 | 0.0003 | 0.0006 | 0.0012 | 0.0022 | 0.0036 | 0.0055 |
| 6 | 0.0000 | 0.0000 | 0.0000 | 0.0000 | 0.0000 | 0.0001 | 0.0001 | 0.0003 | 0.0005 | 0.0008 |
| 7 | 0.0000 | 0.0000 | 0.0000 | 0.0000 | 0.0000 | 0.0000 | 0.0000 | 0.0000 | 0.0000 | 0.0001 |

| | 0.11 | 0.12 | 0.13 | 0.14 | 0.15 | 0.16 | 0.17 | 0.18 | 0.19 | 0.20 |
|---|---|---|---|---|---|---|---|---|---|---|
| 0 | 0.2198 | 0.1898 | 0.1636 | 0.1408 | 0.1209 | 0.1037 | 0.0887 | 0.0758 | 0.0646 | 0.0550 |
| 1 | 0.3532 | 0.3364 | 0.3178 | 0.2979 | 0.2774 | 0.2567 | 0.2362 | 0.2163 | 0.1970 | 0.1787 |
| 2 | 0.2619 | 0.2753 | 0.2849 | 0.2910 | 0.2937 | 0.2934 | 0.2903 | 0.2848 | 0.2773 | 0.2680 |

*(Continued)*

**Table A.7** Binomial probability distribution

| | | | | | | | | | | |
|---|---|---|---|---|---|---|---|---|---|---|
| 3 | 0.1187 | 0.1376 | 0.1561 | 0.1737 | 0.1900 | 0.2049 | 0.2180 | 0.2293 | 0.2385 | 0.2457 |
| 4 | 0.0367 | 0.0469 | 0.0583 | 0.0707 | 0.0838 | 0.0976 | 0.1116 | 0.1258 | 0.1399 | 0.1535 |
| 5 | 0.0082 | 0.0115 | 0.0157 | 0.0207 | 0.0266 | 0.0335 | 0.0412 | 0.0497 | 0.0591 | 0.0691 |
| 6 | 0.0013 | 0.0021 | 0.0031 | 0.0045 | 0.0063 | 0.0085 | 0.0112 | 0.0145 | 0.0185 | 0.0230 |
| 7 | 0.0002 | 0.0003 | 0.0005 | 0.0007 | 0.0011 | 0.0016 | 0.0023 | 0.0032 | 0.0043 | 0.0058 |
| 8 | 0.0000 | 0.0000 | 0.0001 | 0.0001 | 0.0001 | 0.0002 | 0.0004 | 0.0005 | 0.0008 | 0.0011 |
| 9 | 0.0000 | 0.0000 | 0.0000 | 0.0000 | 0.0000 | 0.0000 | 0.0000 | 0.0001 | 0.0001 | 0.0001 |

| | 0.21 | 0.22 | 0.23 | 0.24 | 0.25 | 0.26 | 0.27 | 0.28 | 0.29 | 0.30 |
|---|---|---|---|---|---|---|---|---|---|---|
| 0 | 0.0467 | 0.0396 | 0.0334 | 0.0282 | 0.0238 | 0.0200 | 0.0167 | 0.0140 | 0.0117 | 0.0097 |
| 1 | 0.1613 | 0.1450 | 0.1299 | 0.1159 | 0.1029 | 0.0911 | 0.0804 | 0.0706 | 0.0619 | 0.0540 |
| 2 | 0.2573 | 0.2455 | 0.2328 | 0.2195 | 0.2059 | 0.1921 | 0.1784 | 0.1648 | 0.1516 | 0.1388 |
| 3 | 0.2508 | 0.2539 | 0.2550 | 0.2542 | 0.2517 | 0.2475 | 0.2419 | 0.2351 | 0.2271 | 0.2181 |
| 4 | 0.1667 | 0.1790 | 0.1904 | 0.2007 | 0.2097 | 0.2174 | 0.2237 | 0.2285 | 0.2319 | 0.2337 |
| 5 | 0.0797 | 0.0909 | 0.1024 | 0.1141 | 0.1258 | 0.1375 | 0.1489 | 0.1600 | 0.1705 | 0.1803 |
| 6 | 0.0283 | 0.0342 | 0.0408 | 0.0480 | 0.0559 | 0.0644 | 0.0734 | 0.0829 | 0.0928 | 0.1030 |
| 7 | 0.0075 | 0.0096 | 0.0122 | 0.0152 | 0.0186 | 0.0226 | 0.0272 | 0.0323 | 0.0379 | 0.0442 |
| 8 | 0.0015 | 0.0020 | 0.0027 | 0.0036 | 0.0047 | 0.0060 | 0.0075 | 0.0094 | 0.0116 | 0.0142 |
| 9 | 0.0002 | 0.0003 | 0.0005 | 0.0006 | 0.0009 | 0.0012 | 0.0015 | 0.0020 | 0.0026 | 0.0034 |
| 10 | 0.0000 | 0.0000 | 0.0001 | 0.0001 | 0.0001 | 0.0002 | 0.0002 | 0.0003 | 0.0004 | 0.0006 |
| 11 | 0.0000 | 0.0000 | 0.0000 | 0.0000 | 0.0000 | 0.0000 | 0.0000 | 0.0000 | 0.0000 | 0.0001 |

| | 0.31 | 0.32 | 0.33 | 0.34 | 0.35 | 0.36 | 0.37 | 0.38 | 0.39 | 0.40 |
|---|---|---|---|---|---|---|---|---|---|---|
| 0 | 0.0080 | 0.0066 | 0.0055 | 0.0045 | 0.0037 | 0.0030 | 0.0025 | 0.0020 | 0.0016 | 0.0013 |
| 1 | 0.0469 | 0.0407 | 0.0351 | 0.0302 | 0.0259 | 0.0221 | 0.0188 | 0.0159 | 0.0135 | 0.0113 |
| 2 | 0.1265 | 0.1148 | 0.1037 | 0.0933 | 0.0836 | 0.0746 | 0.0663 | 0.0586 | 0.0516 | 0.0453 |
| 3 | 0.2084 | 0.1981 | 0.1874 | 0.1763 | 0.1651 | 0.1538 | 0.1427 | 0.1317 | 0.1210 | 0.1107 |
| 4 | 0.2341 | 0.2331 | 0.2307 | 0.2270 | 0.2222 | 0.2163 | 0.2095 | 0.2018 | 0.1934 | 0.1845 |
| 5 | 0.1893 | 0.1974 | 0.2045 | 0.2105 | 0.2154 | 0.2190 | 0.2215 | 0.2227 | 0.2226 | 0.2214 |
| 6 | 0.1134 | 0.1239 | 0.1343 | 0.1446 | 0.1546 | 0.1643 | 0.1734 | 0.1820 | 0.1898 | 0.1968 |
| 7 | 0.0509 | 0.0583 | 0.0662 | 0.0745 | 0.0833 | 0.0924 | 0.1019 | 0.1115 | 0.1213 | 0.1312 |
| 8 | 0.0172 | 0.0206 | 0.0244 | 0.0288 | 0.0336 | 0.0390 | 0.0449 | 0.0513 | 0.0582 | 0.0656 |
| 9 | 0.0043 | 0.0054 | 0.0067 | 0.0082 | 0.0101 | 0.0122 | 0.0146 | 0.0175 | 0.0207 | 0.0243 |
| 10 | 0.0008 | 0.0010 | 0.0013 | 0.0017 | 0.0022 | 0.0027 | 0.0034 | 0.0043 | 0.0053 | 0.0065 |
| 11 | 0.0001 | 0.0001 | 0.0002 | 0.0002 | 0.0003 | 0.0004 | 0.0006 | 0.0007 | 0.0009 | 0.0012 |
| 12 | 0.0000 | 0.0000 | 0.0000 | 0.0000 | 0.0000 | 0.0000 | 0.0001 | 0.0001 | 0.0001 | 0.0001 |

| | 0.41 | 0.42 | 0.43 | 0.44 | 0.45 | 0.46 | 0.47 | 0.48 | 0.49 | 0.50 |
|---|---|---|---|---|---|---|---|---|---|---|
| 0 | 0.0010 | 0.0008 | 0.0007 | 0.0005 | 0.0004 | 0.0003 | 0.0003 | 0.0002 | 0.0002 | 0.0001 |
| 1 | 0.0095 | 0.0079 | 0.0066 | 0.0054 | 0.0045 | 0.0037 | 0.0030 | 0.0024 | 0.0020 | 0.0016 |
| 2 | 0.0395 | 0.0344 | 0.0298 | 0.0256 | 0.0220 | 0.0188 | 0.0160 | 0.0135 | 0.0114 | 0.0095 |
| 3 | 0.1007 | 0.0913 | 0.0823 | 0.0739 | 0.0660 | 0.0587 | 0.0519 | 0.0457 | 0.0401 | 0.0349 |
| 4 | 0.1750 | 0.1653 | 0.1553 | 0.1451 | 0.1350 | 0.1250 | 0.1151 | 0.1055 | 0.0962 | 0.0873 |
| 5 | 0.2189 | 0.2154 | 0.2108 | 0.2053 | 0.1989 | 0.1917 | 0.1838 | 0.1753 | 0.1664 | 0.1571 |
| 6 | 0.2029 | 0.2080 | 0.2121 | 0.2151 | 0.2169 | 0.2177 | 0.2173 | 0.2158 | 0.2131 | 0.2095 |

(*Continued*)

**Table A.7** Binomial probability distribution

| | | | | | | | | | | |
|---|---|---|---|---|---|---|---|---|---|---|
| 7 | 0.1410 | 0.1506 | 0.1600 | 0.1690 | 0.1775 | 0.1854 | 0.1927 | 0.1992 | 0.2048 | 0.2095 |
| 8 | 0.0735 | 0.0818 | 0.0905 | 0.0996 | 0.1089 | 0.1185 | 0.1282 | 0.1379 | 0.1476 | 0.1571 |
| 9 | 0.0284 | 0.0329 | 0.0379 | 0.0435 | 0.0495 | 0.0561 | 0.0631 | 0.0707 | 0.0788 | 0.0873 |
| 10 | 0.0079 | 0.0095 | 0.0114 | 0.0137 | 0.0162 | 0.0191 | 0.0224 | 0.0261 | 0.0303 | 0.0349 |
| 11 | 0.0015 | 0.0019 | 0.0024 | 0.0029 | 0.0036 | 0.0044 | 0.0054 | 0.0066 | 0.0079 | 0.0095 |
| 12 | 0.0002 | 0.0002 | 0.0003 | 0.0004 | 0.0005 | 0.0006 | 0.0008 | 0.0010 | 0.0013 | 0.0016 |
| 13 | 0.0000 | 0.0000 | 0.0000 | 0.0000 | 0.0000 | 0.0000 | 0.0001 | 0.0001 | 0.0001 | 0.0001 |

$n = 14$

| $t$ \ $p$ | 0.01 | 0.02 | 0.03 | 0.04 | 0.05 | 0.06 | 0.07 | 0.08 | 0.09 | 0.10 |
|---|---|---|---|---|---|---|---|---|---|---|
| 0 | 0.8687 | 0.7536 | 0.6528 | 0.5647 | 0.4877 | 0.4205 | 0.3620 | 0.3112 | 0.2670 | 0.2288 |
| 1 | 0.1229 | 0.2153 | 0.2827 | 0.3294 | 0.3593 | 0.3758 | 0.3815 | 0.3788 | 0.3698 | 0.3559 |
| 2 | 0.0081 | 0.0286 | 0.0568 | 0.0892 | 0.1229 | 0.1559 | 0.1867 | 0.2141 | 0.2377 | 0.2570 |
| 3 | 0.0003 | 0.0023 | 0.0070 | 0.0149 | 0.0259 | 0.0398 | 0.0562 | 0.0745 | 0.0940 | 0.1142 |
| 4 | 0.0000 | 0.0001 | 0.0006 | 0.0017 | 0.0037 | 0.0070 | 0.0116 | 0.0178 | 0.0256 | 0.0349 |
| 5 | 0.0000 | 0.0000 | 0.0000 | 0.0001 | 0.0004 | 0.0009 | 0.0018 | 0.0031 | 0.0051 | 0.0078 |
| 6 | 0.0000 | 0.0000 | 0.0000 | 0.0000 | 0.0000 | 0.0001 | 0.0002 | 0.0004 | 0.0008 | 0.0013 |
| 7 | 0.0000 | 0.0000 | 0.0000 | 0.0000 | 0.0000 | 0.0000 | 0.0000 | 0.0000 | 0.0001 | 0.0002 |

| | 0.11 | 0.12 | 0.13 | 0.14 | 0.15 | 0.16 | 0.17 | 0.18 | 0.19 | 0.20 |
|---|---|---|---|---|---|---|---|---|---|---|
| 0 | 0.1956 | 0.1670 | 0.1423 | 0.1211 | 0.1028 | 0.0871 | 0.0736 | 0.0621 | 0.0523 | 0.0440 |
| 1 | 0.3385 | 0.3188 | 0.2977 | 0.2759 | 0.2539 | 0.2322 | 0.2112 | 0.1910 | 0.1719 | 0.1539 |
| 2 | 0.2720 | 0.2826 | 0.2892 | 0.2919 | 0.2912 | 0.2875 | 0.2811 | 0.2725 | 0.2620 | 0.2501 |
| 3 | 0.1345 | 0.1542 | 0.1728 | 0.1901 | 0.2056 | 0.2190 | 0.2303 | 0.2393 | 0.2459 | 0.2501 |
| 4 | 0.0457 | 0.0578 | 0.0710 | 0.0851 | 0.0998 | 0.1147 | 0.1297 | 0.1444 | 0.1586 | 0.1720 |
| 5 | 0.0113 | 0.0158 | 0.0212 | 0.0277 | 0.0352 | 0.0437 | 0.0531 | 0.0634 | 0.0744 | 0.0860 |
| 6 | 0.0021 | 0.0032 | 0.0048 | 0.0068 | 0.0093 | 0.0125 | 0.0163 | 0.0209 | 0.0262 | 0.0322 |
| 7 | 0.0003 | 0.0005 | 0.0008 | 0.0013 | 0.0019 | 0.0027 | 0.0038 | 0.0052 | 0.0070 | 0.0092 |
| 8 | 0.0000 | 0.0001 | 0.0001 | 0.0002 | 0.0003 | 0.0005 | 0.0007 | 0.0010 | 0.0014 | 0.0020 |
| 9 | 0.0000 | 0.0000 | 0.0000 | 0.0000 | 0.0000 | 0.0001 | 0.0001 | 0.0001 | 0.0002 | 0.0003 |

| | 0.21 | 0.22 | 0.23 | 0.24 | 0.25 | 0.26 | 0.27 | 0.28 | 0.29 | 0.30 |
|---|---|---|---|---|---|---|---|---|---|---|
| 0 | 0.0369 | 0.0309 | 0.0258 | 0.0214 | 0.0178 | 0.0148 | 0.0122 | 0.0101 | 0.0083 | 0.0068 |
| 1 | 0.1372 | 0.1218 | 0.1077 | 0.0948 | 0.0832 | 0.0726 | 0.0632 | 0.0548 | 0.0473 | 0.0407 |
| 2 | 0.2371 | 0.2234 | 0.2091 | 0.1946 | 0.1802 | 0.1659 | 0.1519 | 0.1385 | 0.1256 | 0.1134 |
| 3 | 0.2521 | 0.2520 | 0.2499 | 0.2459 | 0.2402 | 0.2331 | 0.2248 | 0.2154 | 0.2052 | 0.1943 |
| 4 | 0.1843 | 0.1955 | 0.2052 | 0.2135 | 0.2202 | 0.2252 | 0.2286 | 0.2304 | 0.2305 | 0.2290 |
| 5 | 0.0980 | 0.1103 | 0.1226 | 0.1348 | 0.1468 | 0.1583 | 0.1691 | 0.1792 | 0.1883 | 0.1963 |
| 6 | 0.0391 | 0.0466 | 0.0549 | 0.0639 | 0.0734 | 0.0834 | 0.0938 | 0.1045 | 0.1153 | 0.1262 |
| 7 | 0.0119 | 0.0150 | 0.0188 | 0.0231 | 0.0280 | 0.0335 | 0.0397 | 0.0464 | 0.0538 | 0.0618 |
| 8 | 0.0028 | 0.0037 | 0.0049 | 0.0064 | 0.0082 | 0.0103 | 0.0128 | 0.0158 | 0.0192 | 0.0232 |
| 9 | 0.0005 | 0.0007 | 0.0010 | 0.0013 | 0.0018 | 0.0024 | 0.0032 | 0.0041 | 0.0052 | 0.0066 |
| 10 | 0.0001 | 0.0001 | 0.0001 | 0.0002 | 0.0003 | 0.0004 | 0.0006 | 0.0008 | 0.0011 | 0.0014 |
| 11 | 0.0000 | 0.0000 | 0.0000 | 0.0000 | 0.0000 | 0.0001 | 0.0001 | 0.0001 | 0.0002 | 0.0002 |

*(Continued)*

**Table A.7** Binomial probability distribution

|    | 0.31 | 0.32 | 0.33 | 0.34 | 0.35 | 0.36 | 0.37 | 0.38 | 0.39 | 0.40 |
|----|------|------|------|------|------|------|------|------|------|------|
| 0  | 0.0055 | 0.0045 | 0.0037 | 0.0030 | 0.0024 | 0.0019 | 0.0016 | 0.0012 | 0.0010 | 0.0008 |
| 1  | 0.0349 | 0.0298 | 0.0253 | 0.0215 | 0.0181 | 0.0152 | 0.0128 | 0.0106 | 0.0088 | 0.0073 |
| 2  | 0.1018 | 0.0911 | 0.0811 | 0.0719 | 0.0634 | 0.0557 | 0.0487 | 0.0424 | 0.0367 | 0.0317 |
| 3  | 0.1830 | 0.1715 | 0.1598 | 0.1481 | 0.1366 | 0.1253 | 0.1144 | 0.1039 | 0.0940 | 0.0845 |
| 4  | 0.2261 | 0.2219 | 0.2164 | 0.2098 | 0.2022 | 0.1938 | 0.1848 | 0.1752 | 0.1652 | 0.1549 |
| 5  | 0.2032 | 0.2088 | 0.2132 | 0.2161 | 0.2178 | 0.2181 | 0.2170 | 0.2147 | 0.2112 | 0.2066 |
| 6  | 0.1369 | 0.1474 | 0.1575 | 0.1670 | 0.1759 | 0.1840 | 0.1912 | 0.1974 | 0.2026 | 0.2066 |
| 7  | 0.0703 | 0.0793 | 0.0886 | 0.0983 | 0.1082 | 0.1183 | 0.1283 | 0.1383 | 0.1480 | 0.1574 |
| 8  | 0.0276 | 0.0326 | 0.0382 | 0.0443 | 0.0510 | 0.0582 | 0.0659 | 0.0742 | 0.0828 | 0.0918 |
| 9  | 0.0083 | 0.0102 | 0.0125 | 0.0152 | 0.0183 | 0.0218 | 0.0258 | 0.0303 | 0.0353 | 0.0408 |
| 10 | 0.0019 | 0.0024 | 0.0031 | 0.0039 | 0.0049 | 0.0061 | 0.0076 | 0.0093 | 0.0113 | 0.0136 |
| 11 | 0.0003 | 0.0004 | 0.0006 | 0.0007 | 0.0010 | 0.0013 | 0.0016 | 0.0021 | 0.0026 | 0.0033 |
| 12 | 0.0000 | 0.0000 | 0.0001 | 0.0001 | 0.0001 | 0.0002 | 0.0002 | 0.0003 | 0.0004 | 0.0005 |
| 13 | 0.0000 | 0.0000 | 0.0000 | 0.0000 | 0.0000 | 0.0000 | 0.0000 | 0.0000 | 0.0000 | 0.0001 |

|    | 0.41 | 0.42 | 0.43 | 0.44 | 0.45 | 0.46 | 0.47 | 0.48 | 0.49 | 0.50 |
|----|------|------|------|------|------|------|------|------|------|------|
| 0  | 0.0006 | 0.0005 | 0.0004 | 0.0003 | 0.0002 | 0.0002 | 0.0001 | 0.0001 | 0.0001 | 0.0001 |
| 1  | 0.0060 | 0.0049 | 0.0040 | 0.0033 | 0.0027 | 0.0021 | 0.0017 | 0.0014 | 0.0011 | 0.0009 |
| 2  | 0.0272 | 0.0233 | 0.0198 | 0.0168 | 0.0141 | 0.0118 | 0.0099 | 0.0082 | 0.0068 | 0.0056 |
| 3  | 0.0757 | 0.0674 | 0.0597 | 0.0527 | 0.0462 | 0.0403 | 0.0350 | 0.0303 | 0.0260 | 0.0222 |
| 4  | 0.1446 | 0.1342 | 0.1239 | 0.1138 | 0.1040 | 0.0945 | 0.0854 | 0.0768 | 0.0687 | 0.0611 |
| 5  | 0.2009 | 0.1943 | 0.1869 | 0.1788 | 0.1701 | 0.1610 | 0.1515 | 0.1418 | 0.1320 | 0.1222 |
| 6  | 0.2094 | 0.2111 | 0.2115 | 0.2108 | 0.2088 | 0.2057 | 0.2015 | 0.1963 | 0.1902 | 0.1833 |
| 7  | 0.1663 | 0.1747 | 0.1824 | 0.1892 | 0.1952 | 0.2003 | 0.2043 | 0.2071 | 0.2089 | 0.2095 |
| 8  | 0.1011 | 0.1107 | 0.1204 | 0.1301 | 0.1398 | 0.1493 | 0.1585 | 0.1673 | 0.1756 | 0.1833 |
| 9  | 0.0469 | 0.0534 | 0.0605 | 0.0682 | 0.0762 | 0.0848 | 0.0937 | 0.1030 | 0.1125 | 0.1222 |
| 10 | 0.0163 | 0.0193 | 0.0228 | 0.0268 | 0.0312 | 0.0361 | 0.0415 | 0.0475 | 0.0540 | 0.0611 |
| 11 | 0.0041 | 0.0051 | 0.0063 | 0.0076 | 0.0093 | 0.0112 | 0.0134 | 0.0160 | 0.0189 | 0.0222 |
| 12 | 0.0007 | 0.0009 | 0.0012 | 0.0015 | 0.0019 | 0.0024 | 0.0030 | 0.0037 | 0.0045 | 0.0056 |
| 13 | 0.0001 | 0.0001 | 0.0001 | 0.0002 | 0.0002 | 0.0003 | 0.0004 | 0.0005 | 0.0007 | 0.0009 |
| 14 | 0.0000 | 0.0000 | 0.0000 | 0.0000 | 0.0000 | 0.0000 | 0.0000 | 0.0000 | 0.0000 | 0.0001 |

$n = 15$

| $t$ \ $p$ | 0.01 | 0.02 | 0.03 | 0.04 | 0.05 | 0.06 | 0.07 | 0.08 | 0.09 | 0.10 |
|-----------|------|------|------|------|------|------|------|------|------|------|
| 0 | 0.8601 | 0.7386 | 0.6333 | 0.5421 | 0.4633 | 0.3953 | 0.3367 | 0.2863 | 0.2430 | 0.2059 |
| 1 | 0.1303 | 0.2261 | 0.2938 | 0.3388 | 0.3658 | 0.3785 | 0.3801 | 0.3734 | 0.3605 | 0.3432 |
| 2 | 0.0092 | 0.0323 | 0.0636 | 0.0988 | 0.1348 | 0.1691 | 0.2003 | 0.2273 | 0.2496 | 0.2669 |
| 3 | 0.0004 | 0.0029 | 0.0085 | 0.0178 | 0.0307 | 0.0468 | 0.0653 | 0.0857 | 0.1070 | 0.1285 |
| 4 | 0.0000 | 0.0002 | 0.0008 | 0.0022 | 0.0049 | 0.0090 | 0.0148 | 0.0223 | 0.0317 | 0.0428 |
| 5 | 0.0000 | 0.0000 | 0.0001 | 0.0002 | 0.0006 | 0.0013 | 0.0024 | 0.0043 | 0.0069 | 0.0105 |
| 6 | 0.0000 | 0.0000 | 0.0000 | 0.0000 | 0.0000 | 0.0001 | 0.0003 | 0.0006 | 0.0011 | 0.0019 |
| 7 | 0.0000 | 0.0000 | 0.0000 | 0.0000 | 0.0000 | 0.0000 | 0.0000 | 0.0001 | 0.0001 | 0.0003 |

*(Continued)*

**Table A.7** Binomial probability distribution

|     | 0.11 | 0.12 | 0.13 | 0.14 | 0.15 | 0.16 | 0.17 | 0.18 | 0.19 | 0.20 |
|-----|------|------|------|------|------|------|------|------|------|------|
| 0 | 0.1741 | 0.1470 | 0.1238 | 0.1041 | 0.0874 | 0.0731 | 0.0611 | 0.0510 | 0.0424 | 0.0352 |
| 1 | 0.3228 | 0.3006 | 0.2775 | 0.2542 | 0.2312 | 0.2090 | 0.1878 | 0.1678 | 0.1492 | 0.1319 |
| 2 | 0.2793 | 0.2870 | 0.2903 | 0.2897 | 0.2856 | 0.2787 | 0.2692 | 0.2578 | 0.2449 | 0.2309 |
| 3 | 0.1496 | 0.1696 | 0.1880 | 0.2044 | 0.2184 | 0.2300 | 0.2389 | 0.2452 | 0.2489 | 0.2501 |
| 4 | 0.0555 | 0.0694 | 0.0843 | 0.0998 | 0.1156 | 0.1314 | 0.1468 | 0.1615 | 0.1752 | 0.1876 |
| 5 | 0.0151 | 0.0208 | 0.0277 | 0.0357 | 0.0449 | 0.0551 | 0.0662 | 0.0780 | 0.0904 | 0.1032 |
| 6 | 0.0031 | 0.0047 | 0.0069 | 0.0097 | 0.0132 | 0.0175 | 0.0226 | 0.0285 | 0.0353 | 0.0430 |
| 7 | 0.0005 | 0.0008 | 0.0013 | 0.0020 | 0.0030 | 0.0043 | 0.0059 | 0.0081 | 0.0107 | 0.0138 |
| 8 | 0.0001 | 0.0001 | 0.0002 | 0.0003 | 0.0005 | 0.0008 | 0.0012 | 0.0018 | 0.0025 | 0.0035 |
| 9 | 0.0000 | 0.0000 | 0.0000 | 0.0000 | 0.0001 | 0.0001 | 0.0002 | 0.0003 | 0.0005 | 0.0007 |
| 10 | 0.0000 | 0.0000 | 0.0000 | 0.0000 | 0.0000 | 0.0000 | 0.0000 | 0.0000 | 0.0001 | 0.0001 |

|     | 0.21 | 0.22 | 0.23 | 0.24 | 0.25 | 0.26 | 0.27 | 0.28 | 0.29 | 0.30 |
|-----|------|------|------|------|------|------|------|------|------|------|
| 0 | 0.0291 | 0.0241 | 0.0198 | 0.0163 | 0.0134 | 0.0109 | 0.0089 | 0.0072 | 0.0059 | 0.0047 |
| 1 | 0.1162 | 0.1018 | 0.0889 | 0.0772 | 0.0668 | 0.0576 | 0.0494 | 0.0423 | 0.0360 | 0.0305 |
| 2 | 0.2162 | 0.2010 | 0.1858 | 0.1707 | 0.1559 | 0.1416 | 0.1280 | 0.1150 | 0.1029 | 0.0916 |
| 3 | 0.2490 | 0.2457 | 0.2405 | 0.2336 | 0.2252 | 0.2156 | 0.2051 | 0.1939 | 0.1812 | 0.1700 |
| 4 | 0.1986 | 0.2079 | 0.2155 | 0.2213 | 0.2252 | 0.2273 | 0.2276 | 0.2262 | 0.2231 | 0.2186 |
| 5 | 0.1161 | 0.1290 | 0.1416 | 0.1537 | 0.1651 | 0.1757 | 0.1852 | 0.1935 | 0.2005 | 0.2061 |
| 6 | 0.0514 | 0.0606 | 0.0705 | 0.0809 | 0.0917 | 0.1029 | 0.1142 | 0.1254 | 0.1365 | 0.1472 |
| 7 | 0.0176 | 0.0220 | 0.0271 | 0.0329 | 0.0393 | 0.0465 | 0.0543 | 0.0627 | 0.0717 | 0.0811 |
| 8 | 0.0047 | 0.0062 | 0.0081 | 0.0104 | 0.0131 | 0.0163 | 0.0201 | 0.0244 | 0.0293 | 0.0348 |
| 9 | 0.0010 | 0.0014 | 0.0019 | 0.0025 | 0.0034 | 0.0045 | 0.0058 | 0.0074 | 0.0093 | 0.0116 |
| 10 | 0.0002 | 0.0002 | 0.0003 | 0.0005 | 0.0007 | 0.0009 | 0.0013 | 0.0017 | 0.0023 | 0.0030 |
| 11 | 0.0000 | 0.0000 | 0.0000 | 0.0001 | 0.0001 | 0.0002 | 0.0002 | 0.0003 | 0.0004 | 0.0006 |
| 12 | 0.0000 | 0.0000 | 0.0000 | 0.0000 | 0.0000 | 0.0000 | 0.0000 | 0.0000 | 0.0001 | 0.0001 |

|     | 0.31 | 0.32 | 0.33 | 0.34 | 0.35 | 0.36 | 0.37 | 0.38 | 0.39 | 0.40 |
|-----|------|------|------|------|------|------|------|------|------|------|
| 0 | 0.0038 | 0.0031 | 0.0025 | 0.0020 | 0.0016 | 0.0012 | 0.0010 | 0.0008 | 0.0006 | 0.0005 |
| 1 | 0.0258 | 0.0217 | 0.0182 | 0.0152 | 0.0126 | 0.0104 | 0.0086 | 0.0071 | 0.0058 | 0.0047 |
| 2 | 0.0811 | 0.0715 | 0.0627 | 0.0547 | 0.0476 | 0.0411 | 0.0354 | 0.0303 | 0.0259 | 0.0219 |
| 3 | 0.1579 | 0.1457 | 0.1338 | 0.1222 | 0.1110 | 0.1002 | 0.0901 | 0.0805 | 0.0716 | 0.0634 |
| 4 | 0.2128 | 0.2057 | 0.1977 | 0.1888 | 0.1792 | 0.1692 | 0.1587 | 0.1481 | 0.1374 | 0.1268 |
| 5 | 0.0210 | 0.2130 | 0.2142 | 0.2140 | 0.2123 | 0.2093 | 0.2051 | 0.1997 | 0.1933 | 0.1859 |
| 6 | 0.1575 | 0.1671 | 0.1759 | 0.1837 | 0.1906 | 0.1963 | 0.2008 | 0.2040 | 0.2059 | 0.2066 |
| 7 | 0.0910 | 0.1011 | 0.1114 | 0.1217 | 0.1319 | 0.1419 | 0.1516 | 0.1608 | 0.1693 | 0.1771 |
| 8 | 0.0409 | 0.0476 | 0.0549 | 0.0627 | 0.0710 | 0.0798 | 0.0890 | 0.0985 | 0.1082 | 0.1181 |
| 9 | 0.0143 | 0.0174 | 0.0210 | 0.0251 | 0.0298 | 0.0349 | 0.0407 | 0.0470 | 0.0538 | 0.0612 |
| 10 | 0.0038 | 0.0049 | 0.0062 | 0.0078 | 0.0096 | 0.0118 | 0.0143 | 0.0173 | 0.0206 | 0.0245 |
| 11 | 0.0008 | 0.0011 | 0.0014 | 0.0018 | 0.0024 | 0.0030 | 0.0038 | 0.0048 | 0.0060 | 0.0074 |
| 12 | 0.0001 | 0.0002 | 0.0002 | 0.0003 | 0.0004 | 0.0006 | 0.0007 | 0.0010 | 0.0013 | 0.0016 |
| 13 | 0.0000 | 0.0000 | 0.0000 | 0.0000 | 0.0001 | 0.0001 | 0.0001 | 0.0001 | 0.0002 | 0.0003 |

(*Continued*)

**Table A.7** Binomial probability distribution

| | 0.41 | 0.42 | 0.43 | 0.44 | 0.45 | 0.46 | 0.47 | 0.48 | 0.49 | 0.50 |
|---|---|---|---|---|---|---|---|---|---|---|
| 0 | 0.0004 | 0.0003 | 0.0002 | 0.0002 | 0.0001 | 0.0001 | 0.0001 | 0.0001 | 0.0000 | 0.0000 |
| 1 | 0.0038 | 0.0031 | 0.0025 | 0.0020 | 0.0016 | 0.0012 | 0.0010 | 0.0008 | 0.0006 | 0.0005 |
| 2 | 0.0185 | 0.0156 | 0.0130 | 0.0108 | 0.0090 | 0.0074 | 0.0060 | 0.0049 | 0.0040 | 0.0032 |
| 3 | 0.0558 | 0.0489 | 0.0426 | 0.0369 | 0.0318 | 0.0272 | 0.0232 | 0.0197 | 0.0166 | 0.0139 |
| 4 | 0.1163 | 0.1061 | 0.0963 | 0.0869 | 0.0780 | 0.0696 | 0.0617 | 0.0545 | 0.0478 | 0.0417 |
| 5 | 0.1778 | 0.1691 | 0.1598 | 0.1502 | 0.1404 | 0.1304 | 0.1204 | 0.1106 | 0.1010 | 0.0916 |
| 6 | 0.2060 | 0.2041 | 0.2010 | 0.1967 | 0.1914 | 0.1851 | 0.1780 | 0.1702 | 0.1617 | 0.1527 |
| 7 | 0.1840 | 0.1900 | 0.1949 | 0.1987 | 0.2013 | 0.2028 | 0.2030 | 0.2020 | 0.1997 | 0.1964 |
| 8 | 0.1279 | 0.1376 | 0.1470 | 0.1561 | 0.1647 | 0.1727 | 0.1800 | 0.1864 | 0.1919 | 0.1964 |
| 9 | 0.0691 | 0.0775 | 0.0863 | 0.0954 | 0.1048 | 0.1144 | 0.1241 | 0.1338 | 0.1434 | 0.1527 |
| 10 | 0.0288 | 0.0337 | 0.0390 | 0.0450 | 0.0515 | 0.0585 | 0.0661 | 0.0741 | 0.0827 | 0.0916 |
| 11 | 0.0091 | 0.0111 | 0.0134 | 0.0161 | 0.0191 | 0.0226 | 0.0266 | 0.0311 | 0.0361 | 0.0417 |
| 12 | 0.0021 | 0.0027 | 0.0034 | 0.0042 | 0.0052 | 0.0064 | 0.0079 | 0.0096 | 0.0116 | 0.0139 |
| 13 | 0.0003 | 0.0004 | 0.0006 | 0.0008 | 0.0010 | 0.0013 | 0.0016 | 0.0020 | 0.0026 | 0.0032 |
| 14 | 0.0000 | 0.0000 | 0.0001 | 0.0001 | 0.0001 | 0.0002 | 0.0002 | 0.0003 | 0.0004 | 0.0005 |

$n = 16$

| $t$ \ $p$ | 0.01 | 0.02 | 0.03 | 0.04 | 0.05 | 0.06 | 0.07 | 0.08 | 0.09 | 0.10 |
|---|---|---|---|---|---|---|---|---|---|---|
| 0 | 0.8515 | 0.7238 | 0.6143 | 0.5204 | 0.4401 | 0.3716 | 0.3131 | 0.2634 | 0.2211 | 0.1853 |
| 1 | 0.1376 | 0.2363 | 0.3040 | 0.3469 | 0.3706 | 0.3795 | 0.3771 | 0.3665 | 0.3499 | 0.3294 |
| 2 | 0.0104 | 0.0362 | 0.0705 | 0.1084 | 0.1463 | 0.1817 | 0.2129 | 0.2390 | 0.2596 | 0.2745 |
| 3 | 0.0005 | 0.0034 | 0.0102 | 0.0211 | 0.0359 | 0.0541 | 0.0748 | 0.0970 | 0.1198 | 0.1423 |
| 4 | 0.0000 | 0.0002 | 0.0010 | 0.0029 | 0.0061 | 0.0112 | 0.0183 | 0.0274 | 0.0385 | 0.0514 |
| 5 | 0.0000 | 0.0000 | 0.0001 | 0.0003 | 0.0008 | 0.0017 | 0.0033 | 0.0057 | 0.0091 | 0.0137 |
| 6 | 0.0000 | 0.0000 | 0.0000 | 0.0000 | 0.0001 | 0.0002 | 0.0005 | 0.0009 | 0.0017 | 0.0028 |
| 7 | 0.0000 | 0.0000 | 0.0000 | 0.0000 | 0.0000 | 0.0000 | 0.0000 | 0.0001 | 0.0002 | 0.0004 |
| 8 | 0.0000 | 0.0000 | 0.0000 | 0.0000 | 0.0000 | 0.0000 | 0.0000 | 0.0000 | 0.0000 | 0.0001 |

| | 0.11 | 0.12 | 0.13 | 0.14 | 0.15 | 0.16 | 0.17 | 0.18 | 0.19 | 0.20 |
|---|---|---|---|---|---|---|---|---|---|---|
| 0 | 0.1550 | 0.1293 | 0.1077 | 0.0895 | 0.0743 | 0.0614 | 0.0507 | 0.0418 | 0.0343 | 0.0281 |
| 1 | 0.3065 | 0.2822 | 0.2575 | 0.2332 | 0.2097 | 0.1873 | 0.1662 | 0.1468 | 0.1289 | 0.1126 |
| 2 | 0.2841 | 0.2886 | 0.2886 | 0.2847 | 0.2775 | 0.2675 | 0.2554 | 0.2416 | 0.2267 | 0.2111 |
| 3 | 0.1638 | 0.1837 | 0.2013 | 0.2163 | 0.2285 | 0.2378 | 0.2441 | 0.2475 | 0.2482 | 0.2463 |
| 4 | 0.0658 | 0.0814 | 0.0977 | 0.1144 | 0.1311 | 0.1472 | 0.1625 | 0.1766 | 0.1892 | 0.2001 |
| 5 | 0.0195 | 0.0266 | 0.0351 | 0.0447 | 0.0555 | 0.0673 | 0.0799 | 0.0930 | 0.1065 | 0.1201 |
| 6 | 0.0044 | 0.0067 | 0.0096 | 0.0133 | 0.0180 | 0.0235 | 0.0300 | 0.0374 | 0.0458 | 0.0550 |
| 7 | 0.0008 | 0.0013 | 0.0020 | 0.0031 | 0.0045 | 0.0064 | 0.0088 | 0.0117 | 0.0153 | 0.0197 |
| 8 | 0.0001 | 0.0002 | 0.0003 | 0.0006 | 0.0009 | 0.0014 | 0.0020 | 0.0029 | 0.0041 | 0.0055 |
| 9 | 0.0000 | 0.0000 | 0.0000 | 0.0001 | 0.0001 | 0.0002 | 0.0004 | 0.0006 | 0.0008 | 0.0012 |
| 10 | 0.0000 | 0.0000 | 0.0000 | 0.0000 | 0.0000 | 0.0000 | 0.0001 | 0.0001 | 0.0001 | 0.0002 |

(*Continued*)

**Table A.7** Binomial probability distribution

|    | 0.21 | 0.22 | 0.23 | 0.24 | 0.25 | 0.26 | 0.27 | 0.28 | 0.29 | 0.30 |
|----|------|------|------|------|------|------|------|------|------|------|
| 0  | 0.0230 | 0.0188 | 0.0153 | 0.0124 | 0.0100 | 0.0081 | 0.0065 | 0.0052 | 0.0042 | 0.0033 |
| 1  | 0.0979 | 0.0847 | 0.0730 | 0.0626 | 0.0535 | 0.0455 | 0.0385 | 0.0325 | 0.0273 | 0.0228 |
| 2  | 0.1952 | 0.1792 | 0.1635 | 0.1482 | 0.1336 | 0.1198 | 0.1068 | 0.0947 | 0.0835 | 0.0732 |
| 3  | 0.2421 | 0.2359 | 0.2279 | 0.2185 | 0.2079 | 0.1964 | 0.1843 | 0.1718 | 0.1591 | 0.1465 |
| 4  | 0.2092 | 0.2162 | 0.2212 | 0.2242 | 0.2252 | 0.2243 | 0.2215 | 0.2171 | 0.2112 | 0.2040 |
| 5  | 0.1334 | 0.1464 | 0.1586 | 0.1699 | 0.1802 | 0.1891 | 0.1966 | 0.2026 | 0.2071 | 0.2099 |
| 6  | 0.0650 | 0.0757 | 0.0869 | 0.0984 | 0.1101 | 0.1218 | 0.1333 | 0.1445 | 0.1551 | 0.1649 |
| 7  | 0.0247 | 0.0305 | 0.0371 | 0.0444 | 0.0524 | 0.0611 | 0.0704 | 0.0803 | 0.0905 | 0.1010 |
| 8  | 0.0074 | 0.0097 | 0.0125 | 0.0158 | 0.0197 | 0.0242 | 0.0293 | 0.0351 | 0.0416 | 0.0487 |
| 9  | 0.0017 | 0.0024 | 0.0033 | 0.0044 | 0.0058 | 0.0075 | 0.0096 | 0.0121 | 0.0151 | 0.0185 |
| 10 | 0.0003 | 0.0005 | 0.0007 | 0.0010 | 0.0014 | 0.0019 | 0.0025 | 0.0033 | 0.0043 | 0.0056 |
| 11 | 0.0000 | 0.0001 | 0.0001 | 0.0002 | 0.0002 | 0.0004 | 0.0005 | 0.0007 | 0.0010 | 0.0013 |
| 12 | 0.0000 | 0.0000 | 0.0000 | 0.0000 | 0.0000 | 0.0001 | 0.0001 | 0.0001 | 0.0002 | 0.0002 |

|    | 0.31 | 0.32 | 0.33 | 0.34 | 0.35 | 0.36 | 0.37 | 0.38 | 0.39 | 0.40 |
|----|------|------|------|------|------|------|------|------|------|------|
| 0  | 0.0026 | 0.0021 | 0.0016 | 0.0013 | 0.0010 | 0.0008 | 0.0006 | 0.0005 | 0.0004 | 0.0003 |
| 1  | 0.0190 | 0.0157 | 0.0130 | 0.0107 | 0.0087 | 0.0071 | 0.0058 | 0.0047 | 0.0038 | 0.0030 |
| 2  | 0.0639 | 0.0555 | 0.0480 | 0.0413 | 0.0353 | 0.0301 | 0.0255 | 0.0215 | 0.0180 | 0.0150 |
| 3  | 0.1341 | 0.1220 | 0.1103 | 0.0992 | 0.0888 | 0.0790 | 0.0699 | 0.0615 | 0.0538 | 0.0468 |
| 4  | 0.1958 | 0.1865 | 0.1766 | 0.1662 | 0.1553 | 0.1444 | 0.1333 | 0.1224 | 0.1118 | 0.1014 |
| 5  | 0.2111 | 0.2107 | 0.2088 | 0.2054 | 0.2008 | 0.1949 | 0.1879 | 0.1801 | 0.1715 | 0.1623 |
| 6  | 0.1739 | 0.1818 | 0.1885 | 0.1940 | 0.1982 | 0.2010 | 0.2024 | 0.2024 | 0.2010 | 0.1983 |
| 7  | 0.1116 | 0.1222 | 0.1326 | 0.1428 | 0.1524 | 0.1615 | 0.1698 | 0.1772 | 0.1836 | 0.1889 |
| 8  | 0.0564 | 0.0647 | 0.0735 | 0.0827 | 0.0923 | 0.1022 | 0.1122 | 0.1222 | 0.1320 | 0.1417 |
| 9  | 0.0225 | 0.0271 | 0.0322 | 0.0379 | 0.0442 | 0.1511 | 0.0586 | 0.0666 | 0.0750 | 0.0840 |
| 10 | 0.0071 | 0.0089 | 0.0111 | 0.0137 | 0.0167 | 0.0201 | 0.0241 | 0.0286 | 0.0336 | 0.0392 |
| 11 | 0.0017 | 0.0023 | 0.0030 | 0.0038 | 0.0049 | 0.0062 | 0.0077 | 0.0095 | 0.0117 | 0.0142 |
| 12 | 0.0003 | 0.0004 | 0.0006 | 0.0008 | 0.0011 | 0.0014 | 0.0019 | 0.0024 | 0.0031 | 0.0040 |
| 13 | 0.0000 | 0.0001 | 0.0001 | 0.0001 | 0.0002 | 0.0003 | 0.0003 | 0.0005 | 0.0006 | 0.0008 |
| 14 | 0.0000 | 0.0000 | 0.0000 | 0.0000 | 0.0000 | 0.0000 | 0.0000 | 0.0001 | 0.0001 | 0.0001 |

|    | 0.41 | 0.42 | 0.43 | 0.44 | 0.45 | 0.46 | 0.47 | 0.48 | 0.49 | 0.50 |
|----|------|------|------|------|------|------|------|------|------|------|
| 0  | 0.0002 | 0.0002 | 0.0001 | 0.0001 | 0.0001 | 0.0001 | 0.0000 | 0.0000 | 0.0000 | 0.0000 |
| 1  | 0.0024 | 0.0019 | 0.0015 | 0.0012 | 0.0009 | 0.0007 | 0.0005 | 0.0004 | 0.0003 | 0.0002 |
| 2  | 0.0125 | 0.0103 | 0.0085 | 0.0069 | 0.0056 | 0.0046 | 0.0037 | 0.0029 | 0.0023 | 0.0018 |
| 3  | 0.0405 | 0.0349 | 0.0299 | 0.0254 | 0.0215 | 0.0181 | 0.0151 | 0.0126 | 0.0104 | 0.0085 |
| 4  | 0.0915 | 0.0821 | 0.0732 | 0.0649 | 0.0572 | 0.0501 | 0.0436 | 0.0378 | 0.0325 | 0.0278 |
| 5  | 0.1526 | 0.1426 | 0.1325 | 0.1224 | 0.1123 | 0.1024 | 0.0929 | 0.0837 | 0.0749 | 0.0667 |
| 6  | 0.1944 | 0.1894 | 0.1833 | 0.1762 | 0.1684 | 0.1600 | 0.1510 | 0.1416 | 0.1319 | 0.1222 |
| 7  | 0.1930 | 0.1959 | 0.1975 | 0.1978 | 0.1969 | 0.1947 | 0.1912 | 0.1867 | 0.1811 | 0.1746 |
| 8  | 0.1509 | 0.1596 | 0.1676 | 0.1749 | 0.1812 | 0.1865 | 0.1908 | 0.1939 | 0.1958 | 0.1964 |
| 9  | 0.0932 | 0.1027 | 0.1124 | 0.1221 | 0.1318 | 0.1413 | 0.1504 | 0.1591 | 0.1672 | 0.1746 |

*(Continued)*

**Table A.7** Binomial probability distribution

| 10 | 0.0453 | 0.0521 | 0.0594 | 0.0672 | 0.0755 | 0.0842 | 0.0934 | 0.1028 | 0.1124 | 0.1222 |
|----|--------|--------|--------|--------|--------|--------|--------|--------|--------|--------|
| 11 | 0.0172 | 0.0206 | 0.0244 | 0.0288 | 0.0337 | 0.0391 | 0.0452 | 0.0518 | 0.0589 | 0.0667 |
| 12 | 0.0050 | 0.0062 | 0.0077 | 0.0094 | 0.0115 | 0.0139 | 0.0167 | 0.0199 | 0.0236 | 0.0278 |
| 13 | 0.0011 | 0.0014 | 0.0018 | 0.0023 | 0.0029 | 0.0036 | 0.0046 | 0.0057 | 0.0070 | 0.0085 |
| 14 | 0.0002 | 0.0002 | 0.0003 | 0.0004 | 0.0005 | 0.0007 | 0.0009 | 0.0011 | 0.0014 | 0.0018 |
| 15 | 0.0000 | 0.0000 | 0.0000 | 0.0000 | 0.0001 | 0.0001 | 0.0001 | 0.0001 | 0.0002 | 0.0002 |

$$n = 17$$

| $t$ \ $p$ | 0.01 | 0.02 | 0.03 | 0.04 | 0.05 | 0.06 | 0.07 | 0.08 | 0.09 | 0.10 |
|----|--------|--------|--------|--------|--------|--------|--------|--------|--------|--------|
| 0 | 0.8429 | 0.7093 | 0.5958 | 0.4996 | 0.4181 | 0.3493 | 0.2912 | 0.2423 | 0.2012 | 0.1668 |
| 1 | 0.1447 | 0.2461 | 0.3133 | 0.3539 | 0.3741 | 0.3790 | 0.3726 | 0.3582 | 0.3383 | 0.3150 |
| 2 | 0.0117 | 0.0402 | 0.0775 | 0.1180 | 0.1575 | 0.1935 | 0.2244 | 0.2492 | 0.2677 | 0.2800 |
| 3 | 0.0006 | 0.0041 | 0.0120 | 0.0246 | 0.0415 | 0.0618 | 0.0844 | 0.1083 | 0.1324 | 0.1556 |
| 4 | 0.0000 | 0.0003 | 0.0013 | 0.0036 | 0.0076 | 0.0138 | 0.0222 | 0.0330 | 0.0458 | 0.0605 |
| 5 | 0.0000 | 0.0000 | 0.0001 | 0.0004 | 0.0010 | 0.0023 | 0.0044 | 0.0075 | 0.0118 | 0.0175 |
| 6 | 0.0000 | 0.0000 | 0.0000 | 0.0000 | 0.0001 | 0.0003 | 0.0007 | 0.0013 | 0.0023 | 0.0039 |
| 7 | 0.0000 | 0.0000 | 0.0000 | 0.0000 | 0.0000 | 0.0000 | 0.0001 | 0.0002 | 0.0004 | 0.0007 |
| 8 | 0.0000 | 0.0000 | 0.0000 | 0.0000 | 0.0000 | 0.0000 | 0.0000 | 0.0000 | 0.0000 | 0.0001 |

|  | 0.11 | 0.12 | 0.13 | 0.14 | 0.15 | 0.16 | 0.17 | 0.18 | 0.19 | 0.20 |
|----|--------|--------|--------|--------|--------|--------|--------|--------|--------|--------|
| 0 | 0.1379 | 0.1138 | 0.0937 | 0.0770 | 0.0631 | 0.0516 | 0.0421 | 0.0343 | 0.0278 | 0.0225 |
| 1 | 0.2898 | 0.2638 | 0.2381 | 0.2131 | 0.1893 | 0.1671 | 0.1466 | 0.1279 | 0.1109 | 0.0957 |
| 2 | 0.2865 | 0.2878 | 0.2846 | 0.2775 | 0.2673 | 0.2547 | 0.2402 | 0.2245 | 0.2081 | 0.1914 |
| 3 | 0.1771 | 0.1963 | 0.2126 | 0.2259 | 0.2359 | 0.2425 | 0.2460 | 0.2464 | 0.2441 | 0.2393 |
| 4 | 0.0766 | 0.0937 | 0.1112 | 0.1287 | 0.1457 | 0.1617 | 0.1764 | 0.1893 | 0.2004 | 0.2093 |
| 5 | 0.0246 | 0.0332 | 0.0432 | 0.0545 | 0.0668 | 0.0801 | 0.0939 | 0.1081 | 0.1222 | 0.1361 |
| 6 | 0.0061 | 0.0091 | 0.0129 | 0.0177 | 0.0236 | 0.0305 | 0.0385 | 0.0474 | 0.0573 | 0.0680 |
| 7 | 0.0012 | 0.0019 | 0.0030 | 0.0045 | 0.0065 | 0.0091 | 0.0124 | 0.0164 | 0.0211 | 0.0267 |
| 8 | 0.0002 | 0.0003 | 0.0006 | 0.0009 | 0.0014 | 0.0022 | 0.0032 | 0.0045 | 0.0062 | 0.0084 |
| 9 | 0.0000 | 0.0000 | 0.0001 | 0.0002 | 0.0003 | 0.0004 | 0.0006 | 0.0010 | 0.0015 | 0.0021 |
| 10 | 0.0000 | 0.0000 | 0.0000 | 0.0000 | 0.0000 | 0.0001 | 0.0001 | 0.0002 | 0.0003 | 0.0004 |
| 11 | 0.0000 | 0.0000 | 0.0000 | 0.0000 | 0.0000 | 0.0000 | 0.0000 | 0.0000 | 0.0000 | 0.0001 |

|  | 0.21 | 0.22 | 0.23 | 0.24 | 0.25 | 0.26 | 0.27 | 0.28 | 0.29 | 0.30 |
|----|--------|--------|--------|--------|--------|--------|--------|--------|--------|--------|
| 0 | 0.0182 | 0.0146 | 0.0118 | 0.0094 | 0.0075 | 0.0060 | 0.0047 | 0.0038 | 0.0030 | 0.0023 |
| 1 | 0.0822 | 0.0702 | 0.0597 | 0.0505 | 0.0426 | 0.0357 | 0.0299 | 0.0248 | 0.0206 | 0.0169 |
| 2 | 0.1747 | 0.1584 | 0.1427 | 0.1277 | 0.1136 | 0.1005 | 0.0883 | 0.0772 | 0.0672 | 0.0581 |
| 3 | 0.2322 | 0.2234 | 0.2131 | 0.2016 | 0.1893 | 0.1765 | 0.1634 | 0.1502 | 0.1372 | 0.1245 |
| 4 | 0.2161 | 0.2205 | 0.2228 | 0.2228 | 0.2209 | 0.2170 | 0.2115 | 0.2044 | 0.1961 | 0.1868 |
| 5 | 0.1493 | 0.1617 | 0.1730 | 0.1830 | 0.1914 | 0.1982 | 0.2033 | 0.2067 | 0.2083 | 0.2081 |
| 6 | 0.0794 | 0.0912 | 0.1034 | 0.1156 | 0.1276 | 0.1393 | 0.1504 | 0.1608 | 0.1701 | 0.1784 |
| 7 | 0.0332 | 0.0404 | 0.0485 | 0.0573 | 0.0668 | 0.0769 | 0.0874 | 0.0982 | 0.1092 | 0.1201 |
| 8 | 0.0110 | 0.0143 | 0.0181 | 0.0226 | 0.0279 | 0.0338 | 0.0404 | 0.0478 | 0.0558 | 0.0644 |
| 9 | 0.0029 | 0.0040 | 0.0054 | 0.0071 | 0.0093 | 0.0119 | 0.0150 | 0.0186 | 0.0228 | 0.0276 |

(*Continued*)

**Table A.7** Binomial probability distribution

| | | | | | | | | | | |
|---|---|---|---|---|---|---|---|---|---|---|
| **10** | 0.0006 | 0.0009 | 0.0013 | 0.0018 | 0.0025 | 0.0033 | 0.0044 | 0.0058 | 0.0074 | 0.0095 |
| **11** | 0.0001 | 0.0002 | 0.0002 | 0.0004 | 0.0005 | 0.0007 | 0.0010 | 0.0014 | 0.0019 | 0.0026 |
| **12** | 0.0000 | 0.0000 | 0.0000 | 0.0001 | 0.0001 | 0.0001 | 0.0002 | 0.0003 | 0.0004 | 0.0006 |
| **13** | 0.0000 | 0.0000 | 0.0000 | 0.0000 | 0.0000 | 0.0000 | 0.0000 | 0.0000 | 0.0001 | 0.0001 |

| | **0.31** | **0.32** | **0.33** | **0.34** | **0.35** | **0.36** | **0.37** | **0.38** | **0.39** | **0.40** |
|---|---|---|---|---|---|---|---|---|---|---|
| **0** | 0.0018 | 0.0014 | 0.0011 | 0.0009 | 0.0007 | 0.0005 | 0.0004 | 0.0003 | 0.0002 | 0.0002 |
| **1** | 0.0139 | 0.0114 | 0.0093 | 0.0075 | 0.0060 | 0.0048 | 0.0039 | 0.0031 | 0.0024 | 0.0019 |
| **2** | 0.0500 | 0.0428 | 0.0364 | 0.0309 | 0.0260 | 0.0218 | 0.0182 | 0.0151 | 0.0125 | 0.0102 |
| **3** | 0.1123 | 0.1007 | 0.0898 | 0.0795 | 0.0701 | 0.0614 | 0.0534 | 0.0463 | 0.0398 | 0.0341 |
| **4** | 0.1766 | 0.1659 | 0.1547 | 0.1434 | 0.1320 | 0.1208 | 0.1099 | 0.0993 | 0.0892 | 0.0796 |
| **5** | 0.2063 | 0.2030 | 0.1982 | 0.1921 | 0.1849 | 0.1767 | 0.1677 | 0.1582 | 0.1482 | 0.1379 |
| **6** | 0.1854 | 0.1910 | 0.1952 | 0.1979 | 0.1991 | 0.1988 | 0.1970 | 0.1939 | 0.1895 | 0.1839 |
| **7** | 0.1309 | 0.1413 | 0.1511 | 0.1602 | 0.1685 | 0.1757 | 0.1818 | 0.1868 | 0.1904 | 0.1927 |
| **8** | 0.0735 | 0.0831 | 0.0930 | 0.1032 | 0.1134 | 0.1235 | 0.1335 | 0.1431 | 0.1521 | 0.1606 |
| **9** | 0.0330 | 0.0391 | 0.0458 | 0.0531 | 0.0611 | 0.0695 | 0.0784 | 0.0877 | 0.0973 | 0.1070 |
| **10** | 0.0119 | 0.0147 | 0.0181 | 0.0219 | 0.0263 | 0.0313 | 0.0368 | 0.0430 | 0.0498 | 0.0571 |
| **11** | 0.0034 | 0.0044 | 0.0057 | 0.0072 | 0.0090 | 0.0112 | 0.0138 | 0.0168 | 0.0202 | 0.0242 |
| **12** | 0.0008 | 0.0010 | 0.0014 | 0.0018 | 0.0024 | 0.0031 | 0.0040 | 0.0051 | 0.0065 | 0.0081 |
| **13** | 0.0001 | 0.0002 | 0.0003 | 0.0004 | 0.0005 | 0.0007 | 0.0009 | 0.0012 | 0.0016 | 0.0021 |
| **14** | 0.0000 | 0.0000 | 0.0000 | 0.0001 | 0.0001 | 0.0001 | 0.0002 | 0.0002 | 0.0003 | 0.0004 |
| **15** | 0.0000 | 0.0000 | 0.0000 | 0.0000 | 0.0000 | 0.0000 | 0.0000 | 0.0000 | 0.0000 | 0.0001 |

| | **0.41** | **0.42** | **0.43** | **0.44** | **0.45** | **0.46** | **0.47** | **0.48** | **0.49** | **0.50** |
|---|---|---|---|---|---|---|---|---|---|---|
| **0** | 0.0001 | 0.0001 | 0.0001 | 0.0001 | 0.0000 | 0.0000 | 0.0000 | 0.0000 | 0.0000 | 0.0000 |
| **1** | 0.0015 | 0.0012 | 0.0009 | 0.0007 | 0.0005 | 0.0004 | 0.0003 | 0.0002 | 0.0002 | 0.0001 |
| **2** | 0.0084 | 0.0068 | 0.0055 | 0.0044 | 0.0035 | 0.0028 | 0.0022 | 0.0017 | 0.0013 | 0.0010 |
| **3** | 0.0290 | 0.0246 | 0.0207 | 0.0173 | 0.0144 | 0.0119 | 0.0097 | 0.0079 | 0.0064 | 0.0052 |
| **4** | 0.0706 | 0.0622 | 0.0546 | 0.0475 | 0.0411 | 0.0354 | 0.0302 | 0.0257 | 0.0217 | 0.0182 |
| **5** | 0.1276 | 0.1172 | 0.1070 | 0.0971 | 0.0875 | 0.0784 | 0.0697 | 0.0616 | 0.0541 | 0.0472 |
| **6** | 0.1773 | 0.1697 | 0.1614 | 0.1525 | 0.1432 | 0.1335 | 0.1237 | 0.1138 | 0.1040 | 0.0944 |
| **7** | 0.1936 | 0.1932 | 0.1914 | 0.1883 | 0.1841 | 0.1787 | 0.1723 | 0.1650 | 0.1570 | 0.1484 |
| **8** | 0.1682 | 0.1748 | 0.1805 | 0.1850 | 0.1883 | 0.1903 | 0.1910 | 0.1904 | 0.1886 | 0.1855 |
| **9** | 0.1169 | 0.1266 | 0.1361 | 0.1453 | 0.1540 | 0.1621 | 0.1694 | 0.1758 | 0.1812 | 0.1855 |
| **10** | 0.0650 | 0.0733 | 0.0822 | 0.0914 | 0.1008 | 0.1105 | 0.1202 | 0.1298 | 0.1393 | 0.1484 |
| **11** | 0.0287 | 0.0338 | 0.0394 | 0.0457 | 0.0525 | 0.0599 | 0.0678 | 0.0763 | 0.0851 | 0.0944 |
| **12** | 0.0100 | 0.0122 | 0.0149 | 0.0179 | 0.0215 | 0.0255 | 0.0301 | 0.0352 | 0.0409 | 0.0472 |
| **13** | 0.0027 | 0.0034 | 0.0043 | 0.0054 | 0.0068 | 0.0084 | 0.0103 | 0.0125 | 0.0151 | 0.0182 |
| **14** | 0.0005 | 0.0007 | 0.0009 | 0.0012 | 0.0016 | 0.0020 | 0.0026 | 0.0033 | 0.0041 | 0.0052 |
| **15** | 0.0001 | 0.0001 | 0.0001 | 0.0002 | 0.0003 | 0.0003 | 0.0005 | 0.0006 | 0.0008 | 0.0010 |
| **16** | 0.0000 | 0.0000 | 0.0000 | 0.0000 | 0.0000 | 0.0000 | 0.0001 | 0.0001 | 0.0001 | 0.0001 |

*(Continued)*

**Table A.7** Binomial probability distribution

| | | | | | $n = 18$ | | | | | |
|---|---|---|---|---|---|---|---|---|---|---|
| $p$<br>$t$ | 0.01 | 0.02 | 0.03 | 0.04 | 0.05 | 0.06 | 0.07 | 0.08 | 0.09 | 0.10 |
| 0 | 0.8345 | 0.6951 | 0.5780 | 0.4796 | 0.3972 | 0.3283 | 0.2708 | 0.2229 | 0.1831 | 0.1501 |
| 1 | 0.1517 | 0.2554 | 0.3217 | 0.3597 | 0.3763 | 0.3772 | 0.3669 | 0.3489 | 0.3260 | 0.3002 |
| 2 | 0.0130 | 0.0443 | 0.0846 | 0.1274 | 0.1683 | 0.2047 | 0.2348 | 0.2579 | 0.2741 | 0.2835 |
| 3 | 0.0007 | 0.0048 | 0.0140 | 0.0283 | 0.0473 | 0.0697 | 0.0942 | 0.1196 | 0.1446 | 0.1680 |
| 4 | 0.0000 | 0.0004 | 0.0016 | 0.0044 | 0.0093 | 0.0167 | 0.0266 | 0.0390 | 0.0536 | 0.0700 |
| 5 | 0.0000 | 0.0000 | 0.0001 | 0.0005 | 0.0014 | 0.0030 | 0.0056 | 0.0095 | 0.0148 | 0.0218 |
| 6 | 0.0000 | 0.0000 | 0.0000 | 0.0000 | 0.0002 | 0.0004 | 0.0009 | 0.0018 | 0.0032 | 0.0052 |
| 7 | 0.0000 | 0.0000 | 0.0000 | 0.0000 | 0.0000 | 0.0000 | 0.0001 | 0.0003 | 0.0005 | 0.0010 |
| 8 | 0.0000 | 0.0000 | 0.0000 | 0.0000 | 0.0000 | 0.0000 | 0.0000 | 0.0000 | 0.0001 | 0.0002 |

| | 0.11 | 0.12 | 0.13 | 0.14 | 0.15 | 0.16 | 0.17 | 0.18 | 0.19 | 0.20 |
|---|---|---|---|---|---|---|---|---|---|---|
| 0 | 0.1227 | 0.1002 | 0.0815 | 0.0662 | 0.0536 | 0.0434 | 0.0349 | 0.0281 | 0.0225 | 0.0180 |
| 1 | 0.2731 | 0.2458 | 0.2193 | 0.1940 | 0.1704 | 0.1486 | 0.1288 | 0.1110 | 0.0951 | 0.0811 |
| 2 | 0.2869 | 0.2850 | 0.2785 | 0.2685 | 0.2556 | 0.2407 | 0.2243 | 0.2071 | 0.1897 | 0.1723 |
| 3 | 0.1891 | 0.2072 | 0.2220 | 0.2331 | 0.2406 | 0.2445 | 0.2450 | 0.2425 | 0.2373 | 0.2297 |
| 4 | 0.0877 | 0.1060 | 0.1244 | 0.1423 | 0.1592 | 0.1746 | 0.1882 | 0.1996 | 0.2087 | 0.2153 |
| 5 | 0.0303 | 0.0405 | 0.0520 | 0.0649 | 0.0787 | 0.0931 | 0.1079 | 0.1227 | 0.1371 | 0.1507 |
| 6 | 0.0081 | 0.0120 | 0.0168 | 0.0229 | 0.0301 | 0.0384 | 0.0479 | 0.0584 | 0.0697 | 0.0816 |
| 7 | 0.0017 | 0.0028 | 0.0043 | 0.0064 | 0.0091 | 0.0126 | 0.0168 | 0.0220 | 0.0280 | 0.0350 |
| 8 | 0.0003 | 0.0005 | 0.0009 | 0.0014 | 0.0022 | 0.0033 | 0.0047 | 0.0066 | 0.0090 | 0.0120 |
| 9 | 0.0000 | 0.0001 | 0.0001 | 0.0003 | 0.0004 | 0.0007 | 0.0011 | 0.0016 | 0.0024 | 0.0033 |
| 10 | 0.0000 | 0.0000 | 0.0000 | 0.0000 | 0.0001 | 0.0001 | 0.0002 | 0.0003 | 0.0005 | 0.0008 |
| 11 | 0.0000 | 0.0000 | 0.0000 | 0.0000 | 0.0000 | 0.0000 | 0.0000 | 0.0001 | 0.0001 | 0.0001 |

| | 0.21 | 0.22 | 0.23 | 0.24 | 0.25 | 0.26 | 0.27 | 0.28 | 0.29 | 0.30 |
|---|---|---|---|---|---|---|---|---|---|---|
| 0 | 0.0144 | 0.0114 | 0.0091 | 0.0072 | 0.0056 | 0.0044 | 0.0035 | 0.0027 | 0.0021 | 0.0016 |
| 1 | 0.0687 | 0.0580 | 0.0487 | 0.0407 | 0.0338 | 0.0280 | 0.0231 | 0.0189 | 0.0155 | 0.0126 |
| 2 | 0.1553 | 0.1390 | 0.1236 | 0.1092 | 0.0958 | 0.0836 | 0.0725 | 0.0626 | 0.0537 | 0.0458 |
| 3 | 0.2202 | 0.2091 | 0.1969 | 0.1839 | 0.1704 | 0.1567 | 0.1431 | 0.1298 | 0.1169 | 0.1046 |
| 4 | 0.2195 | 0.2212 | 0.2205 | 0.2177 | 0.2130 | 0.2065 | 0.1985 | 0.1892 | 0.1790 | 0.1681 |
| 5 | 0.1634 | 0.1747 | 0.1845 | 0.1925 | 0.1988 | 0.2031 | 0.2055 | 0.2061 | 0.2048 | 0.2017 |
| 6 | 0.0941 | 0.1067 | 0.1194 | 0.1317 | 0.1436 | 0.1546 | 0.1647 | 0.1736 | 0.1812 | 0.1873 |
| 7 | 0.0429 | 0.0516 | 0.0611 | 0.0713 | 0.0820 | 0.0931 | 0.1044 | 0.1157 | 0.1269 | 0.1376 |
| 8 | 0.0157 | 0.0200 | 0.0251 | 0.0310 | 0.0376 | 0.0450 | 0.0531 | 0.0619 | 0.0713 | 0.0811 |
| 9 | 0.0046 | 0.0063 | 0.0083 | 0.0109 | 0.0139 | 0.0176 | 0.0218 | 0.0267 | 0.0323 | 0.0386 |
| 10 | 0.0011 | 0.0016 | 0.0022 | 0.0031 | 0.0042 | 0.0056 | 0.0073 | 0.0094 | 0.0119 | 0.0149 |
| 11 | 0.0002 | 0.0003 | 0.0005 | 0.0007 | 0.0010 | 0.0014 | 0.0020 | 0.0026 | 0.0035 | 0.0046 |
| 12 | 0.0000 | 0.0001 | 0.0001 | 0.0001 | 0.0002 | 0.0003 | 0.0004 | 0.0006 | 0.0008 | 0.0012 |
| 13 | 0.0000 | 0.0000 | 0.0000 | 0.0000 | 0.0000 | 0.0000 | 0.0001 | 0.0001 | 0.0002 | 0.0002 |

(*Continued*)

**Table A.7** Binomial probability distribution

| | 0.31 | 0.32 | 0.33 | 0.34 | 0.35 | 0.36 | 0.37 | 0.38 | 0.39 | 0.40 |
|---|---|---|---|---|---|---|---|---|---|---|
| 0 | 0.0013 | 0.0010 | 0.0007 | 0.0006 | 0.0004 | 0.0003 | 0.0002 | 0.0002 | 0.0001 | 0.0001 |
| 1 | 0.0102 | 0.0082 | 0.0066 | 0.0052 | 0.0042 | 0.0033 | 0.0026 | 0.0020 | 0.0016 | 0.0012 |
| 2 | 0.0388 | 0.0327 | 0.0275 | 0.0229 | 0.0190 | 0.0157 | 0.0129 | 0.0105 | 0.0086 | 0.0069 |
| 3 | 0.0930 | 0.0822 | 0.0722 | 0.0630 | 0.0547 | 0.0471 | 0.0404 | 0.0344 | 0.0292 | 0.0246 |
| 4 | 0.1567 | 0.1450 | 0.1333 | 0.1217 | 0.1104 | 0.0994 | 0.0890 | 0.0791 | 0.0699 | 0.0614 |
| 5 | 0.1971 | 0.1911 | 0.1838 | 0.1755 | 0.1664 | 0.1566 | 0.1463 | 0.1358 | 0.1252 | 0.1146 |
| 6 | 0.1919 | 0.1948 | 0.1962 | 0.1959 | 0.1941 | 0.1908 | 0.1862 | 0.1803 | 0.1734 | 0.1655 |
| 7 | 0.1478 | 0.1572 | 0.1656 | 0.1730 | 0.1792 | 0.1840 | 0.1875 | 0.1895 | 0.1900 | 0.1892 |
| 8 | 0.0913 | 0.1017 | 0.1122 | 0.1226 | 0.1327 | 0.1423 | 0.1514 | 0.1597 | 0.1671 | 0.1734 |
| 9 | 0.0456 | 0.0532 | 0.0614 | 0.0701 | 0.0794 | 0.0890 | 0.0988 | 0.1087 | 0.1187 | 0.1284 |
| 10 | 0.0184 | 0.0225 | 0.0272 | 0.0325 | 0.0385 | 0.0450 | 0.0522 | 0.0600 | 0.0683 | 0.0771 |
| 11 | 0.0060 | 0.0077 | 0.0097 | 0.0122 | 0.0151 | 0.0184 | 0.0223 | 0.0267 | 0.0318 | 0.0374 |
| 12 | 0.0016 | 0.0021 | 0.0028 | 0.0037 | 0.0047 | 0.0060 | 0.0076 | 0.0096 | 0.0118 | 0.0145 |
| 13 | 0.0003 | 0.0005 | 0.0006 | 0.0009 | 0.0012 | 0.0016 | 0.0021 | 0.0027 | 0.0035 | 0.0045 |
| 14 | 0.0001 | 0.0001 | 0.0001 | 0.0002 | 0.0002 | 0.0003 | 0.0004 | 0.0006 | 0.0008 | 0.0011 |
| 15 | 0.0000 | 0.0000 | 0.0000 | 0.0000 | 0.0000 | 0.0000 | 0.0001 | 0.0001 | 0.0001 | 0.0002 |

| | 0.41 | 0.42 | 0.43 | 0.44 | 0.45 | 0.46 | 0.47 | 0.48 | 0.49 | 0.50 |
|---|---|---|---|---|---|---|---|---|---|---|
| 0 | 0.0001 | 0.0001 | 0.0000 | 0.0000 | 0.0000 | 0.0000 | 0.0000 | 0.0000 | 0.0000 | 0.0000 |
| 1 | 0.0009 | 0.0007 | 0.0005 | 0.0004 | 0.0003 | 0.0002 | 0.0002 | 0.0001 | 0.0001 | 0.0001 |
| 2 | 0.0055 | 0.0044 | 0.0035 | 0.0028 | 0.0022 | 0.0017 | 0.0013 | 0.0010 | 0.0008 | 0.0006 |
| 3 | 0.0206 | 0.0171 | 0.0141 | 0.0116 | 0.0095 | 0.0077 | 0.0062 | 0.0050 | 0.0039 | 0.0031 |
| 4 | 0.0536 | 0.0464 | 0.0400 | 0.0342 | 0.0291 | 0.0246 | 0.0206 | 0.0172 | 0.0142 | 0.0117 |
| 5 | 0.1042 | 0.0941 | 0.0844 | 0.0753 | 0.0666 | 0.0586 | 0.0512 | 0.0444 | 0.0382 | 0.0327 |
| 6 | 0.1569 | 0.1477 | 0.1380 | 0.1281 | 0.1181 | 0.1081 | 0.0983 | 0.0887 | 0.0796 | 0.0708 |
| 7 | 0.1869 | 0.1833 | 0.1785 | 0.1726 | 0.1657 | 0.1579 | 0.1494 | 0.1404 | 0.1310 | 0.1214 |
| 8 | 0.1786 | 0.1825 | 0.1852 | 0.1864 | 0.1864 | 0.1850 | 0.1822 | 0.1782 | 0.1731 | 0.1669 |
| 9 | 0.1379 | 0.1469 | 0.1552 | 0.1628 | 0.1694 | 0.1751 | 0.1795 | 0.1828 | 0.1848 | 0.1855 |
| 10 | 0.0862 | 0.0957 | 0.1054 | 0.1151 | 0.1248 | 0.1342 | 0.1433 | 0.1519 | 0.1598 | 0.1669 |
| 11 | 0.0436 | 0.0504 | 0.0578 | 0.0658 | 0.0742 | 0.0831 | 0.0924 | 0.1020 | 0.1117 | 0.1214 |
| 12 | 0.0177 | 0.0213 | 0.0254 | 0.0301 | 0.0354 | 0.0413 | 0.0478 | 0.1549 | 0.0626 | 0.0708 |
| 13 | 0.0057 | 0.0071 | 0.0089 | 0.0109 | 0.0134 | 0.0162 | 0.0196 | 0.0234 | 0.0278 | 0.0327 |
| 14 | 0.0014 | 0.0018 | 0.0024 | 0.0031 | 0.0039 | 0.0049 | 0.0062 | 0.0077 | 0.0095 | 0.0117 |
| 15 | 0.0003 | 0.0004 | 0.0005 | 0.0006 | 0.0009 | 0.0011 | 0.0015 | 0.0019 | 0.0024 | 0.0031 |
| 16 | 0.0000 | 0.0000 | 0.0001 | 0.0001 | 0.0001 | 0.0002 | 0.0002 | 0.0003 | 0.0004 | 0.0006 |
| 17 | 0.0000 | 0.0000 | 0.0000 | 0.0000 | 0.0000 | 0.0000 | 0.0000 | 0.0000 | 0.0000 | 0.0001 |

$n = 19$

| $p$ / $t$ | 0.01 | 0.02 | 0.03 | 0.04 | 0.05 | 0.06 | 0.07 | 0.08 | 0.09 | 0.10 |
|---|---|---|---|---|---|---|---|---|---|---|
| 0 | 0.8262 | 0.6812 | 0.5606 | 0.4604 | 0.3774 | 0.3086 | 0.2519 | 0.2051 | 0.1666 | 0.1351 |
| 1 | 0.1586 | 0.2642 | 0.3294 | 0.3645 | 0.3774 | 0.3743 | 0.3602 | 0.3389 | 0.3131 | 0.2852 |
| 2 | 0.0144 | 0.0485 | 0.0917 | 0.1367 | 0.1787 | 0.2150 | 0.2440 | 0.2652 | 0.2787 | 0.2852 |
| 3 | 0.0008 | 0.0056 | 0.0161 | 0.0323 | 0.0533 | 0.0778 | 0.1041 | 0.1307 | 0.1562 | 0.1796 |

(*Continued*)

**Table A.7** Binomial probability distribution

| | | | | | | | | | | |
|---|---|---|---|---|---|---|---|---|---|---|
| 4 | 0.0000 | 0.0005 | 0.0020 | 0.0054 | 0.0112 | 0.0199 | 0.0313 | 0.0455 | 0.0618 | 0.0798 |
| 5 | 0.0000 | 0.0000 | 0.0002 | 0.0007 | 0.0018 | 0.0038 | 0.0071 | 0.0119 | 0.0183 | 0.0266 |
| 6 | 0.0000 | 0.0000 | 0.0000 | 0.0001 | 0.0002 | 0.0006 | 0.0012 | 0.0024 | 0.0042 | 0.0069 |
| 7 | 0.0000 | 0.0000 | 0.0000 | 0.0000 | 0.0000 | 0.0001 | 0.0002 | 0.0004 | 0.0008 | 0.0014 |
| 8 | 0.0000 | 0.0000 | 0.0000 | 0.0000 | 0.0000 | 0.0000 | 0.0000 | 0.0001 | 0.0001 | 0.0002 |

| | 0.11 | 0.12 | 0.13 | 0.14 | 0.15 | 0.16 | 0.17 | 0.18 | 0.19 | 0.20 |
|---|---|---|---|---|---|---|---|---|---|---|
| 0 | 0.1092 | 0.0881 | 0.0709 | 0.0569 | 0.0456 | 0.0364 | 0.0290 | 0.0230 | 0.0182 | 0.0144 |
| 1 | 0.2565 | 0.2284 | 0.2014 | 0.1761 | 0.1529 | 0.1318 | 0.1129 | 0.0961 | 0.0813 | 0.0685 |
| 2 | 0.2854 | 0.2803 | 0.2708 | 0.2581 | 0.2428 | 0.2259 | 0.2081 | 0.1898 | 0.1717 | 0.1540 |
| 3 | 0.1999 | 0.2166 | 0.2293 | 0.2381 | 0.2428 | 0.2439 | 0.2415 | 0.2361 | 0.2282 | 0.2182 |
| 4 | 0.0988 | 0.1181 | 0.1371 | 0.1550 | 0.1714 | 0.1858 | 0.1979 | 0.2073 | 0.2141 | 0.2182 |
| 5 | 0.0366 | 0.0483 | 0.0614 | 0.0757 | 0.0907 | 0.1062 | 0.1216 | 0.1365 | 0.1507 | 0.1636 |
| 6 | 0.0106 | 0.0154 | 0.0214 | 0.0288 | 0.0374 | 0.0472 | 0.0581 | 0.0699 | 0.0825 | 0.0955 |
| 7 | 0.0024 | 0.0039 | 0.0059 | 0.0087 | 0.0122 | 0.0167 | 0.0221 | 0.0285 | 0.0359 | 0.0443 |
| 8 | 0.0004 | 0.0008 | 0.0013 | 0.0021 | 0.0032 | 0.0048 | 0.0068 | 0.0094 | 0.0126 | 0.0166 |
| 9 | 0.0001 | 0.0001 | 0.0002 | 0.0004 | 0.0007 | 0.0011 | 0.0017 | 0.0025 | 0.0036 | 0.0051 |
| 10 | 0.0000 | 0.0000 | 0.0000 | 0.0001 | 0.0001 | 0.0002 | 0.0003 | 0.0006 | 0.0009 | 0.0013 |
| 11 | 0.0000 | 0.0000 | 0.0000 | 0.0000 | 0.0000 | 0.0000 | 0.0001 | 0.0001 | 0.0002 | 0.0003 |

| | 0.21 | 0.22 | 0.23 | 0.24 | 0.25 | 0.26 | 0.27 | 0.28 | 0.29 | 0.30 |
|---|---|---|---|---|---|---|---|---|---|---|
| 0 | 0.0113 | 0.0089 | 0.0070 | 0.0054 | 0.0042 | 0.0033 | 0.0025 | 0.0019 | 0.0015 | 0.0011 |
| 1 | 0.0573 | 0.0477 | 0.0396 | 0.0326 | 0.0268 | 0.0219 | 0.0178 | 0.0144 | 0.0116 | 0.0093 |
| 2 | 0.1371 | 0.1212 | 0.1064 | 0.0927 | 0.0803 | 0.0692 | 0.0592 | 0.0503 | 0.0426 | 0.0358 |
| 3 | 0.2065 | 0.1937 | 0.1800 | 0.1659 | 0.1517 | 0.1377 | 0.1240 | 0.1109 | 0.0985 | 0.0869 |
| 4 | 0.2196 | 0.2185 | 0.2151 | 0.2096 | 0.2023 | 0.1935 | 0.1835 | 0.1726 | 0.1610 | 0.1491 |
| 5 | 0.1751 | 0.1849 | 0.1928 | 0.1986 | 0.2023 | 0.2040 | 0.2036 | 0.2013 | 0.1973 | 0.1916 |
| 6 | 0.1086 | 0.1217 | 0.1343 | 0.1463 | 0.1574 | 0.1672 | 0.1757 | 0.1827 | 0.1880 | 0.1916 |
| 7 | 0.0536 | 0.0637 | 0.0745 | 0.0858 | 0.0974 | 0.1091 | 0.1207 | 0.1320 | 0.1426 | 0.1525 |
| 8 | 0.0214 | 0.0270 | 0.0334 | 0.0406 | 0.0487 | 0.0575 | 0.0670 | 0.0770 | 0.0874 | 0.0981 |
| 9 | 0.0069 | 0.0093 | 0.0122 | 0.0157 | 0.0198 | 0.0247 | 0.0303 | 0.0366 | 0.0436 | 0.0514 |
| 10 | 0.0018 | 0.0026 | 0.0036 | 0.0050 | 0.0066 | 0.0087 | 0.0112 | 0.0142 | 0.0178 | 0.0220 |
| 11 | 0.0004 | 0.0006 | 0.0009 | 0.0013 | 0.0018 | 0.0025 | 0.0034 | 0.0045 | 0.0060 | 0.0077 |
| 12 | 0.0001 | 0.0001 | 0.0002 | 0.0003 | 0.0004 | 0.0006 | 0.0008 | 0.0012 | 0.0016 | 0.0022 |
| 13 | 0.0000 | 0.0000 | 0.0000 | 0.0000 | 0.0001 | 0.0001 | 0.0002 | 0.0002 | 0.0004 | 0.0005 |
| 14 | 0.0000 | 0.0000 | 0.0000 | 0.0000 | 0.0000 | 0.0000 | 0.0000 | 0.0000 | 0.0001 | 0.0001 |

| | 0.31 | 0.32 | 0.33 | 0.34 | 0.35 | 0.36 | 0.37 | 0.38 | 0.39 | 0.40 |
|---|---|---|---|---|---|---|---|---|---|---|
| 0 | 0.0009 | 0.0007 | 0.0005 | 0.0004 | 0.0003 | 0.0002 | 0.0002 | 0.0001 | 0.0001 | 0.0001 |
| 1 | 0.0074 | 0.0059 | 0.0046 | 0.0036 | 0.0029 | 0.0022 | 0.0017 | 0.0013 | 0.0010 | 0.0008 |
| 2 | 0.0299 | 0.0249 | 0.0206 | 0.0169 | 0.0138 | 0.0112 | 0.0091 | 0.0073 | 0.0058 | 0.0046 |
| 3 | 0.0762 | 0.0664 | 0.0574 | 0.0494 | 0.0422 | 0.0358 | 0.0302 | 0.0253 | 0.0211 | 0.0175 |
| 4 | 0.1370 | 0.1249 | 0.1131 | 0.1017 | 0.0909 | 0.0806 | 0.0710 | 0.0621 | 0.0540 | 0.0467 |
| 5 | 0.1846 | 0.1764 | 0.1672 | 0.1572 | 0.1468 | 0.1360 | 0.1251 | 0.1143 | 0.1036 | 0.0933 |
| 6 | 0.1935 | 0.1936 | 0.1921 | 0.1890 | 0.1844 | 0.1785 | 0.1714 | 0.1634 | 0.1546 | 0.1451 |

(*Continued*)

**Table A.7** Binomial probability distribution

| | | | | | | | | | | |
|---|---|---|---|---|---|---|---|---|---|---|
| 7 | 0.1615 | 0.1692 | 0.1757 | 0.1808 | 0.1844 | 0.1865 | 0.1870 | 0.1860 | 0.1835 | 0.1797 |
| 8 | 0.1088 | 0.1195 | 0.1298 | 0.1397 | 0.1489 | 0.1573 | 0.1647 | 0.1710 | 0.1760 | 0.1797 |
| 9 | 0.0597 | 0.0687 | 0.0782 | 0.0880 | 0.0980 | 0.1082 | 0.1182 | 0.1281 | 0.1375 | 0.1464 |
| 10 | 0.0268 | 0.0323 | 0.0385 | 0.0453 | 0.0528 | 0.0608 | 0.0694 | 0.0785 | 0.0879 | 0.0976 |
| 11 | 0.0099 | 0.0124 | 0.0155 | 0.0191 | 0.0233 | 0.0280 | 0.0334 | 0.0394 | 0.0460 | 0.0532 |
| 12 | 0.0030 | 0.0039 | 0.0051 | 0.0066 | 0.0083 | 0.0105 | 0.0131 | 0.0161 | 0.0196 | 0.0237 |
| 13 | 0.0007 | 0.0010 | 0.0014 | 0.0018 | 0.0024 | 0.0032 | 0.0041 | 0.0053 | 0.0067 | 0.0085 |
| 14 | 0.0001 | 0.0002 | 0.0003 | 0.0004 | 0.0006 | 0.0008 | 0.0010 | 0.0014 | 0.0018 | 0.0024 |
| 15 | 0.0000 | 0.0000 | 0.0000 | 0.0001 | 0.0001 | 0.0001 | 0.0002 | 0.0003 | 0.0004 | 0.0005 |
| 16 | 0.0000 | 0.0000 | 0.0000 | 0.0000 | 0.0000 | 0.0000 | 0.0000 | 0.0000 | 0.0001 | 0.0001 |

| | 0.41 | 0.42 | 0.43 | 0.44 | 0.45 | 0.46 | 0.47 | 0.48 | 0.49 | 0.50 |
|---|---|---|---|---|---|---|---|---|---|---|
| 0 | 0.0000 | 0.0000 | 0.0000 | 0.0000 | 0.0000 | 0.0000 | 0.0000 | 0.0000 | 0.0000 | 0.0000 |
| 1 | 0.0006 | 0.0004 | 0.0003 | 0.0002 | 0.0002 | 0.0001 | 0.0001 | 0.0001 | 0.0001 | 0.0000 |
| 2 | 0.0037 | 0.0029 | 0.0022 | 0.0017 | 0.0013 | 0.0010 | 0.0008 | 0.0006 | 0.0004 | 0.0003 |
| 3 | 0.0144 | 0.0118 | 0.0096 | 0.0077 | 0.0062 | 0.0049 | 0.0039 | 0.0031 | 0.0024 | 0.0018 |
| 4 | 0.0400 | 0.0341 | 0.0289 | 0.0243 | 0.0203 | 0.0168 | 0.0138 | 0.0113 | 0.0092 | 0.0074 |
| 5 | 0.0834 | 0.0741 | 0.0653 | 0.0572 | 0.0497 | 0.0429 | 0.0368 | 0.0313 | 0.0265 | 0.0222 |
| 6 | 0.1353 | 0.1252 | 0.1150 | 0.1049 | 0.0949 | 0.0853 | 0.0751 | 0.0674 | 0.0593 | 0.0518 |
| 7 | 0.1746 | 0.1683 | 0.1611 | 0.1530 | 0.1443 | 0.1350 | 0.1254 | 0.1156 | 0.1058 | 0.0961 |
| 8 | 0.1820 | 0.1829 | 0.1823 | 0.1803 | 0.1771 | 0.1725 | 0.1668 | 0.1601 | 0.1525 | 0.1442 |
| 9 | 0.1546 | 0.1618 | 0.1681 | 0.1732 | 0.1771 | 0.1796 | 0.1808 | 0.1806 | 0.1791 | 0.1762 |
| 10 | 0.1074 | 0.1172 | 0.1268 | 0.1361 | 0.1449 | 0.1530 | 0.1603 | 0.1667 | 0.1721 | 0.1762 |
| 11 | 0.0611 | 0.0694 | 0.0783 | 0.0875 | 0.0970 | 0.1066 | 0.1163 | 0.1259 | 0.1352 | 0.1442 |
| 12 | 0.0283 | 0.0335 | 0.0394 | 0.0458 | 0.0529 | 0.0606 | 0.0688 | 0.0775 | 0.0866 | 0.0961 |
| 13 | 0.0106 | 0.0131 | 0.0160 | 0.0194 | 0.0233 | 0.0278 | 0.0328 | 0.0385 | 0.0448 | 0.0518 |
| 14 | 0.0032 | 0.0041 | 0.0052 | 0.0065 | 0.0082 | 0.0101 | 0.0125 | 0.0152 | 0.0185 | 0.0222 |
| 15 | 0.0007 | 0.0010 | 0.0013 | 0.0017 | 0.0022 | 0.0029 | 0.0037 | 0.0047 | 0.0059 | 0.0074 |
| 16 | 0.0001 | 0.0002 | 0.0002 | 0.0003 | 0.0005 | 0.0006 | 0.0008 | 0.0011 | 0.0014 | 0.0018 |
| 17 | 0.0000 | 0.0000 | 0.0000 | 0.0000 | 0.0001 | 0.0001 | 0.0001 | 0.0002 | 0.0002 | 0.0003 |

$n = 20$

| $t$ \ $p$ | 0.01 | 0.02 | 0.03 | 0.04 | 0.05 | 0.06 | 0.07 | 0.08 | 0.09 | 0.10 |
|---|---|---|---|---|---|---|---|---|---|---|
| 0 | 0.8179 | 0.6676 | 0.5438 | 0.4420 | 0.3585 | 0.2901 | 0.2342 | 0.1887 | 0.1516 | 0.1216 |
| 1 | 0.1652 | 0.2725 | 0.3364 | 0.3683 | 0.3774 | 0.3703 | 0.3526 | 0.3282 | 0.3000 | 0.2702 |
| 2 | 0.0159 | 0.0528 | 0.0988 | 0.1458 | 0.1887 | 0.2246 | 0.2521 | 0.2711 | 0.2818 | 0.2852 |
| 3 | 0.0010 | 0.0065 | 0.0183 | 0.0364 | 0.0596 | 0.0860 | 0.1139 | 0.1414 | 0.1672 | 0.1901 |
| 4 | 0.0000 | 0.0006 | 0.0024 | 0.0065 | 0.0133 | 0.0233 | 0.0364 | 0.0523 | 0.0703 | 0.0898 |
| 5 | 0.0000 | 0.0000 | 0.0002 | 0.0009 | 0.0022 | 0.0048 | 0.0088 | 0.0145 | 0.0222 | 0.0319 |
| 6 | 0.0000 | 0.0000 | 0.0000 | 0.0001 | 0.0003 | 0.0008 | 0.0017 | 0.0032 | 0.0055 | 0.0089 |
| 7 | 0.0000 | 0.0000 | 0.0000 | 0.0000 | 0.0000 | 0.0001 | 0.0002 | 0.0005 | 0.0011 | 0.0020 |
| 8 | 0.0000 | 0.0000 | 0.0000 | 0.0000 | 0.0000 | 0.0000 | 0.0000 | 0.0001 | 0.0002 | 0.0004 |
| 9 | 0.0000 | 0.0000 | 0.0000 | 0.0000 | 0.0000 | 0.0000 | 0.0000 | 0.0000 | 0.0000 | 0.0001 |

(*Continued*)

**Table A.7** Binomial probability distribution

| | 0.11 | 0.12 | 0.13 | 0.14 | 0.15 | 0.16 | 0.17 | 0.18 | 0.19 | 0.20 |
|---|---|---|---|---|---|---|---|---|---|---|
| 0 | 0.0972 | 0.0776 | 0.0617 | 0.0490 | 0.0388 | 0.0306 | 0.0241 | 0.0189 | 0.0148 | 0.0115 |
| 1 | 0.2403 | 0.2115 | 0.1844 | 0.1595 | 0.1368 | 0.1165 | 0.0986 | 0.0829 | 0.0693 | 0.0576 |
| 2 | 0.2822 | 0.2740 | 0.2618 | 0.2466 | 0.2293 | 0.2109 | 0.1919 | 0.1730 | 0.1545 | 0.1369 |
| 3 | 0.2093 | 0.2242 | 0.2347 | 0.2409 | 0.2428 | 0.2410 | 0.2358 | 0.2278 | 0.2175 | 0.2054 |
| 4 | 0.1099 | 0.1299 | 0.1491 | 0.1666 | 0.1821 | 0.1951 | 0.2053 | 0.2125 | 0.2168 | 0.2182 |
| 5 | 0.0435 | 0.0567 | 0.0713 | 0.1868 | 0.1028 | 0.1189 | 0.1345 | 0.1493 | 0.1627 | 0.1746 |
| 6 | 0.0134 | 0.0193 | 0.0266 | 0.0353 | 0.0454 | 0.0566 | 0.0689 | 0.0819 | 0.0954 | 0.1091 |
| 7 | 0.0033 | 0.0053 | 0.0080 | 0.0115 | 0.0160 | 0.0216 | 0.0282 | 0.0360 | 0.0448 | 0.0545 |
| 8 | 0.0007 | 0.0012 | 0.0019 | 0.0030 | 0.0046 | 0.0067 | 0.0094 | 0.0128 | 0.0171 | 0.0222 |
| 9 | 0.0001 | 0.0002 | 0.0004 | 0.0007 | 0.0011 | 0.0017 | 0.0026 | 0.0038 | 0.0053 | 0.0074 |
| 10 | 0.0000 | 0.0000 | 0.0001 | 0.0001 | 0.0002 | 0.0004 | 0.0006 | 0.0009 | 0.0014 | 0.0020 |
| 11 | 0.0000 | 0.0000 | 0.0000 | 0.0000 | 0.0000 | 0.0001 | 0.0001 | 0.0002 | 0.0003 | 0.0005 |
| 12 | 0.0000 | 0.0000 | 0.0000 | 0.0000 | 0.0000 | 0.0000 | 0.0000 | 0.0000 | 0.0001 | 0.0001 |

| | 0.21 | 0.22 | 0.23 | 0.24 | 0.25 | 0.26 | 0.27 | 0.28 | 0.29 | 0.30 |
|---|---|---|---|---|---|---|---|---|---|---|
| 0 | 0.0090 | 0.0069 | 0.0054 | 0.0041 | 0.0032 | 0.0024 | 0.0018 | 0.0014 | 0.0011 | 0.0008 |
| 1 | 0.0477 | 0.0392 | 0.0321 | 0.0261 | 0.0211 | 0.0170 | 0.0137 | 0.0109 | 0.0087 | 0.0068 |
| 2 | 0.1204 | 0.1050 | 0.0910 | 0.0783 | 0.0669 | 0.0569 | 0.0480 | 0.0403 | 0.0336 | 0.0278 |
| 3 | 0.1920 | 0.1777 | 0.1631 | 0.1484 | 0.1339 | 0.1199 | 0.1065 | 0.0940 | 0.0823 | 0.0716 |
| 4 | 0.2169 | 0.2131 | 0.2070 | 0.1991 | 0.1897 | 0.1790 | 0.1675 | 0.1553 | 0.1429 | 0.1304 |
| 5 | 0.1845 | 0.1923 | 0.1979 | 0.2012 | 0.2023 | 0.2013 | 0.1982 | 0.1933 | 0.1868 | 0.1789 |
| 6 | 0.1226 | 0.1356 | 0.1478 | 0.1589 | 0.1686 | 0.1768 | 0.1833 | 0.1879 | 0.1907 | 0.1916 |
| 7 | 0.0652 | 0.0765 | 0.0883 | 0.1003 | 0.1124 | 0.1242 | 0.1356 | 0.1462 | 0.1558 | 0.1643 |
| 8 | 0.0282 | 0.0351 | 0.0429 | 0.0515 | 0.0609 | 0.0709 | 0.0815 | 0.0924 | 0.1034 | 0.1144 |
| 9 | 0.0100 | 0.0132 | 0.0171 | 0.0217 | 0.0271 | 0.0332 | 0.0402 | 0.0479 | 0.0563 | 0.0654 |
| 10 | 0.0029 | 0.0041 | 0.0056 | 0.0075 | 0.0099 | 0.0128 | 0.0163 | 0.0205 | 0.0253 | 0.0308 |
| 11 | 0.0007 | 0.0010 | 0.0015 | 0.0022 | 0.0030 | 0.0041 | 0.0055 | 0.0072 | 0.0094 | 0.0120 |
| 12 | 0.0001 | 0.0002 | 0.0003 | 0.0005 | 0.0008 | 0.0011 | 0.0015 | 0.0021 | 0.0029 | 0.0039 |
| 13 | 0.0000 | 0.0000 | 0.0001 | 0.0001 | 0.0002 | 0.0002 | 0.0003 | 0.0005 | 0.0007 | 0.0010 |
| 14 | 0.0000 | 0.0000 | 0.0000 | 0.0000 | 0.0000 | 0.0000 | 0.0001 | 0.0001 | 0.0001 | 0.0002 |

| | 0.31 | 0.32 | 0.33 | 0.34 | 0.35 | 0.36 | 0.37 | 0.38 | 0.39 | 0.40 |
|---|---|---|---|---|---|---|---|---|---|---|
| 0 | 0.0006 | 0.0004 | 0.0003 | 0.0002 | 0.0002 | 0.0001 | 0.0001 | 0.0001 | 0.0001 | 0.0000 |
| 1 | 0.0054 | 0.0042 | 0.0033 | 0.0025 | 0.0020 | 0.0015 | 0.0011 | 0.0009 | 0.0007 | 0.0005 |
| 2 | 0.0229 | 0.0188 | 0.0153 | 0.0124 | 0.0100 | 0.0080 | 0.0064 | 0.0050 | 0.0040 | 0.0031 |
| 3 | 0.0619 | 0.0531 | 0.0453 | 0.0383 | 0.0323 | 0.0270 | 0.0224 | 0.0185 | 0.0152 | 0.0123 |
| 4 | 0.1181 | 0.1062 | 0.0947 | 0.0839 | 0.0738 | 0.0645 | 0.0559 | 0.0482 | 0.0412 | 0.0350 |
| 5 | 0.1698 | 0.1599 | 0.1493 | 0.1384 | 0.1272 | 0.1161 | 0.1051 | 0.0945 | 0.0843 | 0.0746 |
| 6 | 0.1907 | 0.1881 | 0.1839 | 0.1782 | 0.1712 | 0.1632 | 0.1543 | 0.1447 | 0.1347 | 0.1244 |
| 7 | 0.1714 | 0.1770 | 0.1811 | 0.1836 | 0.1844 | 0.1836 | 0.1812 | 0.1774 | 0.1722 | 0.1659 |
| 8 | 0.1251 | 0.1354 | 0.1450 | 0.1537 | 0.1614 | 0.1678 | 0.1730 | 0.1767 | 0.1790 | 0.1797 |
| 9 | 0.0750 | 0.0849 | 0.0952 | 0.1056 | 0.1158 | 0.1259 | 0.1354 | 0.1444 | 0.1526 | 0.1597 |
| 10 | 0.0370 | 0.0440 | 0.0516 | 0.0598 | 0.0686 | 0.0779 | 0.0875 | 0.0974 | 0.1073 | 0.1171 |

(*Continued*)

**Table A.7** Binomial probability distribution

| | | | | | | | | | | |
|---|---|---|---|---|---|---|---|---|---|---|
| 11 | 0.0151 | 0.0188 | 0.0231 | 0.0280 | 0.0336 | 0.0398 | 0.0467 | 0.0542 | 0.0624 | 0.0710 |
| 12 | 0.0051 | 0.0066 | 0.0085 | 0.0108 | 0.0136 | 0.0168 | 0.0206 | 0.0249 | 0.0299 | 0.0355 |
| 13 | 0.0014 | 0.0019 | 0.0026 | 0.0034 | 0.0045 | 0.0058 | 0.0074 | 0.0094 | 0.0118 | 0.0146 |
| 14 | 0.0003 | 0.0005 | 0.0006 | 0.0009 | 0.0012 | 0.0016 | 0.0022 | 0.0029 | 0.0038 | 0.0049 |
| 15 | 0.0001 | 0.0001 | 0.0001 | 0.0002 | 0.0003 | 0.0004 | 0.0005 | 0.0007 | 0.0010 | 0.0013 |
| 16 | 0.0000 | 0.0000 | 0.0000 | 0.0000 | 0.0000 | 0.0001 | 0.0001 | 0.0001 | 0.0002 | 0.0003 |

| | 0.41 | 0.42 | 0.43 | 0.44 | 0.45 | 0.46 | 0.47 | 0.48 | 0.49 | 0.50 |
|---|---|---|---|---|---|---|---|---|---|---|
| 0 | 0.0000 | 0.0000 | 0.0000 | 0.0000 | 0.0000 | 0.0000 | 0.0000 | 0.0000 | 0.0000 | 0.0000 |
| 1 | 0.0004 | 0.0003 | 0.0002 | 0.0001 | 0.0001 | 0.0001 | 0.0001 | 0.0000 | 0.0000 | 0.0000 |
| 2 | 0.0024 | 0.0018 | 0.0014 | 0.0011 | 0.0008 | 0.0006 | 0.0005 | 0.0003 | 0.0002 | 0.0002 |
| 3 | 0.0100 | 0.0080 | 0.0064 | 0.0051 | 0.0040 | 0.0031 | 0.0024 | 0.0019 | 0.0014 | 0.0011 |
| 4 | 0.0295 | 0.0247 | 0.0206 | 0.0170 | 0.0139 | 0.0113 | 0.0092 | 0.0074 | 0.0059 | 0.0046 |
| 5 | 0.0656 | 0.0573 | 0.0496 | 0.0427 | 0.0365 | 0.0309 | 0.0260 | 0.0217 | 0.0180 | 0.0148 |
| 6 | 0.1140 | 0.1037 | 0.0936 | 0.0839 | 0.0746 | 0.0658 | 0.0577 | 0.0501 | 0.0432 | 0.0370 |
| 7 | 0.1585 | 0.1502 | 0.1413 | 0.1318 | 0.1221 | 0.1122 | 0.1023 | 0.0925 | 0.0830 | 0.0739 |
| 8 | 0.1790 | 0.1768 | 0.1732 | 0.1683 | 0.1623 | 0.1553 | 0.1474 | 0.1388 | 0.1296 | 0.1201 |
| 9 | 0.1658 | 0.1707 | 0.1742 | 0.1763 | 0.1771 | 0.1763 | 0.1742 | 0.1708 | 0.1661 | 0.1602 |
| 10 | 0.1268 | 0.1359 | 0.1446 | 0.1524 | 0.1593 | 0.1652 | 0.1700 | 0.1734 | 0.1755 | 0.1762 |
| 11 | 0.0801 | 0.0895 | 0.0991 | 0.1089 | 0.1185 | 0.1280 | 0.1370 | 0.1455 | 0.1533 | 0.1602 |
| 12 | 0.0417 | 0.0486 | 0.0561 | 0.0642 | 0.0727 | 0.0818 | 0.0911 | 0.1007 | 0.1105 | 0.1201 |
| 13 | 0.0178 | 0.0217 | 0.0260 | 0.0310 | 0.0366 | 0.0429 | 0.0497 | 0.0572 | 0.0653 | 0.0739 |
| 14 | 0.0062 | 0.0078 | 0.0098 | 0.0122 | 0.0150 | 0.0183 | 0.0221 | 0.0264 | 0.0314 | 0.0370 |
| 15 | 0.0017 | 0.0023 | 0.0030 | 0.0038 | 0.0049 | 0.0062 | 0.0078 | 0.0098 | 0.0121 | 0.0148 |
| 16 | 0.0004 | 0.0005 | 0.0007 | 0.0009 | 0.0013 | 0.0017 | 0.0022 | 0.0028 | 0.0036 | 0.0046 |
| 17 | 0.0001 | 0.0001 | 0.0001 | 0.0002 | 0.0002 | 0.0003 | 0.0005 | 0.0006 | 0.0008 | 0.0011 |
| 18 | 0.0000 | 0.0000 | 0.0000 | 0.0000 | 0.0000 | 0.0000 | 0.0001 | 0.0001 | 0.0001 | 0.0002 |

$n = 25$

| $p$ / $t$ | 0.01 | 0.02 | 0.03 | 0.04 | 0.05 | 0.06 | 0.07 | 0.08 | 0.09 | 0.10 |
|---|---|---|---|---|---|---|---|---|---|---|
| 0 | 0.7778 | 0.6035 | 0.4670 | 0.3604 | 0.2774 | 0.2129 | 0.1630 | 0.1244 | 0.0946 | 0.0718 |
| 1 | 0.1964 | 0.3079 | 0.3611 | 0.3754 | 0.3650 | 0.3398 | 0.3066 | 0.2704 | 0.2340 | 0.1994 |
| 2 | 0.0238 | 0.0754 | 0.1340 | 0.1877 | 0.2305 | 0.2602 | 0.2770 | 0.2821 | 0.2777 | 0.2659 |
| 3 | 0.0018 | 0.0118 | 0.0318 | 0.0600 | 0.0930 | 0.1273 | 0.1598 | 0.1881 | 0.2106 | 0.2265 |
| 4 | 0.0001 | 0.0013 | 0.0054 | 0.0137 | 0.0269 | 0.0447 | 0.0662 | 0.0899 | 0.1145 | 0.1384 |
| 5 | 0.0000 | 0.0001 | 0.0007 | 0.0024 | 0.0060 | 0.0120 | 0.0209 | 0.0329 | 0.0476 | 0.0646 |
| 6 | 0.0000 | 0.0000 | 0.0001 | 0.0003 | 0.0010 | 0.0026 | 0.0052 | 0.0095 | 0.0157 | 0.0239 |
| 7 | 0.0000 | 0.0000 | 0.0000 | 0.0000 | 0.0001 | 0.0004 | 0.0011 | 0.0022 | 0.0042 | 0.0072 |
| 8 | 0.0000 | 0.0000 | 0.0000 | 0.0000 | 0.0000 | 0.0001 | 0.0002 | 0.0004 | 0.0009 | 0.0018 |
| 9 | 0 | 0.0000 | 0.0000 | 0.0000 | 0.0000 | 0.0000 | 0.0000 | 0.0001 | 0.0002 | 0.0004 |
| 10 | 0 | 0.0000 | 0.0000 | 0.0000 | 0.0000 | 0.0000 | 0.0000 | 0.0000 | 0.0000 | 0.0001 |

(*Continued*)

**Table A.7** Binomial probability distribution

| | 0.11 | 0.12 | 0.13 | 0.14 | 0.15 | 0.16 | 0.17 | 0.18 | 0.19 | 0.20 |
|---|---|---|---|---|---|---|---|---|---|---|
| **0** | 0.0543 | 0.0409 | 0.0308 | 0.0230 | 0.0172 | 0.0128 | 0.0095 | 0.0070 | 0.0052 | 0.0038 |
| **1** | 0.1678 | 0.1395 | 0.1149 | 0.0938 | 0.0759 | 0.0609 | 0.0486 | 0.0384 | 0.0302 | 0.0236 |
| **2** | 0.2488 | 0.2283 | 0.2060 | 0.1832 | 0.1607 | 0.1392 | 0.1193 | 0.1012 | 0.0851 | 0.0708 |
| **3** | 0.2358 | 0.2387 | 0.2360 | 0.2286 | 0.2174 | 0.2033 | 0.1874 | 0.1704 | 0.1530 | 0.1358 |
| **4** | 0.1603 | 0.1790 | 0.1940 | 0.2047 | 0.2110 | 0.2130 | 0.2111 | 0.2057 | 0.1974 | 0.1867 |
| **5** | 0.0832 | 0.1025 | 0.1217 | 0.1399 | 0.1564 | 0.1704 | 0.1816 | 0.1897 | 0.1945 | 0.1960 |
| **6** | 0.0343 | 0.0466 | 0.0606 | 0.0759 | 0.0920 | 0.1082 | 0.1240 | 0.1388 | 0.1520 | 0.1633 |
| **7** | 0.0115 | 0.0173 | 0.0246 | 0.0336 | 0.0441 | 0.0559 | 0.0689 | 0.0827 | 0.0968 | 0.1108 |
| **8** | 0.0032 | 0.0053 | 0.0083 | 0.0123 | 0.0175 | 0.0240 | 0.0318 | 0.0408 | 0.0511 | 0.0623 |
| **9** | 0.0007 | 0.0014 | 0.0023 | 0.0038 | 0.0058 | 0.0086 | 0.0123 | 0.0169 | 0.0226 | 0.0294 |
| **10** | 0.0001 | 0.0003 | 0.0006 | 0.0010 | 0.0016 | 0.0026 | 0.0040 | 0.0059 | 0.0085 | 0.0118 |
| **11** | 0.0000 | 0.0001 | 0.0001 | 0.0002 | 0.0004 | 0.0007 | 0.0011 | 0.0018 | 0.0027 | 0.0040 |
| **12** | 0.0000 | 0.0000 | 0.0000 | 0.0000 | 0.0001 | 0.0002 | 0.0003 | 0.0005 | 0.0007 | 0.0012 |
| **13** | 0.0000 | 0.0000 | 0.0000 | 0.0000 | 0.0000 | 0.0000 | 0.0001 | 0.0001 | 0.0002 | 0.0003 |
| **14** | 0.0000 | 0.0000 | 0.0000 | 0.0000 | 0.0000 | 0.0000 | 0.0000 | 0.0000 | 0.0000 | 0.0001 |

| | 0.21 | 0.22 | 0.23 | 0.24 | 0.25 | 0.26 | 0.27 | 0.28 | 0.29 | 0.30 |
|---|---|---|---|---|---|---|---|---|---|---|
| **0** | 0.0028 | 0.0020 | 0.0015 | 0.0010 | 0.0008 | 0.0005 | 0.0004 | 0.0003 | 0.0002 | 0.0001 |
| **1** | 0.0183 | 0.0141 | 0.0109 | 0.0083 | 0.0063 | 0.0047 | 0.0035 | 0.0026 | 0.0020 | 0.0014 |
| **2** | 0.0585 | 0.0479 | 0.0389 | 0.0314 | 0.0251 | 0.0199 | 0.0157 | 0.0123 | 0.0096 | 0.0074 |
| **3** | 0.1192 | 0.1035 | 0.0891 | 0.0759 | 0.0641 | 0.0537 | 0.0446 | 0.0367 | 0.0300 | 0.0243 |
| **4** | 0.1742 | 0.1606 | 0.1463 | 0.1318 | 0.1175 | 0.1037 | 0.0906 | 0.0785 | 0.0673 | 0.0572 |
| **5** | 0.1945 | 0.1903 | 0.1836 | 0.1749 | 0.1645 | 0.1531 | 0.1408 | 0.1282 | 0.1155 | 0.1030 |
| **6** | 0.1724 | 0.1789 | 0.1828 | 0.1841 | 0.1828 | 0.1793 | 0.1736 | 0.1661 | 0.1572 | 0.1472 |
| **7** | 0.1244 | 0.1369 | 0.1482 | 0.1578 | 0.1654 | 0.1709 | 0.1743 | 0.1754 | 0.1743 | 0.1712 |
| **8** | 0.0744 | 0.0869 | 0.0996 | 0.1121 | 0.1241 | 0.1351 | 0.1450 | 0.1535 | 0.1602 | 0.1651 |
| **9** | 0.0373 | 0.0463 | 0.0562 | 0.0669 | 0.0781 | 0.0897 | 0.1013 | 0.1127 | 0.1236 | 0.1336 |
| **10** | 0.0159 | 0.0209 | 0.0269 | 0.0338 | 0.0417 | 0.0504 | 0.0600 | 0.0701 | 0.0808 | 0.0916 |
| **11** | 0.0058 | 0.0080 | 0.0109 | 0.0145 | 0.0189 | 0.0242 | 0.0302 | 0.0372 | 0.0450 | 0.0536 |
| **12** | 0.0018 | 0.0026 | 0.0038 | 0.0054 | 0.0074 | 0.0099 | 0.0130 | 0.0169 | 0.0214 | 0.0268 |
| **13** | 0.0005 | 0.0007 | 0.0011 | 0.0017 | 0.0025 | 0.0035 | 0.0048 | 0.0066 | 0.0088 | 0.0115 |
| **14** | 0.0001 | 0.0002 | 0.0003 | 0.0005 | 0.0007 | 0.0010 | 0.0015 | 0.0022 | 0.0031 | 0.0042 |
| **15** | 0.0000 | 0.0000 | 0.0001 | 0.0001 | 0.0002 | 0.0003 | 0.0004 | 0.0006 | 0.0009 | 0.0013 |
| **16** | 0.0000 | 0.0000 | 0.0000 | 0.0000 | 0.0000 | 0.0001 | 0.0001 | 0.0002 | 0.0002 | 0.0004 |
| **17** | 0.0000 | 0.0000 | 0.0000 | 0.0000 | 0.0000 | 0.0000 | 0.0000 | 0.0000 | 0.0001 | 0.0001 |

| | 0.31 | 0.32 | 0.33 | 0.34 | 0.35 | 0.36 | 0.37 | 0.38 | 0.39 | 0.40 |
|---|---|---|---|---|---|---|---|---|---|---|
| **0** | 0.0001 | 0.0001 | 0.0000 | 0.0000 | 0.0000 | 0.0000 | 0.0000 | 0.0000 | 0.0000 | 0.0000 |
| **1** | 0.0011 | 0.0008 | 0.0006 | 0.0004 | 0.0003 | 0.0002 | 0.0001 | 0.0001 | 0.0001 | 0.0000 |
| **2** | 0.0057 | 0.0043 | 0.0033 | 0.0025 | 0.0018 | 0.0014 | 0.0010 | 0.0007 | 0.0005 | 0.0004 |
| **3** | 0.0195 | 0.0156 | 0.0123 | 0.0097 | 0.0076 | 0.0058 | 0.0045 | 0.0034 | 0.0026 | 0.0019 |
| **4** | 0.0482 | 0.0403 | 0.0334 | 0.0274 | 0.0224 | 0.0181 | 0.0145 | 0.0115 | 0.0091 | 0.0071 |
| **5** | 0.0910 | 0.0797 | 0.0691 | 0.0594 | 0.0506 | 0.0427 | 0.0357 | 0.0297 | 0.0244 | 0.0199 |

*(Continued)*

**Table A.7** Binomial probability distribution

| | | | | | | | | | | |
|---|---|---|---|---|---|---|---|---|---|---|
| 6 | 0.1363 | 0.1250 | 0.1134 | 0.1020 | 0.0908 | 0.0801 | 0.0700 | 0.0606 | 0.0520 | 0.0442 |
| 7 | 0.1662 | 0.1596 | 0.1516 | 0.1426 | 0.1327 | 0.1222 | 0.1115 | 0.1008 | 0.0902 | 0.0800 |
| 8 | 0.1680 | 0.1690 | 0.1681 | 0.1652 | 0.1607 | 0.1547 | 0.1474 | 0.1390 | 0.1298 | 0.1200 |
| 9 | 0.1426 | 0.1502 | 0.1563 | 0.1608 | 0.1635 | 0.1644 | 0.1635 | 0.1609 | 0.1567 | 0.1511 |
| 10 | 0.1025 | 0.1131 | 0.1232 | 0.1325 | 0.1409 | 0.1479 | 0.1536 | 0.1578 | 0.1603 | 0.1612 |
| 11 | 0.0628 | 0.0726 | 0.0828 | 0.0931 | 0.1034 | 0.1135 | 0.1230 | 0.1319 | 0.1398 | 0.1465 |
| 12 | 0.0329 | 0.0399 | 0.0476 | 0.0560 | 0.0650 | 0.0745 | 0.0843 | 0.0943 | 0.1043 | 0.1140 |
| 13 | 0.0148 | 0.0188 | 0.0234 | 0.0288 | 0.0350 | 0.0419 | 0.0495 | 0.0578 | 0.0667 | 0.0760 |
| 14 | 0.0057 | 0.0076 | 0.0099 | 0.0127 | 0.0161 | 0.0202 | 0.0249 | 0.0304 | 0.0365 | 0.0434 |
| 15 | 0.0019 | 0.0026 | 0.0036 | 0.0048 | 0.0064 | 0.0083 | 0.0107 | 0.0136 | 0.0171 | 0.0212 |
| 16 | 0.0005 | 0.0008 | 0.0011 | 0.0015 | 0.0021 | 0.0029 | 0.0039 | 0.0052 | 0.0068 | 0.0088 |
| 17 | 0.0001 | 0.0002 | 0.0003 | 0.0004 | 0.0006 | 0.0009 | 0.0012 | 0.0017 | 0.0023 | 0.0031 |
| 18 | 0.0000 | 0.0000 | 0.0001 | 0.0001 | 0.0001 | 0.0002 | 0.0003 | 0.0005 | 0.0007 | 0.0009 |
| 19 | 0.0000 | 0.0000 | 0.0000 | 0.0000 | 0.0000 | 0.0000 | 0.0001 | 0.0001 | 0.0002 | 0.0002 |

| | 0.41 | 0.42 | 0.43 | 0.44 | 0.45 | .046 | 0.47 | 0.48 | 0.49 | 0.50 |
|---|---|---|---|---|---|---|---|---|---|---|
| 0 | 0.0000 | 0.0000 | 0.0000 | 0.0000 | 0.0000 | 0.0000 | 0.0000 | 0.0000 | 0.0000 | 0.0000 |
| 1 | 0.0000 | 0.0000 | 0.0000 | 0.0000 | 0.0000 | 0.0000 | 0.0000 | 0.0000 | 0.0000 | 0.0000 |
| 2 | 0.0003 | 0.0002 | 0.0001 | 0.0001 | 0.0001 | 0.0000 | 0.0000 | 0.0000 | 0.0000 | 0.0000 |
| 3 | 0.0014 | 0.0011 | 0.0008 | 0.0006 | 0.0004 | 0.0003 | 0.0002 | 0.0001 | 0.0001 | 0.0001 |
| 4 | 0.0055 | 0.0042 | 0.0032 | 0.0024 | 0.0018 | 0.0014 | 0.0010 | 0.0007 | 0.0005 | 0.0004 |
| 5 | 0.0161 | 0.0129 | 0.0102 | 0.0081 | 0.0063 | 0.0049 | 0.0037 | 0.0028 | 0.0021 | 0.0016 |
| 6 | 0.0372 | 0.0311 | 0.0257 | 0.0211 | 0.0172 | 0.0138 | 0.0110 | 0.0087 | 0.0068 | 0.0053 |
| 7 | 0.0703 | 0.0611 | 0.0527 | 0.0450 | 0.0381 | 0.0319 | 0.0265 | 0.0218 | 0.0178 | 0.0143 |
| 8 | 0.1099 | 0.0996 | 0.0895 | 0.0796 | 0.0701 | 0.0612 | 0.0529 | 0.0453 | 0.0384 | 0.0322 |
| 9 | 0.1442 | 0.1363 | 0.1275 | 0.1181 | 0.1084 | 0.0985 | 0.0886 | 0.0790 | 0.0697 | 0.0609 |
| 10 | 0.1603 | 0.1579 | 0.1539 | 0.1485 | 0.1419 | 0.1342 | 0.1257 | 0.1166 | 0.1071 | 0.0974 |
| 11 | 0.1519 | 0.1559 | 0.1583 | 0.1591 | 0.1583 | 0.1559 | 0.1521 | 0.1468 | 0.1404 | 0.1328 |
| 12 | 0.1232 | 0.1317 | 0.1393 | 0.1458 | 0.1511 | 0.1550 | 0.1573 | 0.1581 | 0.1573 | 0.1550 |
| 13 | 0.0856 | 0.0954 | 0.1051 | 0.1146 | 0.1236 | 0.1320 | 0.1395 | 0.1460 | 0.1512 | 0.1550 |
| 14 | 0.0510 | 0.0592 | 0.0680 | 0.0772 | 0.0867 | 0.0964 | 0.1060 | 0.1155 | 0.1245 | 0.1328 |
| 15 | 0.0260 | 0.0314 | 0.0376 | 0.0445 | 0.0520 | 0.0602 | 0.0690 | 0.0782 | 0.0877 | 0.0974 |
| 16 | 0.0113 | 0.0142 | 0.0177 | 0.0218 | 0.0266 | 0.0321 | 0.0382 | 0.0451 | 0.0527 | 0.0609 |
| 17 | 0.0042 | 0.0055 | 0.0071 | 0.0091 | 0.0115 | 0.0145 | 0.0179 | 0.0220 | 0.0268 | 0.0322 |
| 18 | 0.0013 | 0.0018 | 0.0024 | 0.0032 | 0.0042 | 0.0055 | 0.0071 | 0.0090 | 0.0114 | 0.0143 |
| 19 | 0.0003 | 0.0005 | 0.0007 | 0.0009 | 0.0013 | 0.0017 | 0.0023 | 0.0031 | 0.0040 | 0.0053 |
| 20 | 0.0001 | 0.0001 | 0.0001 | 0.0002 | 0.0003 | 0.0004 | 0.0006 | 0.0009 | 0.0012 | 0.0016 |
| 21 | 0.0000 | 0.0000 | 0.0000 | 0.0000 | 0.0001 | 0.0001 | 0.0001 | 0.0002 | 0.0003 | 0.0004 |
| 22 | 0.0000 | 0.0000 | 0.0000 | 0.0000 | 0.0000 | 0.0000 | 0.0000 | 0.0000 | 0.0000 | 0.0001 |

**Table A.8** Critical values of the Kruskal-Wallis test

| $n_1$ | $n_2$ | $n_3$ | Critical value | $\alpha$ | $n_1$ | $n_2$ | $n_3$ | Critical value | $\alpha$ |
|---|---|---|---|---|---|---|---|---|---|
| 2 | 1 | 1 | 2.7000 | 0.500 | | | | 4.7000 | 0.101 |
| 2 | 2 | 1 | 3.6000 | 0.200 | 4 | 4 | 1 | 6.6667 | 0.010 |
| 2 | 2 | 2 | 4.5714 | 0.067 | | | | 6.1667 | 0.022 |
| | | | 3.7143 | 0.200 | | | | 4.9667 | 0.048 |
| 3 | 1 | 1 | 3.2000 | 0.300 | | | | 4.8667 | 0.054 |
| 3 | 2 | 1 | 4.2857 | 0.100 | | | | 4.1667 | 0.082 |
| | | | 3.8571 | 0.133 | | | | 4.0667 | 0.102 |
| 3 | 2 | 2 | 5.3572 | 0.029 | 4 | 4 | 2 | 7.0364 | 0:006 |
| | | | 4.7143 | 0.048 | | | | 6.8727 | 0.011 |
| | | | 4.5000 | 0.067 | | | | 5.4545 | 0.046 |
| | | | 4.4643 | 0.105 | | | | 5.2364 | 0.052 |
| 3 | 3 | 1 | 5.1429 | 0.043 | | | | 4.5545 | 0.098 |
| | | | 4.5714 | 0.100 | | | | 4.4455 | 0.103 |
| | | | 4.0000 | 0.129 | 4 | 4 | 3 | 7.1439 | 0.010 |
| 3 | 3 | 2 | 6.2500 | 0.011 | | | | 7.1364 | 0.011 |
| | | | 5.3611 | 0.032 | | | | 5.5985 | 0.049 |
| | | | 5.1389 | 0.061 | | | | 5.5758 | 0.051 |
| | | | 4.5556 | 0.100 | | | | 4.5455 | 0.099 |
| | | | 4.2500 | 0.121 | | | | 4.4773 | 0.102 |
| 3 | 3 | 3 | 7.2000 | 0.004 | 4 | 4 | 4 | 7.6538 | 0.008 |
| | | | 6.4889 | 0.011 | | | | 7.5385 | 0.011 |
| | | | 5.6889 | 0.029 | | | | 5.6923 | 0.049 |
| | | | 5.6000 | 0.050 | | | | 5.6538 | 0.054 |
| | | | 5.0667 | 0.086 | | | | 4.6539 | 0.097 |
| | | | 4.6222 | 0.100 | | | | 4.5001 | 0.104 |
| 4 | 1 | 1 | 3.5714 | 0.200 | 5 | 1 | 1 | 3.8571 | 0.143 |
| 4 | 2 | 1 | 4.8214 | 0.057 | 5 | 2 | 1 | 5.2500 | 0.036 |
| | | | 4.5000 | 0.076 | | | | 5.0000 | 0.048 |
| | | | 4.0179 | 0.114 | | | | 4.4500 | 0.071 |
| 4 | 2 | 2 | 6.0000 | 0.014 | | | | 4.2000 | 0.095 |
| | | | 5.3333 | 0.033 | | | | 4.0500 | 0.119 |
| | | | 5.1250 | 0.052 | 5 | 2 | 2 | 6.5333 | 0.008 |
| | | | 4.4583 | 0.100 | | | | 6.1333 | 0.013 |
| | | | 4.1667 | 0.105 | | | | 5.1600 | 0.034 |
| 4 | 3 | 1 | 5.8333 | 0.021 | | | | 5.0400 | 0.056 |
| | | | 5.2083 | 0.050 | | | | 4.3733 | 0.090 |
| | | | 5.0000 | 0.057 | | | | 4.2933 | 0.122 |
| | | | 4.0556 | 0.093 | 5 | 3 | 1 | 6.4000 | 0.012 |
| | | | 3.8889 | 0.129 | | | | 4.9600 | 0.048 |
| 4 | 3 | 2 | 6.4444 | 0.008 | | | | 4.8711 | 0.052 |
| | | | 6.3000 | 0.011 | | | | 4.0178 | 0.095 |
| | | | 5.4444 | 0.046 | | | | 3.8400 | 0.123 |

(*Continued*)

**Table A.8** Critical values of the Kruskal-Wallis test

| $n_1$ | $n_2$ | $n_3$ | Critical value | $\alpha$ | $n_1$ | $n_2$ | $n_3$ | Critical value | $\alpha$ |
|---|---|---|---|---|---|---|---|---|---|
| 2 | 1 | 1 | 2.7000 | 0.500 |   |   |   | 4.7000 | 0.101 |
|   |   |   | 4.4444 | 0.102 |   |   |   | 5.2509 | 0.049 |
| 4 | 3 | 3 | 6.7455 | 0.010 |   |   |   | 5.1055 | 0.052 |
|   |   |   | 6.7091 | 0.013 |   |   |   | 4.6509 | 0.091 |
|   |   |   | 5.7909 | 0.046 |   |   |   | 4.4945 | 0.101 |
|   |   |   | 5.7273 | 0.050 | 5 | 3 | 3 | 7.0788 | 0.009 |
|   |   |   | 4.7091 | 0.092 |   |   |   | 6.9818 | 0.011 |
| 5 | 3 | 3 | 5.6485 | 0.049 | 5 | 5 | 1 | 6.8364 | 0.011 |
|   |   |   | 5.5152 | 0.051 |   |   |   | 5.1273 | 0.046 |
|   |   |   | 4.5333 | 0.097 |   |   |   | 4.9091 | 0.053 |
|   |   |   | 4.4121 | 0.109 |   |   |   | 4.1091 | 0.086 |
| 5 | 4 | 1 | 6.9545 | 0.008 |   |   |   | 4.0364 | 0.105 |
|   |   |   | 6.8400 | 0.011 | 5 | 5 | 2 | 7.3385 | 0.010 |
|   |   |   | 4.9855 | 0.044 |   |   |   | 7.2692 | 0.010 |
|   |   |   | 4.8600 | 0.056 |   |   |   | 5.3385 | 0.047 |
|   |   |   | 3.9873 | 0.098 |   |   |   | 5.2462 | 0.051 |
|   |   |   | 3.9600 | 0.102 |   |   |   | 4.6231 | 0.097 |
| 5 | 4 | 2 | 7.2045 | 0.009 |   |   |   | 4.5077 | 0.100 |
|   |   |   | 7.1182 | 0.010 | 5 | 5 | 3 | 7.5780 | 0.010 |
|   |   |   | 5.2727 | 0.049 |   |   |   | 7.5429 | 0.010 |
|   |   |   | 5.2682 | 0.050 |   |   |   | 5.7055 | 0.046 |
|   |   |   | 4.5409 | 0.098 |   |   |   | 5.6264 | 0.051 |
|   |   |   | 4.5182 | 0.101 |   |   |   | 4.5451 | 0.100 |
| 5 | 4 | 3 | 7.4449 | 0.010 |   |   |   | 4.5363 | 0.102 |
|   |   |   | 7.3949 | 0.011 | 5 | 5 | 4 | 7.8229 | 0.010 |
|   |   |   | 5.6564 | 0.049 |   |   |   | 7.7914 | 0.010 |
|   |   |   | 5.6308 | 0.050 |   |   |   | 5.6657 | 0.049 |
|   |   |   | 4.5487 | 0.099 |   |   |   | 5.6429 | 0.050 |
|   |   |   | 4.5231 | 0.103 |   |   |   | 4.5229 | 0.099 |
| 5 | 4 | 4 | 7.7604 | 0.009 |   |   |   | 4.5200 | 0.101 |
|   |   |   | 7.7440 | 0.011 | 5 | 5 | 5 | 8.0000 | 0.009 |
|   |   |   | 5.6571 | 0.049 |   |   |   | 7.9800 | 0.010 |
|   |   |   | 5.6176 | 0.050 |   |   |   | 5.7800 | 0.049 |
|   |   |   | 4.6187 | 0.100 |   |   |   | 5.6600 | 0.051 |
|   |   |   | 4.5527 | 0.102 |   |   |   | 4.5600 | 0.100 |
| 5 | 5 | 1 | 7.3091 | 0.009 |   |   |   | 4.5000 | 0.102 |

**Table A.9** Friedman ANOVA table

(Exact distribution of $\chi_r^2$ for tables with two to nine sets of three ranks [$l = 3$; $k = 2, 3, 4, 5, 6, 7, 8, 9$]). $p$ is the probability of obtaining a value of $\chi_r^2$ as great as or greater than the corresponding value of $\chi_r^2$

| $k = 2$ | | $k = 3$ | | $k = 4$ | | $k = 5$ | |
|---|---|---|---|---|---|---|---|
| $\chi_r^2$ | $p$ | $\chi_r^2$ | $p$ | $\chi_r^2$ | $p$ | $\chi_r^2$ | $p$ |
| 0 | 1.000 | 0.000 | 1.000 | 0.0 | 1.000 | 0.0 | 1.000 |
| 1 | 0.833 | 0.667 | 0.944 | 0.5 | 0.931 | 0.4 | 0.954 |
| 3 | 0.500 | 2.000 | 0.528 | 1.5 | 0.653 | 1.2 | 0.691 |
| 4 | 0.167 | 2.667 | 0.361 | 2.0 | 0.431 | 1.6 | 0.522 |
|  |  | 4.667 | 0.194 | 3.5 | 0.273 | 2.8 | 0.367 |
|  |  | 6.000 | 0.028 | 4.5 | 0.125 | 3.6 | 0.182 |
|  |  |  |  | 6.0 | 0.069 | 4.8 | 0.124 |
|  |  |  |  | 6.5 | 0.042 | 5.2 | 0.093 |
|  |  |  |  | 8.0 | 0.0046 | 6.4 | 0.039 |
|  |  |  |  |  |  | 7.6 | 0.024 |
|  |  |  |  |  |  | 8.4 | 0.0085 |
|  |  |  |  |  |  | 10.0 | 0.00077 |

| $k = 6$ | | $k = 7$ | | $k = 8$ | | $k = 9$ | |
|---|---|---|---|---|---|---|---|
| $\chi_r^2$ | $p$ | $\chi_r^2$ | $p$ | $\chi_r^2$ | $p$ | $\chi_r^2$ | $p$ |
| 0.00 | 1.000 | 0.000 | 1.000 | 0.00 | 1.000 | 0.000 | 1.000 |
| 0.33 | 0.956 | 0.286 | 0.964 | 0.25 | 0.967 | 0.222 | 0.971 |
| 1.00 | 0.740 | 0.857 | 0.768 | 0.75 | 0.794 | 0.667 | 0.814 |
| 1.33 | 0.570 | 1.143 | 0.620 | 1.00 | 0.654 | 0.889 | 0.865 |
| 2.33 | 0.430 | 2.000 | 0.486 | 1.75 | 0.531 | 1.556 | 0.569 |
| 3.00 | 0.252 | 2.571 | 0.305 | 2.25 | 0.355 | 2.000 | 0.398 |
| 4.00 | 0.184 | 3.429 | 0.237 | 3.00 | 0.285 | 2.667 | 0.328 |
| 4.33 | 0.142 | 3.714 | 0.192 | 3.25 | 0.236 | 2.889 | 0.278 |
| 5.33 | 0.072 | 4.571 | 0.112 | 4.00 | 0.149 | 3.556 | 0.187 |
| 6.33 | 0.052 | 5.429 | 0.085 | 4.75 | 0.120 | 4.222 | 0.154 |
| 7.00 | 0.029 | 6.000 | 0.052 | 5.25 | 0.079 | 4.667 | 0.107 |
| 8.33 | 0.012 | 7.143 | 0.027 | 6.25 | 0.047 | 5.556 | 0.069 |
| 9.00 | 0.0081 | 7.714 | 0.021 | 6.75 | 0.038 | 6.000 | 0.057 |
| 9.33 | 0.0055 | 8.000 | 0.016 | 7.00 | 0.030 | 6.222 | 0.048 |
| 10.33 | 0.0017 | 8.857 | 0.0084 | 7.75 | 0.018 | 6.889 | 0.031 |
| 12.00 | 0.00013 | 10.286 | 0.0036 | 9.00 | 0.0099 | 8.000 | 0.019 |
|  |  | 10.571 | 0.0027 | 9.25 | 0.0080 | 8.222 | 0.016 |
|  |  | 11.143 | 0.0012 | 9.75 | 0.0048 | 8.667 | 0.010 |
|  |  | 12.286 | 0.00032 | 10.75 | 0.0024 | 9.556 | 0.0060 |
|  |  | 14.000 | 0.000021 | 12.00 | 0.0011 | 10.667 | 0.0035 |

(*Continued*)

**Table A.9** Friedman ANOVA table

| k = 6 | | k = 7 | | k = 8 | | k = 9 | |
|---|---|---|---|---|---|---|---|
| $\chi_r^2$ | $p$ | $\chi_r^2$ | $p$ | $\chi_r^2$ | $p$ | $\chi_r^2$ | $p$ |
|  |  |  |  | 12.25 | 0.00086 | 10.889 | 0.0029 |
|  |  |  |  | 13.00 | 0.00026 | 11.556 | 0.0013 |
|  |  |  |  | 14.25 | 0.000061 | 12.667 | 0.00066 |
|  |  |  |  | 16.00 | 0.0000036 | 13.556 | 0.00035 |
|  |  |  |  |  |  | 14.000 | 0.00020 |
|  |  |  |  |  |  | 14.222 | 0.000097 |
|  |  |  |  |  |  | 14.889 | 0.000054 |
|  |  |  |  |  |  | 16.222 | 0.000011 |
|  |  |  |  |  |  | 18.000 | 0.0000006 |

| k = 2 | | k = 3 | | k = 4 | | | |
|---|---|---|---|---|---|---|---|
| $\chi_r^2$ | $p$ | $\chi_r^2$ | $p$ | $\chi_r^2$ | $p$ | $\chi_r^2$ | $p$ |
| 0.0 | 1.000 | 0.2 | 1.000 | 0.0 | 1.000 | 5.7 | 0.141 |
| 0.6 | 0.958 | 0.6 | 0.958 | 0.3 | 0.992 | 6.0 | 0.105 |
| 1.2 | 0.834 | 1.0 | 0.910 | 0.6 | 0.928 | 6.3 | 0.094 |
| 1.8 | 0.792 | 1.8 | 0.727 | 0.9 | 0.900 | 6.6 | 0.077 |
| 2.4 | 0.625 | 2.2 | 0.608 | 1.2 | 0.800 | 6.9 | 0.068 |
| 3.0 | 0.542 | 2.6 | 0.524 | 1.5 | 0.754 | 7.2 | 0.054 |
| 3.6 | 0.458 | 3.4 | 0.446 | 1.8 | 0.677 | 7.5 | 0.052 |
| 4.2 | 0.375 | 3.8 | 0.342 | 2.1 | 0.649 | 7.8 | 0.036 |
| 4.8 | 0.208 | 4.2 | 0.300 | 2.4 | 0.524 | 8.1 | 0.033 |
| 5.4 | 0.167 | 5.0 | 0.207 | 2.7 | 0.508 | 8.4 | 0.019 |
| 6.0 | 0.042 | 5.4 | 0.175 | 3.0 | 0.432 | 8.7 | 0.014 |
|  |  | 5.8 | 0.148 | 3.3 | 0.389 | 9.3 | 0.012 |
|  |  | 6.6 | 0.075 | 3.6 | 0.355 | 9.6 | 0.0069 |
|  |  | 70 | 0.054 | 3.9 | 0.324 | 9.9 | 0.0062 |
|  |  | 7.4 | 0.033 | 4.5 | 0.242 | 10.2 | 0.0027 |
|  |  | 8.2 | 0.017 | 4.8 | 0.200 | 10.8 | 0.0016 |
|  |  | 9.0 | 0.0017 | 5.1 | 0.190 | 11.1 | 0.00094 |
|  |  |  |  | 5.4 | 0.158 | 12.0 | 0.000072 |

**Table A.10** Wilcoxon table

(*d*-factors for Wilcoxon signed-rank test and confidence intervals for the median (α' = one-sided significance level, α" = two-sided significance level)

| *n* | *d* | Confidence coefficient | α" | α' | *n* | *d* | Confidence coefficient | α" | α' |
|---|---|---|---|---|---|---|---|---|---|
| 3 | 1 | .750 | .250 | .125 | 14 | 13 | .991 | .009 | .004 |
| 4 | 1 | .875 | .125 | .063 | | 14 | .989 | .011 | .005 |
| 5 | 1 | .938 | .062 | .031 | | 22 | .951 | 0.49 | .025 |
| | 2 | .875 | .125 | .063 | | 23 | .942 | .058 | .029 |
| 6 | 1 | .969 | .031 | .016 | | 26 | .909 | .091 | .045 |
| | 2 | .937 | .063 | .031 | | 27 | .896 | .104 | .052 |
| | 3 | .906 | .094 | .047 | 15 | 16 | .992 | .008 | .004 |
| | 4 | .844 | .156 | .078 | | 17 | .990 | .010 | .005 |
| 7 | 1 | .984 | .016 | .008 | | 26 | .952 | .048 | .024 |
| | 2 | .969 | .031 | .016 | | 27 | .945 | .055 | .028 |
| | 4 | .922 | .078 | .039 | | 31 | .905 | .095 | .047 |
| | 5 | .891 | .109 | .055 | | 32 | .893 | .107 | .054 |
| 8 | 1 | .992 | .008 | .004 | 16 | 20 | .991 | .009 | .005 |
| | 2 | .984 | .016 | .008 | | 21 | .989 | .011 | .006 |
| | 4 | .961 | .039 | .020 | | 30 | .956 | .044 | .022 |
| | 5 | .945 | .055 | .027 | | 31 | .949 | .051 | .025 |
| | 6 | .922 | .078 | .039 | | 36 | .907 | .093 | .047 |
| | 7 | .891 | .109 | .055 | | 37 | .895 | .105 | .052 |
| 9 | 2 | .992 | .008 | .004 | 17 | 24 | .991 | .009 | .005 |
| | 3 | .988 | .012 | .006 | | 25 | .989 | .011 | .006 |
| | 6 | .961 | .039 | .020 | | 35 | .955 | .045 | .022 |
| | 7 | .945 | .055 | .027 | | 36 | .949 | .051 | .025 |
| | 9 | .902 | .098 | .049 | | 42 | .902 | .098 | .049 |
| | 10 | .871 | .129 | .065 | | 43 | .891 | .109 | .054 |
| 10 | 4 | .990 | .010 | .005 | 18 | 28 | .991 | .009 | .005 |
| | 5 | .986 | .014 | .007 | | 29 | .990 | .010 | .005 |
| | 9 | .951 | .049 | .024 | | 41 | .952 | .048 | .024 |
| | 10 | .936 | .064 | .032 | | 42 | .946 | .054 | .027 |
| | 11 | .916 | .084 | .042 | | 48 | .901 | .099 | .049 |
| | 12 | .895 | .105 | .053 | | 49 | .892 | .108 | .054 |
| 11 | 6 | .990 | .010 | .005 | 19 | 33 | .991 | .009 | .005 |
| | 7 | .986 | .014 | .007 | | 34 | .989 | .011 | .005 |
| | 11 | .958 | .042 | .021 | | 47 | .951 | .049 | .025 |
| | 12 | .946 | .054 | .027 | | 48 | .945 | .055 | .027 |
| | 14 | .917 | .083 | .042 | | 54 | .904 | .096 | .048 |
| | 15 | .898 | .102 | .051 | | 55 | .896 | .104 | .052 |

(*Continued*)

**Table A.10** Wilcoxon table

| $n$ | $d$ | Confidence coefficient | $\alpha''$ | $\alpha'$ | $n$ | $d$ | Confidence coefficient | $\alpha''$ | $\alpha'$ |
|---|---|---|---|---|---|---|---|---|---|
| 12 | 8 | .991 | .009 | .005 | 20 | 38 | .991 | .009 | .005 |
| | 9 | .988 | .012 | .006 | | 39 | .989 | .011 | .005 |
| | 14 | .958 | .042 | .021 | | 53 | .952 | .048 | .024 |
| | 15 | .948 | .052 | .026 | | 54 | .947 | .053 | .027 |
| | 18 | .908 | .092 | .046 | | 61 | .903 | .097 | .049 |
| | 19 | .890 | .110 | .055 | | 62 | .895 | .105 | .053 |
| 13 | 10 | .992 | .008 | .004 | 21 | 43 | .991 | .009 | .005 |
| | 11 | .990 | .010 | .005 | | 44 | .990 | .010 | .005 |
| | 18 | .952 | .048 | .024 | | 59 | .954 | .046 | .023 |
| | 19 | .943 | .057 | .029 | | 60 | .950 | .050 | .025 |
| | 22 | .906 | .094 | .047 | | 68 | .904 | .096 | .048 |
| | 23 | .890 | .110 | .055 | | 69 | .897 | .103 | .052 |
| 22 | 49 | .991 | .009 | .005 | 24 | 62 | .990 | .010 | .005 |
| | 50 | .990 | .010 | .005 | | 63 | .989 | .011 | .005 |
| | 66 | .954 | .046 | .023 | | 82 | .951 | .049 | .025 |
| | 67 | .950 | .050 | .025 | | 83 | .947 | .053 | .026 |
| | 76 | .902 | .098 | .049 | | 92 | .905 | .095 | .048 |
| | 77 | .895 | .105 | .053 | | 93 | .899 | .101 | .051 |
| 23 | 55 | .991 | .009 | .005 | 25 | 69 | .990 | .010 | .005 |
| | 56 | .990 | .010 | .005 | | 70 | .989 | .011 | .005 |
| | 74 | .952 | .048 | .024 | | 90 | .952 | .048 | .024 |
| | 75 | .948 | .052 | .026 | | 91 | .948 | .052 | .026 |
| | 84 | .902 | .098 | .049 | | 101 | .904 | .096 | .048 |
| | 85 | .895 | .105 | .052 | | 102 | .899 | .101 | .051 |

# Index

Printed in the United States